AF352922

Application of New Nanoparticle Structures as Catalysts

Application of New Nanoparticle Structures as Catalysts

Editors

Antonio Guerrero Ruiz
Inmaculada Rodríguez-Ramos

MDPI • Basel • Beijing • Wuhan • Barcelona • Belgrade • Manchester • Tokyo • Cluj • Tianjin

Editors
Antonio Guerrero Ruiz
Department of Inorganic and Technical Chemistry
UNED
Spain

Inmaculada Rodríguez-Ramos
Instituto de Catálisis y Petroleoquímica
CSIC
Spain

Editorial Office
MDPI
St. Alban-Anlage 66
4052 Basel, Switzerland

This is a reprint of articles from the Special Issue published online in the open access journal *Nanomaterials* (ISSN 2079-4991) (available at: https://www.mdpi.com/journal/nanomaterials/special_issues/nano_Nanoparticle_Structures_Catalysts).

For citation purposes, cite each article independently as indicated on the article page online and as indicated below:

LastName, A.A.; LastName, B.B.; LastName, C.C. Article Title. *Journal Name* **Year**, *Article Number*, Page Range.

ISBN 978-3-03943-250-9 (Hbk)
ISBN 978-3-03943-251-6 (PDF)

Contents

About the Editors

Antonio Guerrero Ruiz is Full Professor at the Department of Inorganic and Technical Chemistry of the UNED in Madrid, where he heads the Laboratory of Chemistry at Surfaces. His research activities are concentrated in the development of new heterogeneous catalyst materials. For this aim, the preparation and characterization of metallic or bimetallic nanoparticles, functionalized carbon nanotubes, modified graphene composites, etc., has been accomplished. These materials are applied as catalysts or adsorbents for different technical processes.

Inmaculada Rodríguez-Ramos is Research Professor at the Instituto de Catálisis y Petroleoquímica (ICP-CSIC) in Madrid, where she heads the Group for Molecular Design of Heterogeneous Catalysts. Her expertise is Heterogeneous Catalysis, with a particular specialization in carbon materials, their functionalization chemistry, and applications in catalysis. Her research interests involve the design and preparation of (nano)materials for their application in gas/liquid catalytic reactions related to the sustainable production of primary chemicals and energy products.

Editorial

Application of New Nanoparticle Structures as Catalysts

Antonio Guerrero Ruiz [1,*] and Inmaculada Rodríguez-Ramos [2,*]

[1]   Department of Inorganic and Technical Chemistry, Facultad de Ciencias UNED, 28040 Madrid, Spain
[2]   Instituto de Catálisis y Petroleoquímica, CSIC, Campus Cantoblanco, 28049 Madrid, Spain
*   Correspondence: aguerrero@ccia.uned.es (A.G.R.); irodriguez@ccia.uned.es (I.R.-R.)

Received: 3 August 2020; Accepted: 3 August 2020; Published: 27 August 2020

Nanocatalysts, more precisely solids nanomaterials with catalytic properties to be used as heterogeneous catalysts, are an extended and very diverse group of nanostructured materials representing, at present, an active area of research with application in many catalyzed processes. Therefore, this research area not only can lead to significant advances for their potential technological applications, but also must engage a large variety of new materials. For the characterization of the studied nanocatalysts, many different techniques and experimental methods should be used to reveal both the structural and superficial properties of these materials. The scope of the Special Issue "Application of New Nanoparticle Structures as Catalysts" was to provide a non-systematic overview with current research studies in the field of developing nanocatalyts. Practically all the nanomaterials compositions presented as examples in this issue are different from each other, both in their chemical composition of the structured nanomaterials, or in relation to the catalyzed reactions where they are applied. In this way, in this Special Issue, nine selected original research paper and one comprehensive review are collected. More than 40 scientists from universities and research institutions contributed their research studies and expertise for the success of this Special Issue.

The scientific contributions are summarized in the next paragraphs.

An analysis of the recent studies concerning transition metal nitrides applied as heterogeneous catalysts is presented in the review carried out by Dr. Dongil [1]. These materials have a clear interest because they can substitute noble metals in different catalyzed processes. With these nanomaterials, numerous possibilities are opened for new structures for metal nitrides; since chemical ingredients can be combined as mono-, binary and even ternary mixtures or the addition of promoters can be accomplished during the preparation procedures. The description of the most employed synthetic methods is revisited and the application of some transition metal nitrides in different catalyzed reactions, hydrotreatments, oxidations and ammonia synthesis/decomposition is reported.

Two of the contributions of this Special Issue are related to Metal Organic Framework (MOF) nanostructures. The development of these MOF nanomaterials is, at present, one of the major subjects in fundamental research due to, among others, their potential applications as catalytic materials or as selective adsorbents. In the contribution by Zamaro et al. [2], the preparation of nanocatalysts derived from the MOF named UiO-66, when used as support for three transition elements (Cu, Co, and Fe), is described. These materials are evaluated in two CO oxidations: oxidation with air and selective oxidation in a hydrogen-rich stream. The main aim of this research is to find reaction conditions where these new nanostructures can be, for these processes, an alternative to the commercial catalysts based on expensive noble metals.

In the second contribution regarding MOFs by Ordonez et al. [3], a very important aspect when nanomaterials are proposed for a real application is emphasized. Thus, the solid material should be submitted to physic-mechanical treatments, such as grinding or pelleting, in order to transform the original powder into granules, which can be used at industrial scale. Three commercial materials ($[Cu_3(C_9H_3O_6)_2]$, $[C_9H_3FeO_6]$ and $[C_8H_5AlO_5]$) were studied as methane adsorbents in a fixed bed reactor. It was concluded that all these materials suffer structural and textural modifications when subjected to pressure, and consequently their adsorption capacities are largely reduced.

Two other research papers in this issue are related with mixed transition metal oxides. In the contribution by Illán-Gómez et al. [4], the authors synthesized, characterized and tested a series of perovskites ($BaFe_{1-x}Cu_xO_3$ with x = 0, 0.1, 0.3 and 0.4). The target reaction for these nanocatalysts is the soot oxidation, as method for avoiding the atmospheric contamination by exhaust gases of car engines. These perovskites catalyze both the NO_2 oxidation of Diesel soot and, but to a lesser extent, the soot oxidation by O_2 of Gasoline engines. The catalytic activities of these perovskites seem to be related to the amount of oxygen evolved during temperature programmed desorption experiments, which decreases when increasing the copper content.

In the case of the contribution by Cauqui et al. [5], a mixed oxide (ZrO_2 with different loadings of Ce, Ca and Y) is used as a support of Ni nanoparticles. In this study, the synthesized nanomaterials are extensively characterized by complementary methods and techniques, and evaluated as catalysts for the aqueous-phase reforming of methanol. Focusing on the effect of the redox properties of ceria and the basicity properties induced by Ca or Y, it is revealed that the availability of Ni-metallic at the surfaces and the presence of weak basic sites, particularly derived from Ca incorporation, is the key parameter for improving the catalytic performance.

Y. Wang et al. reported on the use of the surface plasmon resonance effect, in this case in a metal nanocomposite, $AuPt/N–TiO_2$, used as a photocatalyst [6]. While the Au nanoparticles were used to obtain energy from visible-light, Pt nanoparticles work as a cocatalyst, trapping the energetic electrons from the semiconductor support. With this material, the selective oxidation of benzyl alcohol under visible-light irradiation can be performed with a markedly enhanced selectivity and yield. An extensive series of irradiation experiments shed light on relevant information concerning the different steps of the photocatalytic mechanism with this material.

Fernández-Morales et al. reported a comparative study of diverse materials, in general solid catalysts with acidic surface properties, when applied to the reaction of isobutene dimerization to C8 olefins [7]. The exposed surface catalytic sites were conveniently characterized in order to interpret catalytic performances. In general, catalytic materials with a higher amount of Brønsted acid sites display improved catalytic performance, but for achieving an optimum selectivity towards C8 compounds, a combination of the nature of acidic sites and structural characteristics of the catalytic materials is required.

In the contribution by Ramirez-Barria et al. [8], an extended series of graphenic materials (doped or not with nitrogen adatoms, with different textural properties, etc.) were prepared and applied as electrocatalysts for the demanding oxygen reduction reaction. The material with nitrogen doping and with smaller grain sizes was demonstrated to be the most efficient electrocatalyst. Moreover, all nitrogen-doped graphenic materials show high tolerance to methanol poisoning and good stability.

The approach of Faroldi et al. [9] for the study of a new heterogeneous catalysts for the dehydrogenation reaction of formic acid, generating high-purity hydrogen, is also of great interest. Instead of noble metal catalysts, they prepared and characterized Ni-based catalysts supported on silica, which were doped with calcium in order to facilitate the adsorption-decomposition of the reactant. From the results of the catalytic performance (100% conversion with a 92% of selectivity to hydrogen) at a moderate reaction temperature, 160 °C, it can be concluded that these materials were very promising for this application. In fact, these results for catalytic behavior are comparable to those reported for noble metals.

Another example of a complex catalytic multicomponent material with an interesting potential application is reported by Ivars-Barcelo et al. [10]. In this case, the composite materials are based on noble metal particles (Pd or bi-metallic Ag/Pd) supported over an iron oxide (Fe_3O_4) with a magnetite structure. The catalytic application of these materials is the direct methane partial oxidation into value-added chemicals as formaldehyde. The presented preliminary catalytic results confirmed the potential of magnetite-supported (Ag)Pd catalysts for CH_4 partial oxidation into formaldehyde, with incipient methane conversion starting at 200 °C, but with very high selectivity above 95%. The prepared nanocomposite materials were investigated by different physicochemical techniques, with the purpose of relating the structural and superficial properties of these nanocatalysts with their detected catalytic performances.

In conclusion, the papers collected in this Special Issue can be described as an impressionistic painting with brushstrokes of different aspects of new developments of catalytic materials. All of them include complementary features involved in the design of special nanocatalysts: preparation-treatments, intensive characterization and evaluation as catalysts in various reactions of applied interest. Although the present Special Issue can cover neither all the research of new structures used as nanocatalysts nor a complete list of application in catalyzed processes, the editors are confident that its contributions to fundamental research will offer new perspectives for the readers.

Author Contributions: All the guest editors wrote and reviewed this Editorial Letter. All authors have read and agreed to the published version of the manuscript.

Funding: This research was funded partially by the Spanish Minister of Science through the projects CTQ2017-89443-C3-1-R and CTQ2017-89443-C3-3-R.

Acknowledgments: We are grateful to all the authors who contributed to this Special Issue. We also acknowledge the referees for reviewing the manuscripts. And especially we have to recognize the immense work developed by Miss Tina Tian, without her support the publication of this Special Issue was impossible.

Conflicts of Interest: The authors declare no conflict of interest.

References

1. Dongil, A.B. Recent Progress on Transition Metal Nitrides Nanoparticles as Heterogeneous Catalysts. *Nanomaterials* **2019**, *9*, 1111. [CrossRef] [PubMed]
2. Lozano, L.A.; Faroldi, B.M.C.; Ulla, M.A.; Zamaro, J.M. Metal—Organic Framework-Based Sustainable Nanocatalysts for CO Oxidation. *Nanomaterials* **2020**, *10*, 165. [CrossRef] [PubMed]
3. Ursueguía, D.; Díaz, E.; Ordóñez, S. Densification-Induced Structure Changes in Basolite MOFs: Effect on Low-Pressure CH_4 Adsorption. *Nanomaterials* **2020**, *10*, 1089. [CrossRef] [PubMed]
4. Torregrosa-Rivero, V.; Moreno-Marcos, C.; Albaladejo-Fuentes, V.; Sánchez-Adsuar, M.S.; Illán-Gómez, M.J. $BaFe_{1-x}Cu_xO_3$ Perovskites as Active Phase for Diesel (DPF) and Gasoline Particle Filters (GPF). *Nanomaterials* **2019**, *9*, 1551. [CrossRef] [PubMed]
5. Goma, D.; Delgado, J.J.; Lefferts, L.; Faria, J.; Calvino, J.J.; Cauqui, M.A. Catalytic Performance of $Ni/CeO_2/X-ZrO_2$ (X = Ca, Y) Catalysts in the Aqueous-Phase Reforming of Methanol. *Nanomaterials* **2019**, *9*, 1582. [CrossRef] [PubMed]
6. Wang, Y.; Chen, Y.; Hou, Q.; Ju, M.; Li, W. Coupling Plasmonic and Cocatalyst Nanoparticles on $N-TiO_2$ for Visible-Light-Driven Catalytic Organic Synthesis. *Nanomaterials* **2019**, *9*, 391. [CrossRef] [PubMed]
7. Fernández-Morales, J.M.; Castillejos, E.; Asedegbega-Nieto, E.; Dongil, A.B.; Rodríguez-Ramos, I.; Guerrero-Ruiz, A. Comparative Study of Different Acidic Surface Structures in Solid Catalysts Applied for the Isobutene Dimerization Reaction. *Nanomaterials* **2020**, *10*, 1235. [CrossRef] [PubMed]
8. Ramirez-Barria, C.S.; Fernandes, D.M.; Freire, C.; Villaro-Abalos, E.; Guerrero-Ruiz, A.; Rodríguez-Ramos, I. Upgrading the Properties of Reduced Graphene Oxide and Nitrogen-Doped Reduced Graphene Oxide Produced by Thermal Reduction toward Efficient ORR Electrocatalysts. *Nanomaterials* **2019**, *9*, 1761. [CrossRef] [PubMed]
9. Faroldi, B.; Paviotti, M.A.; Camino-Manjarrés, M.; González-Carrazán, S.; López-Olmos, C.; Rodríguez-Ramos, I. Hydrogen Production by Formic Acid Decomposition over Ca Promoted Ni/SiO_2 Catalysts: Effect of the Calcium Content. *Nanomaterials* **2019**, *9*, 1516. [CrossRef] [PubMed]
10. Martínez-Navarro, B.; Sanchis, R.; Asedegbega-Nieto, E.; Solsona, B.; Ivars-Barceló, F. $(Ag)Pd-Fe_3O_4$ Nanocomposites as Novel Catalysts for Methane Partial Oxidation at Low Temperature. *Nanomaterials* **2020**, *10*, 988. [CrossRef] [PubMed]

 *nanomaterials*

Review

Recent Progress on Transition Metal Nitrides Nanoparticles as Heterogeneous Catalysts

A.B. Dongil [1,2]

1   Institute of Catalysis and Petrochemistry, CSIC, c/Marie Curie No. 2, Cantoblanco, 28049 Madrid, Spain; a.dongil@csic.es
2   UA UNED-ICP (CSIC) Group Des. Appl. Heter. Catal, 28049 Madrid, Spain

Received: 15 July 2019; Accepted: 23 July 2019; Published: 2 August 2019

Abstract: This short review aims at providing an overview of the most recent literature regarding transition metal nitrides (TMN) applied in heterogeneous catalysis. These materials have received renewed attention in the last decade due to its potential to substitute noble metals mainly in biomass and energy transformations, the decomposition of ammonia being one of the most studied reactions. The reactions considered in this review are limited to thermal catalysis. However the potential of these materials spreads to other key applications as photo- and electrocatalysis in hydrogen and oxygen evolution reactions. Mono, binary and exceptionally ternary metal nitrides have been synthetized and evaluated as catalysts and, in some cases, promoters are added to the structure in an attempt to improve their catalytic performance. The objective of the latest research is finding new synthesis methods that allow to obtain smaller metal nanoparticles and increase the surface area to improve their activity, selectivity and stability under reaction conditions. After a brief introduction and description of the most employed synthetic methods, the review has been divided in the application of transition metal nitrides in the following reactions: hydrotreatment, oxidation and ammonia synthesis and decomposition.

Keywords: heterogeneous catalysis; transition metal nitrides

---

1. Introduction

In the last decade strong interest has emerged in the field of nitrogen-doped catalysts, especially since new carbon nanostructures have been successfully synthetized. In those structures nitrogen can be easily inserted as heteroatoms due to its similarity to carbon, and as a doping agent it offers new possibilities in the field of catalysis [1]. Nitrogen confers basic sites for optimal reactant adsorption and an excellent electron density that may improve the catalytic performance due to the better dispersion of the active phase and the changes in the electronic properties [2].

Transition metal nitrides have been traditionally employed as catalysts in reactions such as hydrodesulfurization or ammonia synthesis [3]. Nowadays, researchers face new challenges in catalytic transformations to find active and selective catalysts in the fields of energy or biomass including reactions such as hydrodeoxygenation, water-gas shift and CO and CO_2 hydrogenation among others [4].

Noble metals have been widely studied on those reactions. However economical catalysts based on abundant materials are more suitable for industrial applications. Hence, the need of searching new catalyst able to replace noble metals and the positive effect of nitrogen doping in several transformations, have brought up renewable attention to transition metal nitrides. These materials can donate electrons from the nitrogen atom and possesses high chemical, mechanical and thermal stability. Also importantly for thermal catalytic applications, when optimal synthesis conditions are employed, nitrides can reach relatively high surface areas of at least 90 m^2/g [5].

Transition metal nitrides are compounds in which nitrogen is incorporated into the interstitial sites of the metal structure. Since the size of the nitrogen atom is small (0.065 nm), it fulfills the Hägg rule (i.e., the ratio of the radii of non-metal to metal is less than 0.59, this allowing the formation of simple typical structures as compiled in Figure 1 [6].

As it can be expected the reactivity of the resulting transition metal nitride is different from that of the parent metal. In general, two main effects can be considered when nitrides are used in catalysis [7,8]:

(a) Ligand effect: Nitrides have a different electronic structure compared to the parent metal. This is due to the charge transfer from the metal to the non-metal, hybridization of the metal d-states with the non-metal sp-states, and expansion of the metal–metal lattice spacing. These changes in the electronic structure, modify the chemical reactivity of the parent metal so that the strenght of adsorption of reactants and products is similar to noble metals, this improving the selectivity of the reactions.

(b) Ensemble effect: When nitrogen is located on the metal surface, it may decrease the number of available metal sites, also creating different adorption sites. This effect can be somehow tuned by changing the metal to nitrogen ratio.

2. Synthesis and Structural Properties

Metal nitrides can be prepared among other methods by thermal treatment of the metal precursor obtained from template methods [9], ball milling [10,11] or temperature programmed reaction/reduction procedures [12].

Template methods are based on the use of a sacrificial template, normally MgO, to which a mixture containing the metal precursor is added followed by thermal treatment at high temperatures and extensive wash to remove the MgO template. In this way Metal-N entities are well dispersed in a carbon based matrix. In the ball milling method, the solid metal precursor either commercial or lab-made are submitted to milling under pressure and connected to a supply of the nitridation agent, N_2 or NH_3.

So far, the most employed method to synthetize transition metal nitrides is a temperature programmed reaction/reduction procedure. The method consists in a reductive nitriding treatment under a gas flow (i.e., NH_3 or a mixture of N_2:H_2) of the corresponding metal oxide precursor to produce the nitride [12]. The characteristics of the final nitride depend on the precursor and preparation conditions such as heating rate and final temperature. It was reported that higher areas are obtained when heating rates below 1 °C/min and/or high gas space velocity (150,000 h^{-1}) are used, as these conditions reduce sintering due to water release [13,14]. Other nitrogen sources such as urea or *m*-phenylenediamine have been employed in order to improve the efficiency of the nitridation [15,16].

The use of nitrogen-rich transition metal nitrides in catalysis is very promising. However, the synthesis of TMN with high nitrogen to metal ratio is energetically demanding. This has prompted researchers to find aternative methodologies and new structures such as polymorphs of Zr_3N_4, Hf_3N_4, Ta_2N_3, and noble metal dinitrides OsN_2, IrN_2, and PtN_2 have been explored [17].

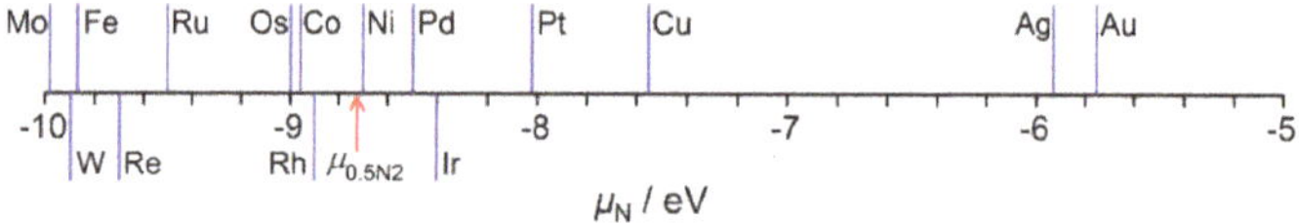

Figure 1. Value of the nitrogen chemical potential (μN) where the nitrogen covered surface becomes more stable than the clean metal surface; Reproduced from [18], with permission from American Chemical Society, 2018.

Sautet et al. [18] recently studied the nitridation of transition metal surfaces and tried to offer some insight into the synthesis and stability of transition metal nitrides under working conditions.

The authors studied fifteen transition metals and using theoretical and experimental techniques evaluated the extent of nitridation (surface vs. bulk) depending on the metal and the shape of nanoparticles. By using the nitrogen chemical potential at which metal covered by nitrogen is more stable than bare metal, the authors were able to stablish a nitridation trend among the studied serie. As depicted in Figure 1, Mo, W, Fe and Re are more easily nitrided, becomign more difficult when moving to the right hand side of the periodic table, i.e., less oxophilic metals, which agrees well with previous experimental results [19,20].

The most studied transition metal nitride in catalysis is molybdenum. In general cubic γ-MoN$_x$ ($0.5 \leq x < 1$) is obtained by NH$_3$ treatment of the oxide precursor, MoO$_3$, while mixtures of H$_2$ + N$_2$ lead to the formation of tetragonal β-MoNx ($x \leq 0.5$) or γ-MoNx. Hexagonal δ-MoNx ($x \geq 1$) is obtained from MoS$_2$ and NH$_3$. Similarly, the content of nitrogen on iron nitrides influences their structure, so upon increasing the nitrogen content, the lattice structure changes from fcc γ'-Fe$_4$N to hcp ε-FexN ($2 < x \leq 3$) and to orthorhombic ζ-Fe$_2$N. In Figure 2, the most common structures of transition metal nitrides are shown.

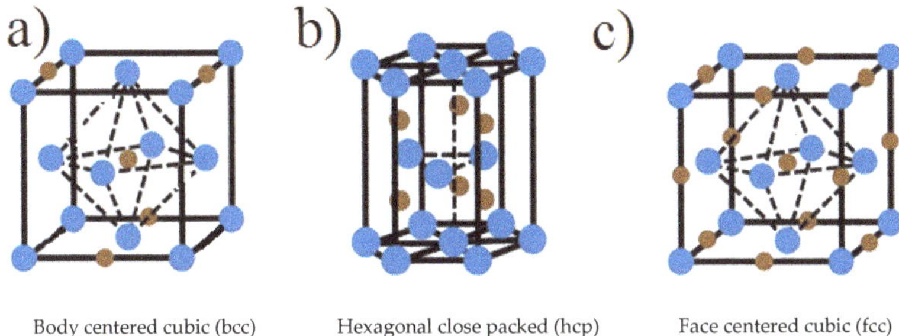

Figure 2. (**a**) bcc: TiN, ZrN, HfN, VN, CrN; (**b**) hcp: Mo$_2$N, W$_2$N; (**c**) fcc: MoN, TaN. Blue points represent transition metal atoms and brown points nitrogen atoms. Adapted from [6], with permission from Wiley, 2013.

3. Transition Metal Nitrides as Catalyst

3.1. Hydrotreatment Reactions

Transition metal nitrides are considered excellent candidates to replace noble metals in hydrogen-treatment reactions since they show similar or even better performance than noble metals. It has been reported that Mo nitrides can easily chemisorb hydrogen due to the contraction of the d-band and the changes in the electron density that result as a consequence of the interstitial incorporation of N in the Mo metal lattice. Moreover, in the Mo$_2$N-based catalysts hydrogenation reaction occurs over nitrogen vacancies, so Mo/N ratio also influences the catalytic performance [21].

Both CO and CO$_2$ hydrogenations are very interesting reactions since they overcome key environmental challenges while providing energy or valuable chemicals. On the one hand CO hydrogenation can be employed to purify H$_2$ gas streams before feeding to fuel cells to avoid poisoning. On the other hand, CO$_2$ hydrogenation has been appointed as a solution to reduce the amount of CO$_2$ evolved to the atmosphere from the industrial activity and convert it to fuels and chemicals [22,23].

However CO and CO$_2$ hydrogenation face several challenges. CO$_2$ activation is difficult due to the inert nature of the molecule and the cleavage of the C−O bonds in CO$_2$ demands high activation energy. Also, methanation of CO and CO$_2$ which provides an efficient alternative to conventional natural gas, is highly exotermic. The produced reaction heat favours metal sintering, which decreases the catalyst activity. Catalyst deactivation can also take place when carbon deposits on the active phase.

Zaman et al. [24,25] have studied the influence of adding alkali promoters to the Mo$_2$N systems in CO hydrogenation. The synthesis in both cases was performed by a simple temperature programmed treatment of the molybdenum and alkali precursors under ammonia flow. This leads to a material containing a mix of different phases: Mo, Mo_2N, Mo oxide and alkali-Mo oxide phase.

The promotion with alkalis favoured the conversion to oxygenates, i.e., methanol, ethanol and propanol. Cesium was found to be better promoter to oxygenates at 5% wt. (28% selectivity to oxygenates vs. 11% with Li promoted and 6.5% with unpromoted). The lower selectivity achieved with Li compared to Cs was attributed to the formation of Li_2MoO_4 phases during nitridation.

On the other hand, bare Mo_2N leads to the preferential conversion of CO to hydrocarbons, which can be ascribed to (i) CO dissociative hydrogenation and (ii) water-gas shift reaction, as shown in Figure 3. However, the presence of alkali hinders CO dissociation, which benefits the molecular insertion of CO into—CHx intermediate and promotes the coupling of alcohols.

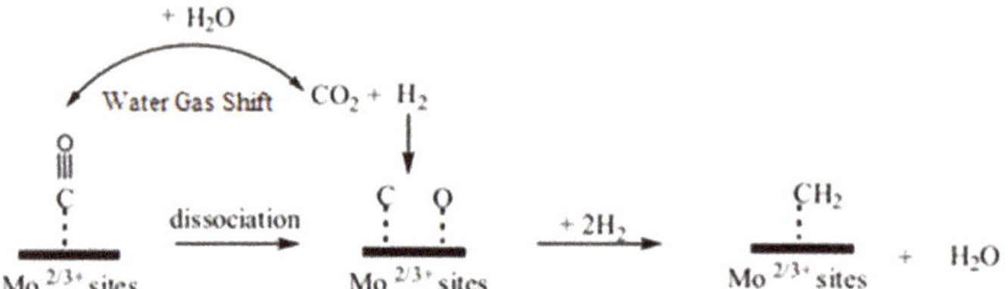

Figure 3. Plausible CO hydrogenation reaction pathways on Mo_2N and $K-Mo_2N$ catalysts. Reproduced from [25] with permission from Elsevier, 2018.

Regarding the effect of alkali loading, the authors performed a thorough study with potassium as promoter using several weight percentages: 0.45, 1.3, 3 and 6.2% [24]. The best selectivity to oxygenates, 44%, was obtained over promoted $K-Mo_2N$ with a K/Mo surface ratio of 0.06 which also corresponded to the best K distribution among the samples. The XRD showed that both $\gamma-Mo_2N$ cubic and monoclinic K_2MoO_4 were formed and this latter phase seems to increase upon K addition further than 3% wt, this being detrimental for K distribution and hence for oxygenates conversion.

Methanation, is the catalytic hydrogenation of carbon oxides (CO and CO_2) to obtain synthetic natural gas. Ru, Rh and Ni catalysts have proven to offer good catalytic performance in terms of activity and selectivity to methane which can be further improved by using bimetallic systems and optimizing catalyst synthesis method [26].

Methanation and hydrodesulfurization of dibenzothiophene was studied over molybdenum nitride by Zhao et al. [27]. The authors were able to synthetize a rich nitrogen molybdenum nitride using high pressure, 3.5 GPa, through a solid-state ion exchange reaction. This new nitride, 3R−MoN$_2$ holds a rhombohedral R3m structure, isotypic with MoS$_2$. However it offered catalytic activities three times higher than MoS$_2$ for the hydrodesulfurization of dibenzothiophene and over twice as high in the sour methanation of syngas at 723 K.

The binary nitride, Ni$_2$Mo$_3$N, was studied by Leybo et al. [28] in the methanation of CO$_2$. Yet, modest selectivites to CH$_4$, ca. 20%, were obtained and the active phase suffered sintering upon reaction conditions, decreasing the stability of the catalyst.

Besides molybdenum other metal nitrides have been tested in methanation reaction. For example co-methanation of CO and CO$_2$ have been evaluated by Li et al. [29] over cobalt nitrides supported on alumina. The authors studied the effect of metal loading on Co$_4$N/γ-Al$_2$O$_3$ and Co/γ-Al$_2$O$_3$ catalysts. According to the characterization, the cobalt nitride favoured stronger interactions with the support, this improving the dispersion of the nanoparticles and also their resistance to coking and metal sintering after 250 h which, as previously mentioned, is critical in such a exotermic reaction. Moreover, it was confirmed that the nitrogen atoms improved the adsorption of reactants due to their basicity, leading to better catalytic performance compared to the Co metal supported catalysts. The better results were also explained by the uniform metal dispersion and superior metal-support interaction.

Fisher-Tropsh is a well-known transformation to convert syngas, CO + H$_2$, into liquid hydrocarbons and that was very relevant in catalytic research in the 70–80s due to the oil crisis. Now, it has revised attention since it can also use syngas from biomass to produce fuels and chemicals. The most studied systems are those based on iron as they offer optimal results under a variety of conditions that allow to tune the selectivity, and are economically interesting. However, under reactions conditions the water produced can oxidize the catalyst with its subsequent deactivation [30].

Bao et al. [31] studied the confination effect of FeN cubic nanoparticles inside carbon nanotubes (FexN-in) (see Figure 4). Firstly, FeN supported on CNT resulted 5–7 times more active than FeN supported on silica and metallic Fe on CNT. This seems to be related to the better stability of the nitrides under CO hydrogenation conditions compared to metallic and carbide iron which are oxidized by water, resulting in catalyst deactivation. Also, the catalyst where FeN nanoparticles were selectively loaded inside the CNT was more active than the catalyst with FeN nanoparticles mainly dispersed on the external walls, FexN-out, (1.4 times). The authors explained the better activity of the confined nanoparticles by the lower particle size and the formation of more FeCxN1x entities on FexN-in than on FexN-out during reaction, leading to stronger retention of nitrogen atoms in the lattice.

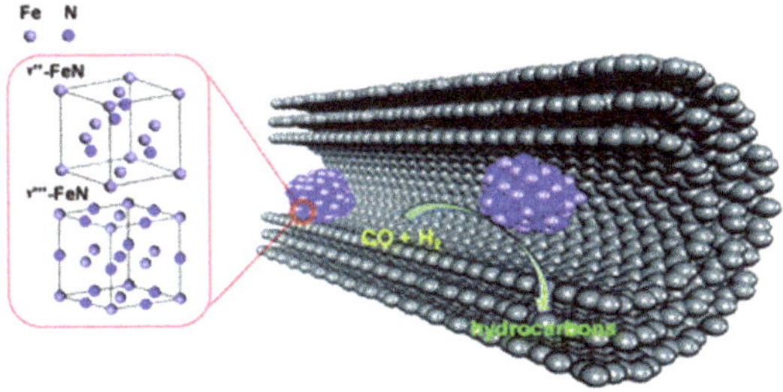

Figure 4. Reproduced from [32], with permission from Royal Society of Chemistry, 2011.

In contrast to the activity and stability enhancement of iron nitride compared to its parent metal, cobalt nitride seems to be a poison for FT synthesis. In literature it was reported that by adding a nitrogen source such as acetonitrile or ammonia to the FT gas feed, cobalt nitride phases are formed which result in catalyst deactivation by deposition on the most active metallic cobalt sites (steps and edges) [32].

So far, theoretical results have predicted that CO adsorption and dissociation over γ-$Mo_2N(111)$ has a similar activation barrier to that of MoS_2, so similar activity for syngas conversion can be expected [33].

CO_2 dry reforming of methane (DRM) has received much attention, as it transforms two greenhouse gases (CH_4 and CO_2) into syngas. The most studied catalysts have been noble metals, Ru, Rh and non-noble metals such as Ni [34]. Some reports appeared using transition metal carbides due to their low cost and similar structure compared to noble metals. However at atmospheric pressure, carbides can easily be deactivated due to oxidation by CO_2 or H_2O [35]. Hence, alternatively nitrides have been tested as potential catalysts for DRM.

Gu et al. [36] studied Mo_2N, Ni_3Mo_3N and Co_3Mo_3N above 550 °C and atmospheric pressures and found that a synergic effect is observed in the bimetallic nitrides that improve the activity and resistance to oxidation and coke deposition on the DRM compared to the monometallic nitride Mo_2N. The most active and stable catalyst was Co_3Mo_3N which, among other factors, was ascribed to the synergistic effect between the Mo and interstitial metal Co.

Owing to environmental concerns and the depletion of fossil resources, in the last decade researchers have focus on the study of biomass derived compounds to obtain fuels and chemicals. Since the starting lignocellulosic biomass owns a high oxygen concentration (>50%), most of the transformations require selective removal of oxygen. More specifically, one of the most studied reactions is hydrodeoxygenation (HDO). Under HDO condititions, the reactants can be also converted through the decarboxylation/decarbonylation (DCO) path, promoting the C−C cleaveage which is undesirable for fuels and chemicals.

For example, Monnier et al. [37] studied the conversion of oleic acid and canola oil with nitrides of Mo, W, and V supported on γ-Al_2O_3. The Mo_2N catalyst exhibited superior activity for oleic acid conversion compared to the other nitride catalysts, and also favored the HDO route vs. DCO. The HDO path produces preferentially n-$C_{18}H_{38}$ (diesel fuel cetane enhancers). Also, Mo_2N/γ-Al_2O_3 was stable in continuous hydrotreatment of canola oil at 400 °C under 83 bar hydrogen, reaching a constant yield of 50% middle distillates.

Murzin et al. [38] studied Ni and Mo_2N-MoO_2 on the HDO of more complex reactants, *Chlorella* algal oil extracted with supercritical hexane and stearic acid, at 300 °C under 30 bar in the presence of hydrogen. The catalysts were selective to fatty acids, indicating deactivation of decarbonylation sites. The catalyst Mo_2N-MoO_2, despite being less active than the Ni based catalysts, was more stable and it showed no deactivation after a 360 min test. These results open new possibilities that should be explored regarding mixtures of nitrides and oxides. Nitrides are known to be deactivated in the presence of water due to oxidation, but these oxides might tolerate better the presence of impurities, enhancing their stability.

Zhang et al. [39] prepared cobalt nitride supported on a nitrogen doped carbon CoNx@NC using cellulose and ammonia as the carbon and nitrogen source respectively at different synthesis temperatures from 500 to 800 °C. The catalysts were tested in the HDO of eugenol at 2 MPa H_2 and 200 °C. According to the reaction results, the reaction follows different paths when using nitrides or metallic cobalt. While the nitrides favour the cleaveage of the C-aryl−OCH_3 bond to form 4-propylphenol, metallic cobalt promotes the hydrogenation of the alkene moiety. The best catalyst was CoNx@NC-650 which displayed the largest surface area and dispersion of the nitrides nanoparticles. The catalyst was also successfully tested to promote the HDO of phenolic compounds.

Supported CoNx on carbon nanotubes on the hydrogenation of nitrobenzene and hydrogenated coupling of nitrobenzene with benzaldehyde was studied by Zhang et al. [40] as schematized in Figure 5. Some catalytic tests verified that catalytic activity was mainly due to the CoNx entities and the authors also suggested that the activity was mainly due to the cobalt chelate complexes bonded to nitrogen atoms of the graphene lattice.

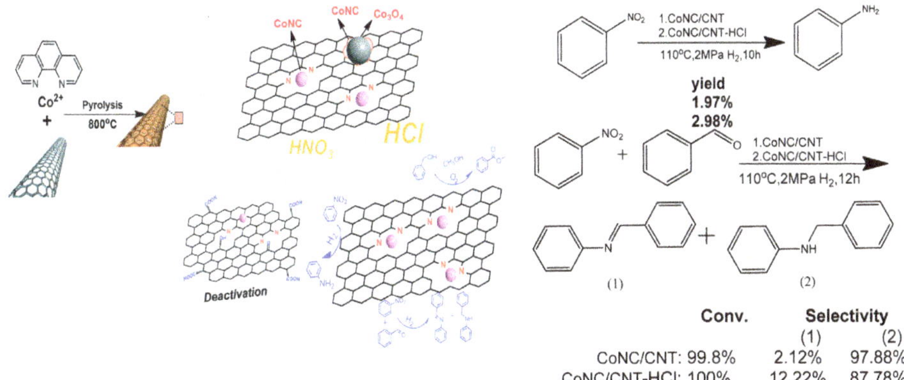

Figure 5. CoNC/CNT active sites on nitro compounds hydrogenation and hydrogenated coupling of nitrobenzene with benzaldehyde. Reproduced from [41], with permission from Royal Society of Chemistry, 2016.

Lodeng et al. [41] compared the activity of molybdenum nitride, carbide, and phosphide supported on TiO_2 on the HDO of phenol at 25 bar and in a temperature range between 350 and 450 °C. All the catalysts were highly active to benzene and only minor amounts of aromatic ring hydrogenation were obtained. Molybdenum nitride displayed lower activity compared to its carbide and phosphide counterparts, but its selectivity to cyclohexene was higher than that of phosphide and similar to carbide.

Hydrogenation of −COH moieties constitutes one of the most interesting and studied reactions in fine chemistry since it allows obtaining a high number of compounds that are used for example in pharmaceuticals and/or fragances. Reactants such as cinnamaldehyde or crotonaldehyde have been widely studied in an attempt to heterogeneized the catalytic system. The catalysts must be selective to the unsaturated alcohols without reducing the C=C bonds. With that aim heterogeneous catalysts based on noble metals mainly Ru, Pd and Pt have been widely investigated with good results in terms of activity and selectivity [42] and the significant role of nitrogen improving the selectivity to the desired products have also been reported [43]. To date, these specific transformations has been tested with bimetallic systems of metal nitrides and noble metals.

Fu et al. [44] have used a previously functionalized support to obtain small nitrides nanoparticles supported on the mesoporous silica SBA-15. The synthesis of Mo_2N over SBA-15 started by functionalizing the support with a monoamine that is located homogeneously into the pores of the support as shown ion Figure 6. This amine is then used as anchoring point for the molybdenum precursors which preferentially adsorbs on the moieties. Then, the procedure follows the previously explained temperature reduction procedure in NH_3. Finally the noble metal is impregnated and reduced with $NaBH_4$ forming bimetallic phases with Mo_2N. In this way, Mo_2N and Pt nanoparticles with a size of about 8.0 and 5-6 nm respectively were obtained with metal loadings of Mo (22% wt.) and Pt (3% wt.).

Then, $Pt/Mo_2N/SBA$-15 with different Pt loadings (1–3% wt.) was tested in the chemoselective hydrogenation of cinnamaldehyde. Both activity and selectivity to the cinnamyl alcohol was higher over Pt/Mo_2N-based catalysts than over monometallic Pt/SBA-15, that the authors ascribed to the synergy between Pt and Mo_2N nanoparticles and the more efficient use of the Pt surface on the bimetallic sample.

In the work of Thomson et al. [45], some more insight was given regarding the reaction mechanism and nature of active phases. To do so, the authors used a thorough experiment by hydrogenating a high surface area γ-Mo_2N to obtain a partially hydrogenated entity γ-Mo_2N-Hx. Then, by combining H_2-TPD experiments and DFT simulations three different hydrogen species were identified: surface nitrogen bound (κ1-NHsurf), surface Mo bound (κ1-MoHsurf) and subsurface Mo-bound (μ6-Mo6Hsub).

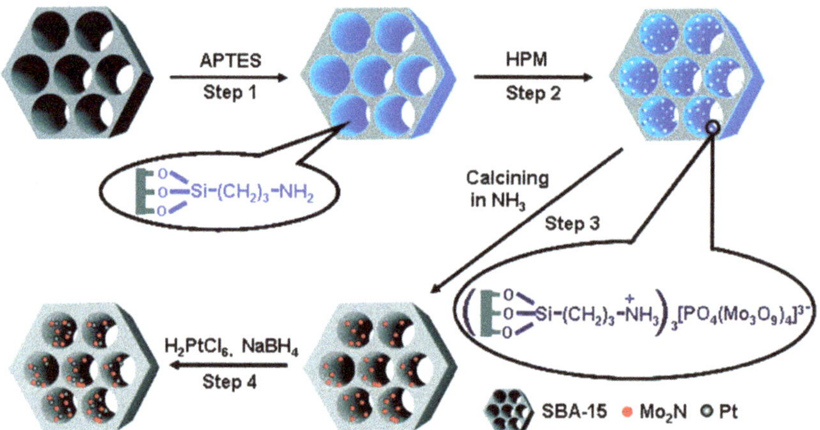

Figure 6. Synthesis of Pt-Mo$_2$N-SAB-15. Reproduced from [45], with permission from American Chemical Society, 2016.

The reactivity of these species was assessed by testing them in the hydrogenation of crotonaldehyde. Accordingly, the authors proposed that reaction starts by a heterolytic dissociation of H$_2$ to form surface NH (κ1-NHsurf) and MoH (κ1-MoHsurf) as schematized in Figure 7. Then, since subsurface interstitial H site (μ6-MoHsub) is more energetically favored than surface κ1-MoHsurf, hydrogen migrates into the lattice. Moreover, based on the catalytic results the authors proposed that surface and subsurface species MoH (κ1-MoHsurf/μ6-MoHsub) are more selective to the hydrogenation of the C=O bond and the surface κ1-NHsurf sites hydrogenate preferentially the C=C bond.

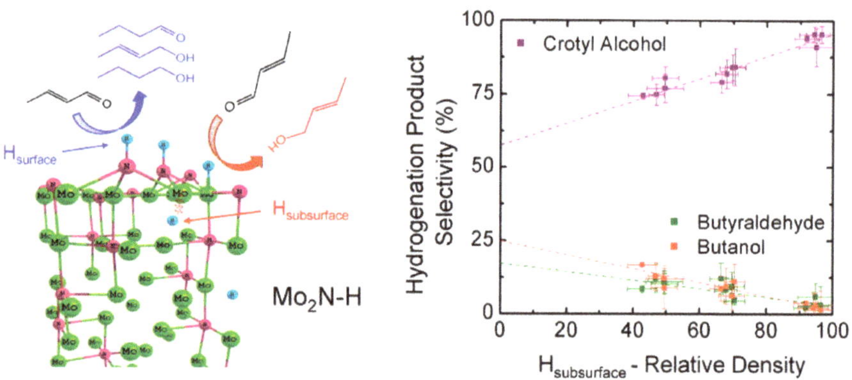

Figure 7. Reproduced from [45], with permission from American Chemical Society, 2016.

The selective hydrogenation of acetylene to ethylene is an important transformation since ethylene is the monomer to produce polyethylene polymers and it has a strategic relevance in refineries, being critical its high purity production.

The catalytic behaviour of β- and γ-Mo$_2$N in the partial hydrogenation of acetylene was evaluated by Lizana et al. [46] that studied the influence of synthesis parameters on textural properties of the nitrides and its effect on the catalytic performance. The results showed that selectivity of both β- and γ-Mo$_2$N was higher than over Pd-based catalysts. Also, β-Mo$_2$N which displays higher surface Mo/N ratio compared to γ-Mo$_2$N, offered lower selectivity to partial hydrogenation and a two-fold higher specific acetylene hydrogenation rate.

Altarawneh et al. [47], used computational methods to study mechanism of the selective hydrogenation C$_2$H$_2$ over γ-Mo$_2$N to C$_2$H$_4$ rather than complete hydrogenation to the corresponding alkane.

Reactions take place through H$_2$ adsorption followed by dissociation. The authors obtained the modes of H$_2$ adsorption as shown in Figure 8: 3-fold hollow fcc (H1) and 4-fold hollow fcc (H3) sites over the (111) and (100) terminations of γ-Mo$_2$N, respectively.

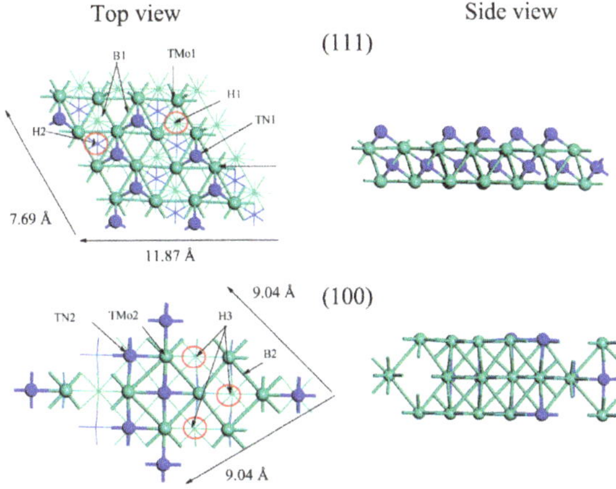

Figure 8. H$_2$ adsorption sites on Mo$_2$N (111) and (100) faces. Reproduced from [48], with permission from American Chemical Society, 2016.

In agreement with experimental results, this work seems to confirm that dissociation of H$_2$ occurs over nitrogen vacancies. It is also proposed that the lower stability of the partial hydrogenated molecule, C$_2$H$_4$ leads the selectivity.

Another interesting reaction within fine chemistry is the hydrogenation of nitroaromatic compounds, since aromatic haloamines are important intermediates in the manufacturing of drugs, pesticides, and pigments among others. The reaction has been successfully performed over Au [48], Ir [49] and Pd [50] catalysts supported over a variety of materials.

Keane et al. [51] demonstrated experimentally that Mo$_2$N improved the performance of Au in the selective hydrogenation of *p*-chloronitrobenzene (p-CNB) to *p*-chloroaniline (p-CAN) reaching 100% selectivity to p-CAN, a four-fold higher hydrogenation rate compared to Au/Al$_2$O$_3$ and showed stability upon several cycles.

The system Pd/Mo$_2$N was an effective catalyst for the hydrogenation of p-nitrophenol (PNP) to p-aminophenol (PAP) [52]. In this study, the authors synthetized Pd/Mo$_2$N nanoparticles of 2–3 nm size over SBA-15. The high dispersion improved the interaction between Pd and the nitride so that 1 wt% Pd–Mo2N/SBA-15 showed better catalytic performance than 1 wt% Pd/SBA-15 and 20 wt% Pd/SBA-15. In Figure 9 the proposed reaction mechanism over the Pd–Mo$_2$N/SBA-15 is shown.

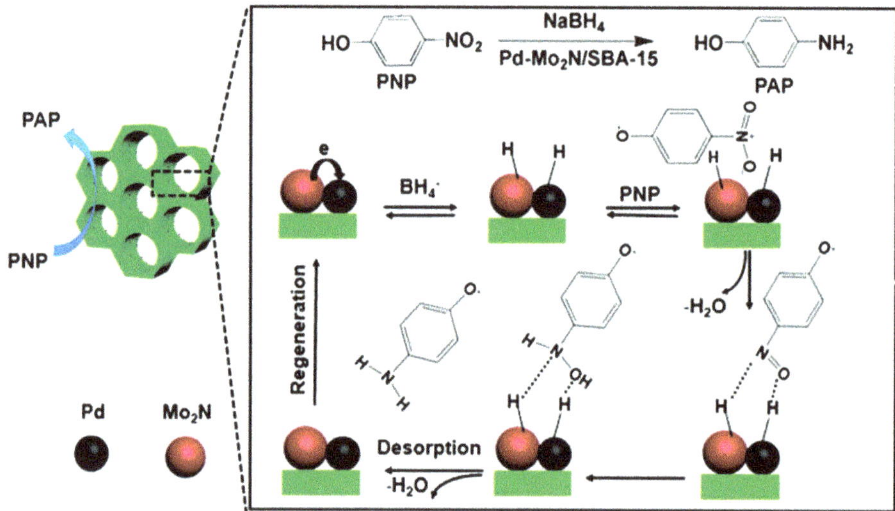

Figure 9. Catalytic conversion mechanism of PNP into PAP over the Pd–Mo$_2$N/SBA-15 hybrids in the presence of NaBH$_4$. Reproduced from [53], with permission from American Chemical Society, 2018.

Wu et al. [53] reported a green solvent-free synthesis method for CoNx entities supported on doped mesoporous carbon materials (CoNx-OMC) with surface areas in the range 678–1250 m^2/g, high N content (4.3–10.8 wt%) and rich in CoNx sites as verified by XPS. The optimized CoNx-OMC, thermally treated at 800 °C) catalyst showed an interesting catalytic performance on the hydrogenation of several nitro compounds, i.e., 100% conversion, almost 100% selectivity and stable upon recycling, under mild conditions (5 bar H$_2$ pressure, 110 °C). According to the catalytic and characterization results, a synergy effect is reached between CoNx sites and the nitrogen doped support. On the one hand, the CoNx entity provides specific sites for the adsorption and activation of nitro groups. On the other hand, the nitrogen heteroatoms of the support act as anchoring sites, increasing the dispersion of the active components and facilitating mass transportation. Also, it was confirmed that cobalt nitride was responsible of the activity and not the metallic Co nanoparticles.

Density functional theory calculations were performed by Altarawneh et al. [54] who studied the hydrogenation of p-CNB to p-CAN over the model γ-Mo$_2$N(111) surface. The results showed that adsorption of p-CNB is thermodynamically favoured over Mo-hollow face-centered cubic (fcc) and N-hollow hexagonal close-packed (hcp) sites with adsorption energies of −32.1 and −38.5 kcal/mol, respectively. Also, the results are in agreement with previous experimental reports that described a high selectivity to p-CAN at low temperatures, the direct path being preferential versus the condensation route [55].

The theoretical results suggest that activated hydrogen, H*, is transferred from both fcc and hcp hollow sites to the NO/−NH groups, the hydrogenation of chloronitrosobenzene being the rate-limiting step with an energetic barrier of 55.8 kcal/mol. Also, the high energy barrier for direct fission of the C−Cl bond excludes the formation of aniline.

Another industrial transformation of acetylene is the hydrochlorination to obtain vinyl chloride monomer (VCM), the basic unit of polyvinyl chloride (PVC). Industrially this reaction is performed using HgCl$_2$ as catalyst, however due to environmental and health concerns, alternative systems have been evaluated [56]. Among them, noble metal, i.e., Au, Pd, Pt and Ru over activated carbon, have been the most studied, gold being the best in terms of activity and selectivity.

Since it has been previously found that nitrogen doping can promote adsorption for HCl [57], which is the VCM synthesis reaction rate-controlling step, a potential catalytic system is that consisting of metal nitrides.

Dai et al. [58] studied V, Mo, and W nitrides (10 wt% metal loading) supported on activated carbon. Their experiments showed that VN/AC offered very low activity in the reaction. Mo_2N was initially the most active, but deactivated with time to give lower conversion values than those reached with W_2N/AC. The selectivity to C_2H_3Cl was similar for all the tested nitrides and reached near 100%. To further explain these results, TPD of the reactants, C_2H_2 and HCl, was performed and showed that W_2N/AC dislayed a stronger and similar interaction with both reactants, compared to the other catalysts. However, Mo_2N/AC which deactivates during reaction, showed an easy desorption of HCl and difficult desorption of C_2H_2, this producing significant amounts of coke which can be responsible of the observed deactivation.

The authors also studied binary Mo and Ti nitrides with different Mo/Ti ratios supported on activated carbon for acetylene hydrochlorination [59]. All the binary nitrides displayed better catalytic performance compared to the mononitrides and a Mo/Ti ratio of 3 was optimal among the studied systems offering 89% conversion and selectivity over 98.5%. Apparently a synergy effect among Mo and Ti occurs so that adsorption of HCl is enhanced while adsorption of acetylene is reduced.

Other chloro compounds have been evaluated with TMN. For example Keane et al. [60] studied the gas phase hydrodechlorination of 1,3-dichlorobenzene (1,3-DCB) using molybdenum and tungsten carbide (Mo_2C, W_2C) and nitride (Mo_2N). fcc-Mo_2N showed better activity (by a factor of 20) compared to pure hcp- and fcc-carbides, that displayed similar activity.

3.2. Oxidation

The selective oxidation of carbon monoxide with low concentration of O_2 (CO-PROX) is used to purify hydrogen rich streams obtained from hydrocarbon reforming. So far, the most studied and active system are the CuO–CeO_2 and Pt based catalyst [61].

With the aim of finding more economical active and selective catalysts, the catalytic performance of several transition metal nitrides have been assessed. For example, Yang et al. [62] studied the effect of Co loading (1 to 10 wt%) on Co_4N supported on γ-Al_2O_3. The cobalt nitrides displayed similar activities compared Pt-group metals in the temperature range 200–220 °C. The sample 3 wt% Co/γ-Al_2O_3 offered the best activity and selectivity in PROX reaction which could be related to the higher concentration nitrogen vacancies in the near-surface that enhance the adsorption of reactants.

Selective oxidation of alcohols plays an important role in many industrial transformations such as energy conversion and storage or the production of fine chemicals. The most studied and active catalysts are supported Au Pt, Pd, Ag and Ru and other non-noble-metal catalysts such as Co and Cu [63].

In order to test more economically viable alternatives Deng et al. [64] evaluated the catalytic activity of iron nitride, FeN_4, supported on graphene in the oxidation of benzene using H_2O_2 as oxidant. The characterization suggested that atomically dispersed FeN_4 were obtained and these entities were also stable after reaction. The catalyst reached 23.4% conversion and 18.7% yield of phenol at room temperature, however no comparison with other catalyst is given.

Later, Yuan et al. [65] studied several metal nitrides (MNx/C-T, M = Fe, Co, Cu, Cr, and Ni) synthetized at different pyrolysis temperatures, T, in the oxidation of unsaturated alcohols. The most active and selective catalyst to the corresponding aldehydes was the iron nitride. Among them, the catalysts prepared by thermal treatment at 900 °C displayed the better catalytic performance in the selective oxidation of HMF to DFF with almost complete conversion and selectivity exceeding 97%. According to the characterization performed the authors suggested that when thermal treatment was performed at lower temperatures, c.a. 600 °C, a higher concentration of nitrogen doped carbon was formed, in detriment of iron nitride; and that nitrogen doped carbon offered lower activity. In contrast, the formation of FeN_4 was favored at higher synthesis temperatures, resulting in materials that offered better activity. Moreover, the activity of the recycled catalysts could be restored by thermal treatment under NH_3/N_2.

With the aim of gaining more insight into the active sites of the Fe−N−C catalysts. Zhang et al. [9] studied atomically dispersed Fe−N−C catalyst synthetized using nano-MgO as a template. The catalysts were tested in the selective oxidation of the C−H bond of a wide range of aromatics, heterocyclic, and aliphatic alkanes at room temperature. The catalysts showed high activity and selectivity of up to 99%, as well as great reusability. The atomical dispersion of FeNx (x = 4 − 6) was verified using sub-Ångström-resolution HAADF-STEM along with XPS, XAS, ESR, and Mössbauer spectroscopy.

The authors also reported that the concentration of each FeNx species depends on the pyrolysis temperature. Among the studied samples, the most active was Fe−N−C-700 which is comprised of high-spin FeN_6 (28.3%), low-spin FeN_6 (53.8%), and medium-spin FeN_5 (17.9%) species, as shown in Figure 10. This latter is over one order of magnitude more active than the other two species. Upon increasing the pyrolysis temperature, the concentration of FeN_5 decreases to less than 10%, leading to lower activity.

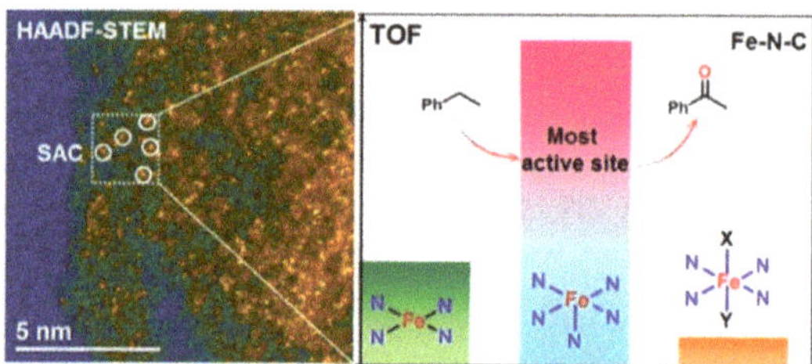

Figure 10. Reproduced from [9], with permission from American Chemical Society, 2017.

3.3. Ammonia Synthesis and Decomposition

The production of ammonia through the Haber−Bosch Process was a technological breakthough of the last century since it is a relatively easy way of obtaining a synthetic fertiliser. The process uses promoted iron catalyst to produce ammonia from N_2 and H_2 at temperatures of ca 400 °C and high pressure of around 100–200 atmospheres. The process of ammonia synthesis is so relevant that it currently consumes near 1–2% of the world's energy demand [66].

This data reveals that more efforts can be made to optimize the process as small changes on the reaction conditions could have a huge impact on global energy consumption. Thus, despite being a well-known process, several attempts have been performed to obtain more active and stable catalysts to work under milder conditions [67].

In the last decades binary and ternary transition metal nitrides of the type Mo, Co and Fe have been tested in both ammonia decomposition and ammonia synthesis. Simulations and experimental tests have shown that cobalt molybdenum nitride, Co_3Mo_3N can be the most active catalyst for ammonia synthesis, superior to the industrial iron catalyst and promoted ruthenium catalyst [68,69].

The synthesis of ammonia catalyzed by molybdenum nitride seems to be a structure-sensitive reaction so that the bigger the particle size, the higher the intrinsic activity of the catalyst [70]. On the other hand, the effect of the nanoparticle morphology does not seem to be so clear. Sun et al. studied the synthesis of ammonia over plate-like γ-Mo_2N and nanorod β-Mo_2N and γ-Mo_2N [71] and did not find significant differences among them. Also, the synthesis conditions limit the comparison since residual sulphur from the molybdenum precursor, Mo_2S, can poison the catalyst.

In order to increase the dispersion of the nitrides and improve the stability, Ding et al. [72] synthetized molybdenum nitride supported on HZSM-5, using MoOx as precursor and NH_3 as nitriding agent at 973 K. In this way molybdenum oxide exchanges with hydroxyl groups on the zeolite surface and obtained a Mo to N ratio close to 2. The authors proved that this supported molybdenum nitrides are more stable against oxidation by comparing with unsupported catalysts.

The resulting catalyst, MoNx/ZSM-5, displayed excellent activity in the ammonia synthesis under ambient pressure. A kinetic approach showed that the apparent activation energy for ammonia synthesis on MoNx/ZSM-5 is around 20% lower than that on bulk Mo_2N (9.8 kcal/mol vs. 12.4 kcal/mol).

Interestingly, the effect of reaction pressure was different on bulk Mo_2N and MoNx/ZSM-5, i.e., higher reaction pressures are more favorable on the MoNx/ZSM-5 catalyst than on γ-Mo_2N. The authors suggested that this difference could be due to the interaction of nitrogen with the zeolite framework, acting as an equivalent pressure which may enhance the pressure effect on the ammonia synthesis. Another significant finding of this work is that upon increasing the Si/Al ratio of the zeolite, the catalytic activity of MoNx increases.

Recently, Catlow et al. [73] used DFT calculations to evaluate the associative and dissociative mechanisms of ammonia synthesis over Co_3Mo_3N. The results showed that in the associative mechanism, Eley-Rideal/Mars van Krevelen, hydrogen reacts with nitrogen adsorbed and activated on the surface to generate ammonia.

Promotion of cobalt molybdenum nitrides with other elements have been also studied. According to Paweł Adamski [74] chromium and potassium were able to generate a well-developed porous structure, increasing the activity of the catalysts in ammonia synthesis by 50% compared to the non-promoted catalyst.

Ammonia decomposition has become a strategic research topic since it allows to obtain CO free hydrogen to feed fuel cells. Ammonia can also be employed directly as fuel in vehicles since it has a high energy density (8.9 kW h/kg), is easily liquified at room temperature and low pressure, i.e., less than 10 bar, and its narrow combustion range allows safe operation.

Many cataytic monometallic and bimetallic systems have been tested based on Fe, Ni, Co, Ru, Ir, Pt or Rh, supported on several materials. Among them, ruthenium catalysts supported on different carbon materials such as active carbon, carbon nanotubes showed the highest activity.

In all cases, researchers observed a low activity for ammonia decomposition reaction at temperatures below 400 °C, since recombinative desorption rate of the adsorbed N atoms from active metals is slow at those conditions [75]. Also, hydrogen molecule seems to cover active sites over Ni- and Ru catalysts, this hindering the ammonia decomposition reaction. Hence, despite being well studied systems, the temperatures required and the use of noble metals should be avoided to make the process technical and economically feasible.

Eguchi et al. [76], studied the effect of a second transition metal on Mo nitride based catalysts, prepared by temperature-programmed reaction under NH_3 flow of the oxides precursors: MoO_3, $CoMoO_4$, $NiMoO_4$, and $FeMoO_4$. Incorporation of the second metal into the Mo nitride resulted in a significant decrease in the surface area (3.1–8.8 vs. 80 of Mo_2N). However, the area of Mo_2N was reduced to 23 m^2/g after reaction, indicating that the material was not stable under the reaction conditions. Despite the lower surface area, the addition of a second metal was beneficial for ammonia decomposition and the activity followed the trend Co_3Mo_3N > Ni_3Mo_3N > Fe_3Mo_3N > Mo_2N. Several transition metal nitrides have been studied for ammonia decomposition, being the binary system metal-cobalt molybdenum nitride the most interesting among them [77–81].

Based on the NH_3-TPSR results, the authors suggested that the addition of Co and Fe favoured the desorption of hydrogen. However, over Mo_2N de desorption of nitrogen was slower since nitrogen atoms tend to interact stronger with the nitride. According to the results the authors concluded that the order in which each metal nitride system dissociates metal nitride–N bond was Co_3Mo_3N ≈ Ni_3Mo_3N > Fe_3Mo_3N > Mo_2N. Moreover, the presence of Co, Ni, and Fe improves the stability against poisoning by H_2 on the active sites and the best results are obtained when doped catalysts are employed.

In order to improve the surface area of nitrides, alternative synthesis paths have been explored. For example Podila et al. [77] studied the use of citric acid as chelating agent to prepare bulk Co_3Mo_3N, with surface areas in the range 93–129 m^2/g. The use of citric acid (CA) as chelating agent afforded better nitride dispersion which resulted in better catalytic performance. An optimal concentration of citric acid was related to a higher surface area, lower particle size and increased proportion of

Mo$_2$N and Co$_3$Mo$_3$N phases, so that when the CA/Mo ratio was changed from 1 to 3 in the synthesis, the conversion increased from 75% to 97% at 550 °C.

Similarly Zaman et al. synthetized nickel [78] and cobalt molybdenum nitrides [79] using citric acid and compared its catalytic performance on ammonia decomposition with that obtained employing γ-Mo$_2$N. Under these synthesis conditions, the surface area of the binary nitrides was increased but still below 20 m^2/g. According to the results both catalyst, Ni$_2$Mo$_3$N and Co$_3$Mo$_3$N offered over 97% conversion, while the use of pure γ-Mo$_2$N resulted in 50–70% conversion under the same experimental conditions.

Zhao et al. [80] used supported binary CoMo nitrides over several porous materials with different physico-chemical characteristics: CNTs, Al$_2$O$_3$, activated carbon and 5A Zolite. The experimental procedure was a simple impregnation of the precursors followed by a temperature-programmed reaction in N$_2$–H$_2$. The activity followed the order: CoMoNx/CNTs > CoMoNx/Zeolite 5 > CoMoNx/AC > CoMoNx/Al2O3. However, despite the clear better performance of CNT, the supports differ in many features such as surface chemical composition and morphology, making the comparison and related conclusions quite difficult.

The effect of synthesis conditions, nitridation temperature and iron loading was also evaluated in iron nitrides supported on carbon nanotubes [81]. A higher synthesis temperature of ca. 500 °C and Fe loading of 10% prepared under NH$_3$ flow resulted in well-dispersed Fe$_2$N nanoparticles which exists along with Fe$_2$O$_3$ entities and offered the best ammonia conversion among the catalysts tested. In contrast the synthesis under a N$_2$/H$_2$ flow resulted in the formation of both Fe$_2$N and Fe$_4$N, this latter species being detrimental for ammonia decomposition.

4. Conclusions

This short review of the most recent literature has shown that transition metal nitrides possess a wide spectrum of catalytic applications with interesting catalytic performance. The review has focused on thermal heterogeneous catalysis which requires high surface areas and good dispersion of the active phase. Despite the significant advances done up to date, there is plenty of room for research on transition metal nitrides to optimize the synthesis conditions in order to obtain higher surface areas and better nanoparticles dispersion, ideally reducing the synthesis temperature.

Also, further work to improve the stability will be required in order to obtain potential industrial catalysts. One of the main reason for deactivation is the oxidation of the nitrides, which is likely to occur in reactions that generate water such as carbon dioxide hydrogenation, or when water is already in the reactants mixtures as it happens with biomass transformations. In this sense, it seems that subsurface hydrogen can delay deactivation and more insight into this reaction mechanism would allow to propose regeneration mechanism.

Again, the use of a high surface area support for transition metal nitrides nanoparticles can improve the dispersion of the active phase and potentially improve their stability upon reaction conditions against sintering and oxidation.

The improved catalytic performance that has been reached with more complex systems that incorporate a second or third metal, should be complemented with deeper understanding of the actual active phase and the chemical structure of the nitrides. Similarly, the use of promoters like alkalis and its effect of the structure need to be further studied since these materials have also demonstrated a significant potential for future catalytic applications. However, there is still no clear correlation mainly due to the complexity of these systems and difficulties to perform in situ investigations.

Funding: A.B.Dongil acknowledges financial support from Fundación General CSIC (Programa ComFuturo).

Acknowledgments: A.B. Dongil acknowledges financial support from Fundación General CSIC (Programa ComFuturo).

Conflicts of Interest: The author declares no conflicts of interest.

References

1. Inagaki, M.; Toyoda, M.; Soneda, Y.; Morishita, T. Nitrogen-doped carbon materials. *Carbon* **2018**, *132*, 104–140. [CrossRef]
2. Thandavarayan, H.W.; Wang, M.X. Review on Recent Progress in Nitrogen-Doped Graphene: Synthesis, Characterization, and Its Potential Applications. *ACS Catal.* **2012**, 25781–25794.
3. Nagai, M. Transition-metal nitrides for hydrotreating catalyst—Synthesis, surface properties, and reactivities. *Appl. Catal. A Gen.* **2007**, *322*, 178–190. [CrossRef]
4. Dinh, K.N.; Liang, Q.; Du, C.F.; Zhao, J.; Tok, A.I.Y.; Mao, H.; Yan, Q. Nanostructured metallic transition metal carbides, nitrides, phosphides, and borides for energy storage and conversion. *Nano Today* **2019**, *25*, 99–121. [CrossRef]
5. Volpe, L.; Boudart, M.J. Compounds of molybdenum and tungsten with high specific surface area: I. Nitrides. *Solid State Chem.* **1985**, *59*, 332–347. [CrossRef]
6. Zhong, Y.; Xia, X.; Shi, F.; Zhan, J.; Tu, J.; Fan, H.J. Transition Metal Carbides and Nitrides in Energy Storage and Conversion. *Adv. Sci.* **2016**, *3*, 1500286. [CrossRef] [PubMed]
7. Hargreaves, J.S.J.; McFarlane, A.R.; Laassiri, S. (Eds.) *Alternative Catalytic Materials: Carbides, Nitrides, Phosphides and Amorphous Boron Alloys*; Metal Nitride Catalysts Chapter 5; Royal Society of Chemistry: London, UK, 2018.
8. Lee, J.S.; Ham, D.J. *Metal Nitrides, Encyclopedia of Catalysis*; Wiley: Hoboken, NJ, USA, 2010.
9. Liu, W.; Zhang, L.; Liu, X.; Liu, X.; Yang, X.; Miao, S.; Wang, W.; Wang, A.; Zhang, T. Discriminating Catalytically Active FeNx Species of Atomically Dispersed Fe–N–C Catalyst for Selective Oxidation of the C–H Bond. *J. Am. Chem. Soc.* **2017**, *139*, 10790–10798. [CrossRef] [PubMed]
10. Rounaghi, S.A.; Vanpoucke, D.E.P.; Esmaeili, E.; Scudino, S.; Eckert, J. Synthesis, characterization and thermodynamic stability of nanostructured ε-iron carbonitride powder prepared by a solid-state mechanochemical route. *J. Alloys Compd.* **2019**, *778*, 327–336. [CrossRef]
11. Roldan, M.A.; López-Flores, V.; Alcala, M.D.; Ortega, A.; Real, C. Mechanochemical synthesis of vanadium nitride. *J. Eur. Ceram. Soc.* **2010**, *30*, 2099–2107. [CrossRef]
12. Rasaki, S.A.; Zhang, B.; Thomas, K.A.T.; Yang, M. Synthesis and application of nano-structured metal nitrides and carbides: A review. *Prog. Solid State Chem.* **2018**, *50*, 1–15. [CrossRef]
13. Markel, E.J.; Burdick, S.E.; Leaphart, M.E.; Roberts, K.L. Synthesis, Characterization, and Thiophene Desulfurization Activity of Unsupported γ-Mo2N Macrocrystalline Catalysts. *J. Catal.* **1999**, *182*, 136–147. [CrossRef]
14. Wise, R.S.; Markel, E.J. Synthesis of High Surface Area Molybdenum Nitride in Mixtures of Nitrogen and Hydrogen. *J. Catal.* **1994**, *145*, 344–355. [CrossRef]
15. Giordano, C.; Erpen, C.; Yao, W.; Milke, B.; Antonietti, M. Metal Nitride and Metal Carbide Nanoparticles by a Soft Urea Pathway. *Chem. Mater.* **2009**, *21*, 5136–5144. [CrossRef]
16. Giordano, C.; Erpen, C.; Yao, W.; Antonietti, M. Synthesis of Mo and W Carbide and Nitride Nanoparticles via a Simple "Urea Glass" Route. *Nano Lett.* **2008**, *8*, 4659–4663. [CrossRef] [PubMed]
17. Salamat, A.; Hector, A.L.; Kroll, P.; McMillan, P.F. Nitrogen-rich transition metal nitrides. *Coord. Chem. Rev.* **2013**, *257*, 2063. [CrossRef]
18. Wang, T.; Yan, Z.; Michel, C.; Pera-Titus, M.; Sautet, P. Trends and Control in the Nitridation of Transition-Metal Surfaces. *ACS Catal.* **2018**, *8*, 63–68. [CrossRef]
19. Crowhurst, J.C.; Goncharov, A.F.; Sadigh, B.; Evans, C.L.; Morrall, P.G.; Ferreira, J.L.; Nelson, A.J. Synthesis and characterization of the nitrides of platinum and iridium. *Science* **2006**, *311*, 1275–1278. [CrossRef] [PubMed]
20. Crowhurst, J.C.; Goncharov, A.F.; Sadigh, B.; Zaug, J.M.; Aberg, D.; Meng, Y.; Prakapenka, V.B. Synthesis and characterization of nitrides of iridium and palladium. *J. Mater. Res.* **2008**, *23*, 1–5. [CrossRef]
21. Hargreaves, J.S.J. Heterogeneous catalysis with metal nitrides Coordination. *Chem. Rev.* **2013**, *257*, 2015–2031.
22. Álvarez, A.; Bansode, A.; Urakawa, A.; Bavykina, A.V.; Wezendonk, T.A.; Makkee, M.; Gascon, J.; Kapteijn, F. Challenges in the Greener Production of Formates/Formic Acid, Methanol, and DME by Heterogeneously Catalyzed CO_2 Hydrogenation Processes. *Chem. Rev.* **2017**, *117*, 9804–9838. [CrossRef]
23. Din, I.U.; Shaharun, M.S.; Alotaibi, M.A.; Alharthi, A.I.; Naeem, A. Recent developments on heterogeneous catalytic CO2 reduction to methanol. *J. CO_2 Utilizat.* **2019**, *34*, 20–33. [CrossRef]

24. Zaman, S.F.; Pasupulety, N.; Al-Zahrania, A.A.; Daousa, M.A.; Al-Shahrania, S.S.; Driss, H.; Petrova, L.A.; Smith, K.J. Carbon monoxide hydrogenation on potassium promoted Mo2N catalysts. *Appl. Catal. A Gener.* **2017**, *532*, 133–145. [CrossRef]

25. Zaman, S.F.; Pasupulety, N.; Al-Zahrani, A.A.; Daous, M.A.; Driss, H.; Al-Shahrani, S.S.; Petrov, L. Influence of alkali metal (Li and Cs) addition to Mo2N catalyst for CO hydrogenation to hydrocarbons and oxygenates. *Can. J. Chem. Eng.* **2018**, *96*, 1770–1779. [CrossRef]

26. Rönscha, S.; Schneidera, J.; Matthischke, S.; Schlütera, M.; Götz, M.; Lefebvre, J.; Prabhakaran, P.; Bajohr, S. Review on methanation—From fundamentals to current projects. *Fuel* **2016**, *166*, 276–296. [CrossRef]

27. Wang, S.; Ge, H.; Sun, S.; Zhang, J.; Liu, F.; Wen, X.; Yu, X.; Wang, L.; Zhang, Y.; Xu, H.; et al. A New Molybdenum Nitride Catalyst with Rhombohedral MoS$_2$ Structure for Hydrogenation Applications. *J. Am. Chem. Soc.* **2015**, *137*, 4815–4822. [CrossRef] [PubMed]

28. Leybo, D.V.; Kosova, N.I.; Chuprunov, K.O.; Kuznetsov, D.V.; Kurzina, I.A. Bimetallic Ni-Mo Nitride as the Carbon Dioxide Hydrogenation Catalyst. *Adv. Mater. Res.* **2014**, *872*, 3–9. [CrossRef]

29. Razzaq, R.; Li, C.; Usman, M.; Suzuki, K.; Zhan, S. A highly active and stable Co4N/γ-Al2O3 catalyst for CO and CO$_2$ methanation to produce synthetic natural gas (SNG). *Chem. Eng. J.* **2015**, *262*, 1090–1098. [CrossRef]

30. Jahangiri, H.; Bennett, J.; Mahjoubi, P.; Wilson, K.; Gua, S. A review of advanced catalyst development for Fischer–Tropsch synthesis of hydrocarbons from biomass derived syn-gas. *Catal. Sci. Technol.* **2014**, *4*, 2210–2229. [CrossRef]

31. Yang, Z.; Guo, S.; Pan, X.; Wang, J.; Bao, X. A review of advanced catalyst development for Fischer–Tropsch synthesis of hydrocarbons from biomass derived syn-gas, FeN nanoparticles confined in carbon nanotubes for CO hydrogenation. *Energy Environ. Sci.* **2011**, *4*, 4500. [CrossRef]

32. Ordomsky, V.V.; Carvalho, A.; Legras, B.; Paul, S.; Khodakov, A.Y. Effects of co-feeding with nitrogen-containing compounds on the performance of supported cobalt and iron catalysts in Fischer–Tropsch synthesis. *Catal. Today* **2016**, *275*, 84–93. [CrossRef]

33. Zaman, S.F. DFT study of CO Adsorption and Dissociation over γ-Mo2N(111) Plane. *Bull. Chem. Commun.* **2015**, *47*, 125–132.

34. Aramouni, N.A.K.; Touma, J.G.; Tarboush, B.A.; Zeaiter, J.; Ahmad, M.N. Catalyst design for dry reforming of methane: Analysis review. *Renew. Sustain. Energy Rev.* **2018**, *82*, 2570–2585. [CrossRef]

35. Du, X.; France, L.J.; Kuznetsov, V.L.; Xiao, T.; Edwards, P.P.; AlMegren, H.; Bagabas, A. Dry reforming of methane over ZrO2-supported Co–Mo carbide catalyst. *Appl. Petrochem. Res.* **2014**, *4*, 137–144. [CrossRef]

36. Fu, X.; Su, H.; Yin, W.; Huang, Y.; Gu, X. Bimetallic molybdenum nitride Co3Mo3N: a new promising catalyst for CO$_2$ reforming of methane. *Catal. Sci. Technol.* **2017**, *7*, 1671–1678. [CrossRef]

37. Monnier, J.; Sulimma, H.; Dalai, A.; Caravaggio, G. Hydrodeoxygenation of oleic acid and canola oil over alumina-supported metal nitrides. *Appl. Catal. A* **2010**, *382*, 176. [CrossRef]

38. Nguyen, H.S.H.; Mäki-Arvela, P.; Akhmetzyanova, U.; Tišler, Z.; Hachemi, I.; Rudnäs, A.; Smeds, A.; Eränen, K.; Aho, A.; Kumar, N.; et al. Direct hydrodeoxygenation of algal lipids extracted from Chlorellaalga. *J. Chem. Technol. Biotechnol.* **2017**, *92*, 741–748. [CrossRef]

39. Liu, X.; Xu, L.; Xu, G.; Jia, W.; Ma, Y.; Zhang, Y. Selective Hydrodeoxygenation of Lignin-Derived Phenols to Cyclohexanols or Cyclohexanes over Magnetic CoNx@NC Catalysts under Mild Conditions. *ACS Catal.* **2016**, *6*, 7611–7620. [CrossRef]

40. Cheng, T.; Yu, H.; Peng, F.; Wang, H.; Zhang, B.; Su, D. Identifying active sites of CoNC/CNT from pyrolysis of molecularly defined complexes for oxidative esterification and hydrogenation reactions. *Catal. Sci. Technol.* **2016**, *6*, 1007. [CrossRef]

41. Boullosa-Eiras, S.; Lødeng, R.; Bergem, H.; Stöcker, M.; Hannevold, L.; Blekkana, E.A. Catalytic hydrodeoxygenation (HDO) of phenol over supported molybdenum carbide, nitride, phosphide and oxide catalysts. *Catal. Today* **2014**, *223*, 44–53. [CrossRef]

42. Junga, A.; Jessa, A.; Schubert, T.; Schütz, W. Performance of carbon nanomaterial (nanotubes and nanofibres) supported platinum and palladium catalysts for the hydrogenation of cinnamaldehyde and of 1-octyne. *Appl. Catal. A Gen.* **2009**, *362*, 95–105. [CrossRef]

43. Dongil, A.B.; Bachiller-Baeza, B.; Guerrero-Ruiz, A.; Rodríguez-Ramos, I. Chemoselective hydrogenation of cinnamaldehyde: A comparison of the immobilization of Ru–phosphine complex on graphite oxide and on graphitic surfaces. *J. Catal.* **2011**, *282*, 299–309. [CrossRef]

44. Wang, D.; Zhu, Y.; Tian, C.; Wang, L.; Zhou, W.; Dong, Y.; Han, Q.; Liu, Y.; Yuana, F.; Fu, H. Synergistic effect of Mo_2N and Pt for promoted selective hydrogenation of cinnamaldehyde over Pt–Mo_2N/SBA-15. *Catal. Sci. Technol.* **2016**, *6*, 2403. [CrossRef]

45. Wyvratt, B.M.; Gaudet, J.R.; Pardue, B.; Marton, A.; Rudić, S.; Mader, E.A.; Cundari, T.R.; Mayer, J.M.; Thompson, L.T. Reactivity of Hydrogen on and in Nanostructured Molybdenum Nitride: Crotonaldehyde Hydrogenation. *ACS Catal.* **2016**, *6*, 5797–5806. [CrossRef]

46. Cárdenas-Lizana, F.; Lamey, D.; Kiwi-Minsker, L.; Keane, M.A. Molybdenum nitrides: a study of synthesis variables and catalytic performance in acetylene hydrogenation. *J. Mater. Sci.* **2018**, *53*, 6707–6718. [CrossRef]

47. Jaf, Z.N.; Altarawneh, M.; Miran, H.A.; Jianga, Z.T.; Dlugogorski, B.Z. Mechanisms governing selective hydrogenation of acetylene over γ-Mo_2N surfaces. *Catal. Sci. Technol.* **2017**, *7*, 943. [CrossRef]

48. Cárdenas-Lizana, F.; de Pedro, Z.M.; Gómez-Quero, S.; Keane, M.A. Gas phase hydrogenation of nitroarenes: A comparison of the catalytic action of titania supported gold and silver. *J. Mol. Catal. A Chem.* **2010**, *326*, 48–54. [CrossRef]

49. Dongil, A.B.; Rivera-Cárcamo, C.; Pastor-Pérez, L.; Sepúlveda-Escribano, A.; Reyes, P. Ir supported over carbon materials for the selective hydrogenation of nitrocompounds. *Catal. Today* **2015**, *249*, 72–78. [CrossRef]

50. Dongil, A.B.; Pastor-Pérez, L.; Fierro, J.L.G.; Escalona, N.; Sepúlveda-Escribano, A. Synthesis of palladium nanoparticles over graphite oxide and carbon nanotubes by reduction in ethylene glycol and their catalytic performance on the chemoselective hydrogenation of para-chloronitrobenzene. *Appl. Catal. A Gen.* **2016**, *513*, 89–97. [CrossRef]

51. Perret, N.; Cárdenas-Lizana, F.; Lamey, D.; Laporte, V.; Kiwi-Minsker, L.; Keane, M.A. Effect of Crystallographic Phase (β vs. γ) and Surface Area on Gas Phase Nitroarene Hydrogenation over Mo2N and Au/Mo2N. *Top. Catal.* **2012**, *55*, 955–968. [CrossRef]

52. Cheng, X.; Wang, D.; Liu, J.; Kang, X.; Yan, H.; Wu, A.; Gu, Y.; Tian, C.; Fu, H. Ultra-small Mo2N on SBA-15 as a highly efficient promoter of low-loading Pd for catalytic hydrogenation. *Nanoscale* **2018**, *10*, 22348. [CrossRef]

53. Wei, X.; Zhang, Z.; Zhou, M.; Zhang, A.; Wu, W.D.; Wu, Z. Solid-state nanocasting synthesis of ordered mesoporous CoNx–carbon catalysts for highly efficient hydrogenation of nitro compounds. *Nanoscale* **2018**, *10*, 16839–16847. [CrossRef]

54. Jaf, Z.N.; Altarawneh, M.; Miran, H.A.; Almatarneh, M.H.; Jiang, Z.-T.; Dlugogorski, B.Z. Catalytic Hydrogenation of p-Chloronitrobenzene to p-Chloroaniline Mediated by γ-Mo2N. *ACS Omega* **2018**, *3*, 14380–14391. [CrossRef]

55. Song, J.; Huang, Z.-F.; Pan, L.; Li, K.; Zhang, X.; Wang, L.; Zou, J.-J. Review on selective hydrogenation of nitroarene by catalytic, photocatalytic and electrocatalytic reactions. *Appl. Catal. B Environ.* **2018**, *227*, 386–408. [CrossRef]

56. Davies, C.; Miedziak, P.J.; Brett, G.L.; Hutchings, G.J. Vinyl chloride monomer production catalysed by gold: A review 2016. *Chin. J. Catal.* **2016**, *37*, 1600–1607. [CrossRef]

57. Dai, B.; Chen, K.; Wang, Y.; Kang, L.; Zhu, M. Boron and Nitrogen Doping in Graphene for the Catalysis of Acetylene Hydrochlorination. *ACS Catal.* **2015**, *5*, 2541–2547. [CrossRef]

58. Dai, H.; Zhu, M.; Zhang, H.; Yu, F.; Wang, C.; Dai, B. Activated carbon supported VN, Mo2N, and W2N as catalysts for acetylene hydrochlorination. *J. Ind. Eng. Chem.* **2017**, *50*, 72–78. [CrossRef]

59. Dai, H.; Zhu, M.; Zhang, H.; Yu, F.; Wang, C.; Dai, B. Activated Carbon Supported Mo-Ti-N Binary Transition Metal Nitride as Catalyst for Acetylene Hydrochlorination. *Catalysts* **2017**, *7*, 200. [CrossRef]

60. Jujjuri, S.; Cárdenas-Lizana, F.; Keane, M.A. Synthesis of group VI carbides and nitrides: Application in catalytic hydrodechlorination. *J. Mater. Sci.* **2014**, *49*, 5406–5417. [CrossRef]

61. Mohammad Al Soubaihi, R.; Saoud, K.M.; Dutta, J. Critical Review of Low-Temperature CO Oxidation and Hysteresis Phenomenon on Heterogeneous Catalysts. *Catalysts* **2018**, *8*, 660. [CrossRef]

62. Yao, Z.; Zhang, X.; Peng, F.; Yu, H.; Wang, H.; Yang, J. Novel highly efficient alumina-supported cobalt nitride catalyst for preferential CO oxidation at high temperatures. *Int. J. Hydrogen Energy* **2011**, *36*, 1955–1959. [CrossRef]

63. Davis, S.E.; Ide, M.S.; Davis, R.J. Selective oxidation of alcohols and aldehydes over supported metal nanoparticles. *Green Chem.* **2013**, *15*, 17–45. [CrossRef]

64. Deng, D.H.; Chen, X.Q.; Yu, L.; Wu, X.; Liu, Q.F.; Liu, Y.X.; Yang, H.; Tian, H.F.; Hu, Y.F.; Du, P.P.; et al. A single iron site confined in a graphene matrix for the catalytic oxidation of benzene at room temperature. *Sci. Adv.* **2015**, *1*, e1500462. [CrossRef]

65. Zhang, J.; Nagamatsu, S.; Du, J.; Tong, C.; Fang, H.; Deng, D.; Liu, X.; Asakura, K.; Yuan, Y. A study of FeNx/C catalysts for the selective oxidation of unsaturated alcohols by molecular oxygen. *J. Catal.* **2018**, *367*, 16–26. [CrossRef]

66. Ammonia—A Renewable Fuel Made from Sun, Air, and Water—Could Power the Globe without Carbon. Available online: https://www.sciencemag.org/news/2018/07/ammonia-renewable-fuel-made-sun-air-and-water-could-power-globe-without-carbon (accessed on 12 July 2018).

67. Hargreaves, J.S.J. Nitrides as ammonia synthesis catalysts and as potential nitrogen transfer reagents. *Appl. Petrochem. Res.* **2014**, *4*, 3–10. [CrossRef]

68. Jacobsen, C.J.H. Novel class of ammonia synthesis catalysts. *Chem. Commun.* **2000**, 1057–1058. [CrossRef]

69. Jacobsen, C.J.H.; Dahl, S.; Clausen, B.S.; Bahn, S.; Logadottir, A.; Norskov, J.K. Catalyst design by interpolation in the periodic table: Bimetallic ammonia synthesis catalysts. *J. Am. Chem. Soc.* **2001**, *123*, 8404–8405. [CrossRef]

70. Kojima, R.; Aika, K. Molybdenum nitride and carbide catalysts for ammonia synthesis. *Appl. Catal. A Gener.* **2001**, *219*, 141–147. [CrossRef]

71. Mckay, D.; Hargreaves, J.S.J.; Rico, J.L.; Rivera, J.L.; Sun, X.L. The influence of phase and morphology of molybdenum nitrides on ammonia synthesis activity and reduction characteristics. *J. Solid State Chem.* **2008**, *181*, 325–333. [CrossRef]

72. Liu, N.; Nie, L.; Xue, N.; Dong, H.; Peng, L.; Guo, X.; Ding, W. Catalytic Ammonia Synthesis over Mo Nitride/ZSM-5. *ChemCatChem* **2010**, *2*, 167–174. [CrossRef]

73. Zeinalipour-Yazdi, C.D.; Hargreaves, J.S.J.; Catlow, C.A. Low-T Mechanisms of Ammonia Synthesis on Co3Mo3N. *Phys. Chem. C* **2018**, *122*, 6078–6082. [CrossRef]

74. Moszyński, D.; Adamski, P.; Nadziejko, M.; Komorowska, A.; Sarnecki, A. Cobalt molybdenum nitrides co-promoted by chromium and potassium as catalysts for ammonia synthesis. *Chem. Pap.* **2018**, *72*, 425–430. [CrossRef]

75. Lamba, K.E.; Dolana, M.D.; Kennedy, D.F. Ammonia for hydrogen storage; A review of catalytic ammonia decomposition and hydrogen separation and purification. *Int. J. Hydrogen Energy* **2019**, *44*, 3580–3593. [CrossRef]

76. Srifa, A.; Okura, K.; Okanishi, T.; Muroyama, H.; Matsuia, T.; Eguch, K. COx-free hydrogen production via ammonia decomposition over molybdenum nitride-based catalysts. *Catal. Sci. Technol.* **2016**, *6*, 7495. [CrossRef]

77. Podila, S.; Zaman, S.F.; Driss, H.; Al-Zahrani, A.A.; Daous, M.A.; Petrov, L.A. High performance of bulk Mo2N and Co3Mo3N catalysts for hydrogen production from ammonia: Role of citric acid to Mo molar ratio in preparation of high surface area nitride. *Int. J. Hydrogen Energy* **2017**, *42*, 8006–8020. [CrossRef]

78. Zaman, S.F.; Jolaoso, L.A.; Podila, S.; Al-Zahrani, A.A.; Alhamed, Y.A.; Driss, H.; Daous, M.M.; Petrov, L. Ammonia decomposition over citric acid chelated γ-Mo2N and Ni2Mo3N catalysts. *Int. J. Hydrogen Energy* **2018**, *43*, 17252–17258. [CrossRef]

79. Jolaoso, L.A.; Zaman, S.F.; Podila, S.; Driss, H.; Al-Zahrani, A.A.; Daous, M.A.; Petrov, L. Ammonia decomposition over citric acid induced γ-Mo2N and Co3Mo3N catalysts. *Int. J. Hydrogen Energy* **2017**, *43*, 4839–4844. [CrossRef]

80. Zhao, Z.H.; Zou, H.B.; Lin, W.M. Effect of Supports on the Properties of Co-Mo Nitride Catalysts for Ammonia Decomposition 1271. *Adv. Mater. Res.* **2012**, *391–392*, 1215–1219.

81. Zhang, H.; Gong, Q.; Ren, S.; Ali Arshid, M.; Chu, W.; Chen, C. Implication of iron nitride species to enhance the catalytic activity and stability of carbon nanotubes supported Fe catalysts for carbon-free hydrogen production via low-temperature ammonia decomposition. *Catal. Sci. Technol.* **2018**, *8*, 907–915. [CrossRef]

 *nanomaterials*

Article

Metal–Organic Framework-Based Sustainable Nanocatalysts for CO Oxidation

Luis A. Lozano, Betina M. C. Faroldi, María A. Ulla and Juan M. Zamaro *

Instituto de Investigaciones en Catálisis y Petroquímica, INCAPE (FIQ, UNL, CONICET), Santiago del Estero 2829 (3000), Santa Fe, Argentina; llozano@fiq.unl.edu.ar (L.A.L.); bfaroldi@gmail.com (B.M.C.F.); mulla@fiq.unl.edu.ar (M.A.U.)
* Correspondence: zamaro@fiq.unl.edu.ar

Received: 17 December 2019; Accepted: 14 January 2020; Published: 17 January 2020

Abstract: The development of new catalytic nanomaterials following sustainability criteria both in their composition and in their synthesis process is a topic of great current interest. The purpose of this work was to investigate the preparation of nanocatalysts derived from the zirconium metal–organic framework UiO-66 obtained under friendly conditions and supporting dispersed species of non-noble transition elements such as Cu, Co, and Fe, incorporated through a simple incipient wetness impregnation technique. The physicochemical properties of the synthesized solids were studied through several characterization techniques and then they were investigated in reactions of relevance for environmental pollution control, such as the oxidation of carbon monoxide in air and in hydrogen-rich streams (COProx). By controlling the atmospheres and pretreatment temperatures, it was possible to obtain active catalysts for the reactions under study, consisting of Cu-based UiO-66-, bimetallic CuCo–UiO-66-, and CuFe–UiO-6-derived materials. These solids represent new alternatives of nanostructured catalysts based on highly dispersed non-noble active metals.

Keywords: UiO-66; copper; iron; cobalt; nanocatalyst; CO oxidation; COProx

---

1. Introduction

The use of metal–organic frameworks (MOFs) as host matrices to disperse metal nanoparticles is a topic of great interest in the ongoing research for new nanocatalysts. MOFs have the advantage of presenting a variety of transition metals and a wide range of organic ligands in their composition, which makes them attractive for applications in catalysis [1,2]. There are several alternatives to obtain catalytic functionality in the structure of MOFs. One is based on the coordination around metal centers or structural defects that are not committed to the material framework, which can act as active sites [3]. Another possibility is to take advantage of the ligand chemistry since terephthalates or tricarboxylates to which acidic or basic functional groups can be added are usually used [4]. Moreover, the large internal volume available in these materials can be used to host active species [5]. In addition, in recent years, the use of MOFs as templates, which according to their construction units can generate nanostructured metal/metal oxide systems [6], became a consolidated strategy. For example, Sun et al. [7] reported a number of catalysts obtained with this concept, such as Co_3O_4 materials from Co-based MOFs or α-Fe_2O_3 and Fe_3O_4 nanomaterials from Fe-MIL-88B. CuO and CuO–CeO_2 nanoparticle-based catalysts derived from the MOF HKUST-1 with high catalytic activity were also reported [8].

On the other hand, the catalytic oxidation of carbon monoxide in gaseous streams is a reaction of great environmental relevance. Carbon monoxide is one of the main pollutants of indoor and industrial environments because, due to its high affinity for hemoglobin, it is extremely toxic to living beings [9] and, therefore, numerous studies are currently being conducted for their catalytic removal [10]. Moreover, the elimination of this gas in concentrated hydrogen streams (COProx) has significance in the field of renewable energy, as it is one of the most accepted alternatives to carry out the final purification of H_2 to be used in fuel cells [11]. For the CO oxidation, numerous catalytic formulations were tested to date, and those composed of supported oxides represent one of the most promising alternatives [12]. For example, catalysts based on supported particles of copper oxides, cobalt oxides, and iron oxides showed good performance [13–15]. These types of solids avoid the use of noble metals traditionally used in this reaction such as Pt, Pd, or Au [12], some of which were also supported in MOFs [16,17]. These elements have a limited abundance and involve much higher costs; thus, efforts are being made for the development of catalysts based on non-noble metals. Currently, it is considered that the use of precious metals is not sustainable compared to earth-abundant metals which are available in orders-of-magnitude higher quantities [18]. Of special interest are the metals of the first transition period, such as Cu, Co, and Fe, since they are not only less expensive but also less toxic compared to those of the second and third period [19]. Since MOFs have high specific surface areas, they represent a new support alternative for the efficient and low-cost obtainment of catalysts based on dispersed non-noble metal species. In this scenario, UiO-66 is an attractive structure for this purpose since it is a microporous zirconium terephthalate forming a three-dimensional arrangement with high specific surface area and good thermal, mechanical, and chemical stability [20]. In addition, it requires low-cost precursors for its synthesis and can be obtained under fairly sustainable conditions. Very recently, active catalysts based on atomically dispersed ionic Cu species on UiO-66 and hybrid nanostructures of CuO nanocrystals encapsulated in UiO-66 crystals were reported [21,22]. In addition, with the concept of a matrix, CuO/CeO_2 active catalysts derived from Ce–UiO-66 were obtained, as well as $CuCe/ZrO_2$ catalysts derived from metal-impregnated UiO-66 [23,24].

In this context, the present work proposes to systematically analyze the use of UiO-66 synthesized through a sustainable protocol [25] as a dispersion matrix of Cu, Co, and Fe species to obtain new nanoparticle structures with potential use in catalysis. In addition, the incorporation of these metallic species is proposed by simple procedures traditionally used to prepare supported catalysts, such as the incipient wetness impregnation with precursors and subsequent thermal decomposition. The physicochemical properties of the obtained nanomaterials were tested in the gas-phase oxidation of carbon monoxide in air streams and in the CO preferential oxidation in hydrogen-rich streams (COProx).

2. Materials and Methods

2.1. Synthesis of UiO-66

Benzenedicarboxylic acid (BDC, Aldrich, 98.0% purity, St. Louis, MO, USA), $ZrCl_4$ (Zr, Aldrich, 98.0% purity, Darmstadt, Germany), and acetone (Cicarelli, 99.0% purity, San Lorenzo, Argentina) were used without further purification, and UiO-66 synthesis was performed employing a sustainable protocol reported elsewhere [25]. Briefly, the procedure consisted of mixing the two solid reactants together with the solvent in the molar proportions $BDC:ZrCl_4:solvent = 1:1:1622$. After obtaining the homogeneous mixture, it was placed under solvothermal treatment at 80 °C for 24 h. At the end of the treatments, the solids were recovered by centrifugation (10,000 rpm, 10 min), washed twice with ethanol, and finally dried at 80 °C overnight.

2.2. Incorporation of Metallic Species

$Cu(NO_3)_2 \cdot 3H_2O$ (Aldrich, 98.0–103% purity, St. Louis, MO, USA), $Co(NO_3)_2 \cdot 6H_2O$ (Alfa Aesar, 98.5% purity, Tewksbury, MA, USA), and $Fe(NO_3)_3 \cdot 9H_2O$ (Aldrich, ≥99.99%, purity, St. Louis, MO, USA) were employed as metal precursors. For their incorporation to MOF, incipient wetness impregnation was used as described in S1 (Supplementary Materials). The nomenclature employed was the following: First the incorporated metal, then its load in wt.% with respect to the catalyst total mass, and finally the support which was MOF (M) or its degradation product (Zr), i.e., Cu_{10}/M or $Cu_{16}Fe_7/Zr$.

2.3. Catalyst Characterization

The crystalline structure of the materials was analyzed trough X-ray diffraction (XRD) with a Shimadzu XD-D1 instrument (Shimadzu Corp., Kyoto, Japan, CuKα radiation, λ = 1.5418 Å, 2 °C·min^{-1}, 30 mV, 40 mA, 2θ = 5° to 65°). In order to analyze the thermal evolution of the MOF and precursor, thermogravimetric analysis (TGA) and single differential thermal analysis (SDTA) were conducted with a Mettler Toledo STARe (version 4.1, Bristol, UK) TGA/SDTA 851e module from 25 to 800 °C at 10 °C·min^{-1} in air or nitrogen flow (50 mL·min^{-1}, standard temperature and pressure (STP)). Laser Raman spectroscopy (LRS) was performed using a LabRam spectrometer (Horiba-Jobin-Yvon, Stanmore, UK) coupled to an Olympus confocal microscope (Olympus Corp., Shinjuku, Tokyo, Japan) equipped with a charge-coupled device (CCD) detector cooled to about 200 K. The excitation wavelength was 532 nm (Spectra Physics argon-ion laser), and the laser power was set at 30 mW. Transmission electron microscopy (TEM) images of the synthesized UiO-66 crystals and metal-based nanocatalysts were acquired using a JEOL 2100 Plus microscope (JEOL Ltd., Tokyo, Japan) equipped with an energy dispersive X-ray (EDX) detector (JEOL Ltd.) and a scanning transmission electron microscope (STEM) (JEOL Ltd.). The samples were milled and suspended in ethanol by ultrasonic treatment, and a drop of the fine suspension was placed on a carbon-coated nickel grid to be loaded into the microscope.

2.4. Catalytic Evaluations

The samples were evaluated in a glass tubular reactor connected to a continuous flow system equipped with mass flow controllers (Brooks 4800, Brooks Instrument, Hatfield, PA, USA) and heated with a furnace controlled with a proportional–integral–derivative (PID) system. Before each evaluation, the reactor was heated at different temperatures (200–650 °C) in He or Air flow (30 mL·min^{-1}) according to the pretreatment required by the sample before its test and maintaining such a temperature for 60 or 120 min. Then, the catalytic tests were performed with a mixture of molar composition 1% CO, 2% O$_2$ (employing synthetic air) in He balance, maintaining a total flow of 30 mL·min^{-1} with an initial mass of the solid of 70 mg. Some tests were also performed by adding 50% of H$_2$ in the reaction stream (COProx), at the same total flow. The catalytic measurements were taken after stabilizing the reactor at different temperatures for 8 min. The CO conversions were determined with an on-line Shimadzu GC-2014 chromatograph (Shimadzu Corp., Kyoto, Japan) equipped with a thermal conductivity detector (TCD) and a 5-Å molecular sieve packed bed column. CO conversions (XCO were calculated as follows:

$$\text{XCO (\%)} = ([CO]^0 - [CO])/[CO]^0 \times 100,$$

where $[CO]^0$ and $[CO]$ are inlet and outlet gas concentrations (ppm), respectively. Moreover, for COProx, the selectivity of O$_2$ to CO$_2$ (S) was calculated as follows [12]:

$$S\ (\%) = \text{XCO (\%)}/(\lambda \times (([O_2]^0 - [O_2])/[O_2]^0)),$$

where $[O_2]^0$ and $[O_2]$ are inlet and outlet oxygen concentrations (ppm), respectively. In our case, the value of the factor λ (excess oxygen in the reaction) was four.

3. Results and Discussion

3.1. Dispersion of Copper Species in UiO-66

3.1.1. Study of the Precursor Decomposition

The obtained MOF corresponded to a crystalline and pure phase of UiO-66 since all its diffraction signals were observed (Figure S1, Supplementary Materials); the most important ones were identified at 2θ = 7.38°, 8.52°, and 25.75°, corresponding to the (111), (200), and (600) planes, respectively [20]. All these signals matched those indexed for this MOF from its crystallographic data (CCDC 733458) and no impurities were detected, such as benzenedicarboxylic acid. Afterward, the compatibility between the decomposition temperatures of the metal precursors and the MOF was analyzed by TGA, both in an inert and in air atmosphere. During the impregnation process, the transition metals were incorporated as nitrate salts onto the UiO-66 surface. Afterward, the nitrate ion decomposition was required to obtain the metal oxides. In this context, it is necessary to analyze the atmosphere and temperature condition in which this decomposition was done, maintaining the MOF structure. Figure 1a, profile 3, presents the TGA of UiO-66 in inert gas, and Figure 1b, profile 3, depicts its corresponding derivative TGA (dTGA). Typical evolutions were observed, with an initial mass loss up to 130 °C due to physisorbed water and/or residual solvent of synthesis. Then, another small mass loss from 180 °C due to the dehydroxylation of the inorganic cluster of the MOF from $Zr_6O_4(OH)_4$ to Zr_6O_6. Finally, the thermal decomposition of the ligands was observed, which was similar to the one corresponding to this MOF synthesized under conventional conditions [26], with a maximum decomposition rate (T^{max}) at 555 °C (Figure 1(b3) and Table 1). Instead, under air atmosphere, the TGA (Figure 1(c3)) and dTGA profiles (Figure 1(d3)) showed a somewhat lower stability with a T^{max} of 520 °C (Table 1). This evolution was more exothermic compared to the previous one as could be seen in the SDTA (Figure S2, Supplementary Materials). Meanwhile, the copper precursor showed a complete decomposition at 290 °C or 265 °C in inert gas (Figure 1(a4)) or air (Figure 1(b4)), respectively. The degradation temperature window of the MOF and the precursor suggested the possibility of obtaining dispersed copper species by applying heat treatments.

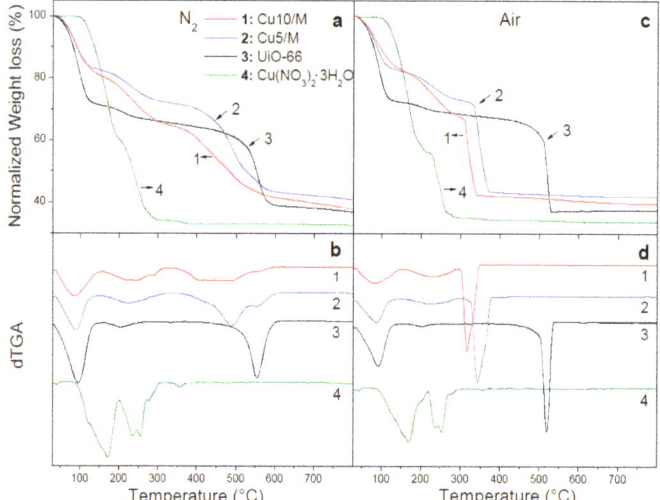

Figure 1. Thermogravimetric analysis (TGA) of UiO-66 (M), Cu/M, and copper precursor: (**a**) N_2 atmosphere; (**c**) air atmosphere. Corresponding derivative TGA (dTGA) curves: (**b**) N_2 atmosphere; (**d**) air atmosphere.

Table 1. Maximum decomposition temperatures (T^{max}) of the fresh metal–organic framework (MOF) and of the MOF impregnated with copper (Cu/M) and that of the respective metal precursors (obtained from the derivative thermogravimetric analysis (dTGA) data).

	T^{max} N_2 [1]	T^{max} Air [2]
$Cu(NO_3)_2 \cdot 3H_2O$	259	255
UiO-66	555	520
Cu_5/UiO-66	490	346
Cu_{10}/UiO-66	460	318

[1] Decomposition temperature (°C) of the maximum in the N_2 dTGA profile. [2] Decomposition temperature (°C) of the maximum in the air dTGA profile.

The MOF impregnated with 5 wt.% copper (Cu$_5$/M) showed a marked decrease in the structural stability of the framework (Figure 1a,b), which was magnified with a higher copper amount (Cu$_{10}$/M). The T^{max} was reduced from 555 °C to 490 °C and 460 °C for fresh UiO-66, Cu$_5$/M, and Cu$_{10}$/M, respectively (Table 1). The same effect was observed in air but with a greater destabilization of the framework. This phenomenon is attributed to the copper species formed after the precursor decomposition, which catalyzed the oxidation of the organic part of the MOF, accelerating its structural collapse. However, a small temperature gap persisted in which it would be possible to decompose the said precursor before the MOF collapsed. The SDTA of Cu/M in both atmospheres (Figure S2, Supplementary Materials) showed only an endothermic process due to the evacuation of host molecules that ended at 160 °C, and an exothermic one near 300 °C due to MOF collapse. From the previous results, pretreatments of Cu$_{10}$/M in inert and combinations with air were carried out at different temperatures prior to carrying out its catalytic test in order to get insight into the catalytic performances and structural stabilities after the different pretreatments.

3.1.2. Copper-Based UiO-66 Catalysts

In Figure 2a it can be observed that UiO-66 by itself presented no activity in the oxidation of CO. On the other hand, Cu$_{10}$/M treated for 1 h at 200 °C in He (1) presented activity, reaching a maximum conversion of 60% at the said temperature, while a treatment at 225 °C (2) yielded an improvement, reaching conversions of 75%. Meanwhile, a pretreatment at 300 °C impaired activity, as observed in solid (3). The diffractograms of these evaluated samples (Figure 2b) indicate that samples (1) and (2) maintained the MOF structure. Nevertheless, a degradation of the MOF took place in sample (3). The absence of definite signals of copper species in all the diffractograms should be highlighted since it indicates their high dispersion. Accordingly, in solids (1) and (2), the catalytic activity was due to copper species highly dispersed in the MOF structure, while, in sample (3), those species were dispersed in an amorphous solid.

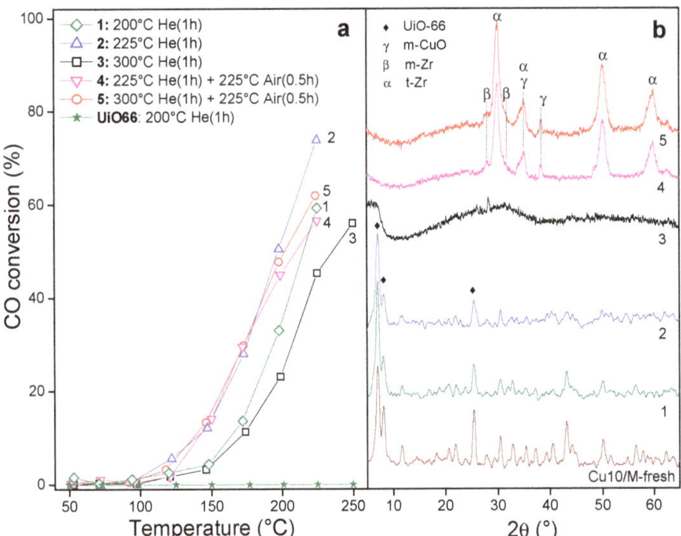

Figure 2. Cu/UiO-66 solids (Cu/M): (**a**) catalytic evaluation in the CO oxidation; (**b**) X-ray diffraction (XRD) patterns after the reaction.

The change in the pretreatment atmosphere was analyzed in terms of the activity of Cu$_{10}$/M by combining the pretreatment in inert atmosphere followed by a brief exposure to air for 0.5 h (sample (4)). The as-treated solid was also active (Figure 2a) even though the treatment promoted the total degradation of the MOF (Figure 2b), favoring its evolution to a crystalline phase of tetragonal zirconia (t-Zr), (JCPDS 17-923). The decomposition of the hydrated nitrate salts in the presence of air can generate various oxidizing agents such as HNO_3 and NO_2, which, added to the treatment in oxygen, can give rise to a hyperoxidizing atmosphere and accelerate the transformation of the MOF into zirconia. This evolution is in agreement with what was reported for the decomposition of UiO-66 in air [27]. A high dispersion of cupric oxide was observed in these solids, characterized by weak signals at $2\theta = 35.5°$ (masked by a signal of t-Zr) and 38.5°, corresponding to the (11−1) and (111) planes of a CuO monoclinic phase (JCPDS 48-1548), respectively.

The TEM image of the synthesized MOF shows nanometric crystals with sizes ranging from 30–100 nm which formed globular aggregates (Figure 3a), making it possible to distinguish the facets of the individual crystals with a polyhedral morphology (Figure 3b) similar to that reported for this MOF but obtained under conventional conditions [20]. Meanwhile, when the said crystals were functionalized with copper following the sequence of impregnation and heat treatment in He to obtain Cu/UiO-66, the porous structure of the MOF was maintained (Figure 3c). The high-resolution (HR) TEM image showed the characteristic aspect of a porous material but no particles could be distinguished inside the MOF porosity (Figure 3d). This highlights the small size of these copper species dispersed in the MOF, consistent with the XRD results.

In brief, a pretreatment of Cu_{10}/M in He for 1 h at 225 °C allowed obtaining an active catalyst in the CO oxidation, based on copper species with high dispersion in the MOF structure which were preserved after the tests in reaction. This nanocatalyst represents a new alternative not only for this reaction but also for other reactions demanding a high dispersion of active copper phases and led at relatively low temperatures (<225 °C). These reactions could be, for example, the reduction of C–C multiple bonds and carbonyl, the hydroxylation of benzene, the reduction of aromatic nitrocompounds or NO_x [28], or the synthesis of methanol from CO and H_2 [29].

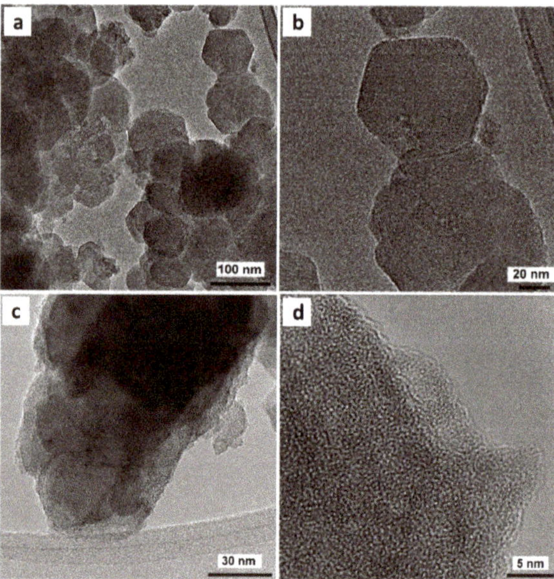

Figure 3. TEM images: (**a**) as-synthesized UiO-66 crystals; (**b**) close view of UiO-66; (**c**) Cu/UiO-66 catalyst with He treatment at 225 °C; (**d**) close view of activated (He treated) Cu/UiO-66.

3.2. Derived Cu/UiO-66 Catalysts

3.2.1. Monitoring of the Thermal Transformation of Cu/MOF

Given the structural changes observed after the thermal pretreatments, the transformation of Cu/M was analyzed through temperature-programmed X-ray diffraction (T-XRD) both in an inert atmosphere and in air. In the first case, from 250 °C, the MOF underwent a reduction in crystallinity (Figure 4a), totally losing itself at 325 °C (red curve). Then, the solid persisted as an amorphous material in which signals at 43.3° and 50.1° emerged (the latter masked with a t-Zr signal), which corresponded to the (111) and (200) planes of a cubic Cu^0 phase, respectively (JCPDS 4-836). At higher temperatures, these species increased the crystallinity, while the support evolved into a tetragonal zirconia (t-Zr) of low crystallinity. This is consistent with what was discussed above. On the other hand, heat treatments in air showed that the structure of the MOF was more unstable (Figure 4b) losing the crystallinity at 275 °C (red curve). In addition, it quickly transformed into a t-Zr system with a high dispersion of copper oxide species. The stable formation of a t-Zr phase from this MOF was attributed to the initial transformation of the small zirconia nuclei from inorganic nodes, which have a low surface energy and facilitate the evolution toward a tetragonal phase instead of a monoclinic (m-Zr) [27]. From about 400 °C a small contribution of m-Zr was detected, characterized by strong signals at 2θ = 27.9°, 31.2°, 34.1°, 40.7°, and 49.1°, corresponding to the (−111), (111), (200), (−112), and (220) planes, respectively (JCPDS 37-1484). An increase in the proportion of the monoclinic phase in copper-doped zirconia was attributed to copper inclusion in the ZrO_2 network, which increased the size of the crystallites, causing a growth in the free surface energy and, thus, promoting the evolution toward m-ZrO_2 [30]. This could be, in our case, due to a migration of part of the copper to the zirconia phase in formation during the heat treatment in air.

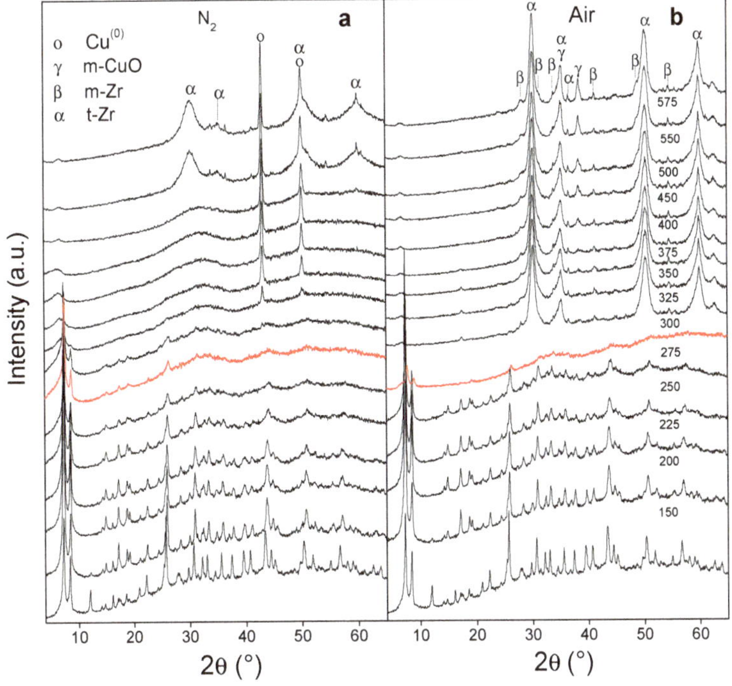

Figure 4. Analyses of X-ray diffraction at programmed temperature (T-XRD) with the Cu/M sample: (**a**) in nitrogen; (**b**) in air.

By T-XRD, it was shown that thermal pretreatments of the Cu/M solid in an inert atmosphere caused a delayed degradation of the MOF toward an amorphous solid in which Cu^0 species evolved. Meanwhile, the MOF degradation was accelerated in air with a fast growth of t-Zr phase with a small contribution of m-Zr and with highly dispersed CuO species. Given the potential of solids derived from Cu/M, their catalytic behavior was analyzed.

3.2.2. Catalytic Behavior of Derived Cu/UiO-66 Catalysts

Cu_{10}/M was pretreated in situ in He flow at 225 °C for 1 h, and its catalytic curve showed an inflection in the profile starting at 250 °C (Figure 5a) due to a fall in the CO conversion. This was caused by a smaller availability of oxygen for the reaction (as observed in the insert in Figure 5a), which was consumed in the MOF degradation. Hence, 250 °C is the maximum temperature at which Cu_{10}/M maintained its structure under reaction conditions. From 375 °C the oxygen was recovered, and the catalyst was taken up to 400 °C for 1 h in reaction, maintaining conversions of 100%; later, the catalyst was cooled and evaluated again. In this case, Cu_{23}/Zr (1), a marked activation was observed (Figure 5a) due to the evolution of the solid to the system of copper oxide dispersed in a developing t-Zr phase (Figure 5b). Since the zirconia mass in the derived solid was around 37.4 wt.% with respect to the initial mass of the MOF, the proportion of copper in these systems was 23 wt.%. It is noticeable that, with this high load, the copper species were highly dispersed in the t-Zr support. Given the good catalytic performance of these solids and taking into account the studies of the transformation of a Cu/M solid into Cu/Zr, a pretreatment of Cu_{10}/M in air was performed at 400 °C for 2 h. In this case, Cu_{23}/Zr (2), a remarkable shift of the catalytic curves was observed, reaching total conversion at 225 °C (Figure 5a) without extra oxygen consumption due to the presence of a stabilized phase of CuO/ZrO_2 (Figure 5b) with a contribution of a m-Zr phase. Compared with classical CuO/ZrO_2 catalysts obtained via other techniques such as sol–gel [31] or urea combustion [32], the use of the MOF as a template allowed minimizing the generation of bulk CuO of low interaction with zirconia, which would generate a lower catalytic activity. In contrast to the MOF-derived zirconia (Zr in Figure 5b), the Cu_{23}/Zr (2) solid showed a contribution of the m-Zr phase. When Cu_{10}/M was pretreated in air at 500 °C, Cu_{23}/Zr (3), a slight catalytic improvement was observed with respect to the former case, reaching total conversion at 175 °C.

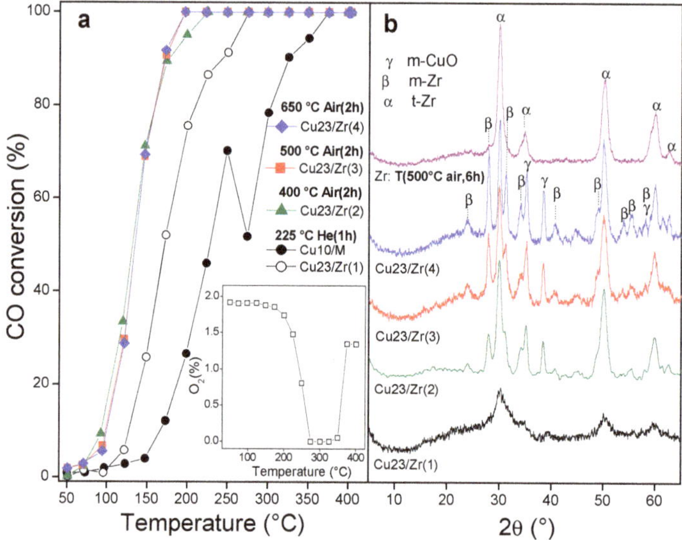

Figure 5. Cu/UiO-66-derived solids: (**a**) catalytic evaluation; (**b**) XRD patterns after reaction.

However, the contribution of m-Zr in this sample was more evident, which could be related to its better catalytic behavior. In this sense, it was demonstrated that the adsorption capacity of CO in m-Zr supports was higher than in t-Zr, which can be explained by a higher Lewis acidity (Zr^{4+}), as well as a higher Lewis basicity (O^{2-}), on the surface of the m-Zr solid [33]. Finally, a pretreatment at 650 °C in air, Cu_{23}/Zr (4), did not improve the conversion (Figure 5a), even though it favored the development of the monoclinic phase, probably due to a sintering of the CuO species (Figure 5b).

3.3. Derived Cobalt and Iron-Based UiO-66 Catalysts

Other non-noble metals of interest that have activity in the CO oxidation reaction are cobalt and iron [13], added to the fact that the latter is a very low-cost metal with high abundance. The decomposition temperature of cobalt and iron nitrate precursors in air was far from the limit of MOF stability (Table 2, Figure S3, Supplementary Materials) while the incorporation of 10 wt.% Co or Fe in the MOF (Co_{10}/M, Fe_{10}/M) decreased the framework stability due to the formed oxides, although the shift was lower than that of Cu_{10}/M. The order of structural stability was as follows: Co/M > Fe/M > Cu/M. The SDTA profiles in air (Figure S3, Supplementary Materials) were very similar to that of Cu_{10}/M, with an endothermic peak due to the evaporation of host molecules and an exothermic one due to structural collapse. Taking into account the similar structure stability of Cu_{10}/M under either He or air atmosphere at temperatures lower than 275 °C, Co_{10}/M and Fe_{10}/M solids were pretreated at 250 °C in air, and their catalytic behavior was analyzed.

Table 2. Maximum decomposition temperatures (T^{max}) of the MOF impregnated with cobalt (Co/UiO-66) and iron (Fe/UiO-66), and that of the respective metal precursors (obtained from the dTGA data in air).

	T^{max} Air [1]
$Co(NO_3)_2 \cdot 6H_2O$	263
$Fe(NO_3)_3 \cdot 9H_2O$	163
$Co_{10}/UiO\text{-}66$	465
$Fe_{10}/UiO\text{-}66$	435

[1] Decomposition temperature (°C) of the maximum in the air dTGA profile.

For the Co_{10}/M solid (Figure S4a, Supplementary Materials), from 200 °C onward, conversion increased proportionally with temperature, and, when it was over 325 °C, both a conversion fall and an abrupt consumption of oxygen were produced due to the MOF degradation. The activity evolved until total conversion but at a higher temperature than the Cu_{10}/M solid, previously analyzed. Subsequently, this solid was taken to 400 °C and was kept 1 h under reaction. When evaluated again, an improvement in activity was observed, Co_{23}/Zr (1). The catalyst consisted of a t-Zr phase evolving with a high dispersion of cobalt species due to the absence of characteristic signals of their oxides (Figure S4b, Supplementary Materials). Since it was observed that an improvement in ZrO_2 crystallinity favored the activity, a pretreatment of the Co_{10}/M sample was performed in air but at 400 °C for 2 h. This effectively improved the catalytic performance (Figure S4a, Supplementary Materials) due to the formation of a stabilized phase of Co_3O_4 in a t-Zr of high crystallinity (Figure S4b, Supplementary Materials). This is in agreement with what was reported regarding the formation of this cubic spinel (JPDS 43-1003) on conventional ZrO_2 supports [15].

On the other hand, Fe_{10}/M showed less activity (Figure S4a, Supplementary Materials), even over 300 °C when the MOF decomposed. After 1 h in reaction at 400 °C, the system was evaluated again, and an improvement was observed even though total conversion was not reached in the temperature range analyzed. This solid consisted of an incipiently formed t-Zr with a high dispersion of iron species (Figure S4b, Supplementary Materials). In this case, a remarkable catalytic improvement was also observed when pretreating at 400 °C in air. This solid consisted of a highly crystalline t-Zr with a small contribution of m-Zr which dispersed a rhombohedral hematite phase (α-Fe_2O_3). The previous confinement of the iron precursor in the pores of UiO-66 facilitated, after degradation, the generation of small Fe_2O_3 crystals that were quite active in CO oxidation, as already observed for iron oxide crystals [34].

The MOF degradation under reaction conditions of CO oxidation started at 325, 300, and 250 °C for Co_{10}/M, Fe_{10}/M, and Cu_{10}/M respectively. Although the thermal stability of the latter was slightly lower, its CO conversion at 250 °C was significantly higher (70%) than that of the other two samples (17% and 5%). The best catalytic performance corresponded to the Cu-based sample after the degradation of the MOF structure in air at 400 °C for 2 h (Cu_{23}/Zr), which reached 100% CO conversion at 225 °C. At that temperature, the conversion for the Co_{23}/Zr solid was 45%, while, for the Fe_{23}/Zr solid, it was only 8%.

3.4. Bimetallic CuCo/UiO-66- and CuFe/UiO-66-Derived Nanocatalysts

3.4.1. CO Oxidation

It was reported that mixed cobalt and copper oxides [35], as well as copper–iron mixed oxides synthesized by low-temperature co-precipitation methods [33,36,37], have synergistic effects on the oxidation of CO; therefore, bimetallic systems were prepared incorporating these metals into UiO-66. The solids obtained by successive impregnation were analyzed, firstly by incorporating the copper precursor followed by their thermal treatment (500 °C, air, 2 h); subsequently, a cobalt or iron precursor was added, followed by a final calcination step in air (400 °C, 2 h) to obtain $Cu_{16}Co_7/Zr$ and $Cu_{16}Fe_7/Zr$ systems. The Cu_{16}/Zr, $Cu_{16}Co_7/Zr$, and $Cu_{16}Fe_7/Zr$ samples exhibited well-developed t-Zr phases with a small contribution of m-Zr (promoted by the presence of copper as discussed above), adding to a CuO phase with high dispersion (Figure 6). Additionally, a Co_3O_4 phase was observed in the solid containing cobalt, while, in the iron-containing bimetallic solid, no iron oxide phases were detected. This accounts for the high dispersion of the FeO phases in this solid. The catalytic assays showed that, among these catalysts, an improvement in the activity of $Cu_{16}Fe_7/Zr$ was found (Figure 6a). This was due to both the initial presence of a very small proportion of m-Zr phase before the incorporation of iron and the subsequent development of a Fe–Cu synergy among these species due to their intimate contact, favored by the high dispersion achieved by these oxides in the solid, as shown by their XRD patterns.

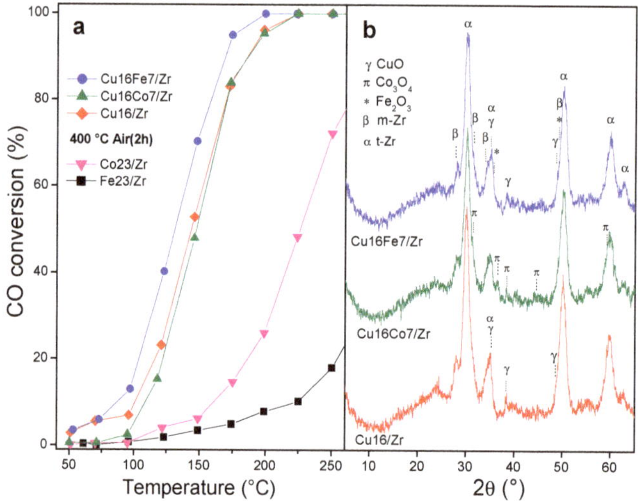

Figure 6. CuCo/Zr and CuFe/Zr nanocatalysts: (**a**) catalytic behavior; (**b**) XRD patterns after reaction.

The catalytically evaluated nanomaterials were analyzed by laser Raman spectroscopy (LRS). These spectra are shown in Figure 7, and the respective spectra of the used monometallic samples are included for comparison. The vibrational signals observed in all of these spectra are consistent with the crystalline phases identified by XRD. The vibrations of the monoclinic ZrO_2 (m-Zr) and tetragonal ZrO_2 (t-Zr) phases were present in the Cu_{16}/Zr spectrum (Figure 7), proving the existence of both zirconia phases. The characteristic narrow vibration signals of m-Zr were at 179, 192, 335, 347, 385, 476, 614, and 635 cm^{-1}, with that at 476 cm^{-1} being the strongest one [15], while the typical broad signals of t-Zr were at 145, 275, 310, 460, and 650 cm^{-1} [27]. In the case of the sample with higher copper content, Cu_{23}/Zr, the monoclinic phase was clearly identified due to its narrow vibration signals (Figure 7). The vibrations of CuO at 280, 335, and 615 cm^{-1} [8] overlapped with those of the zirconia, thus hindering their identification. The spectrum of Co_{23}/Zr pointed out the existence of a $t-ZrO_2$ phase and a Co_3O_4 spinel (485, 523, and 687 cm^{-1}) [15], in clear agreement with the XRD results. The absence of $m-ZrO_2$ was evident, inferring that the UiO-66 degradation in the presence of cobalt hampered the formation of this monoclinic phase. The same outcome was obtained when iron was the impregnated metal. The Fe_{23}/Zr spectrum (Figure 7) revealed the presence of just $t-ZrO_2$ and $\alpha-Fe_2O_3$ (226, 246, 293, 411, and 610 cm^{-1} [36]), and no signals of $m-ZrO_2$ were identified.

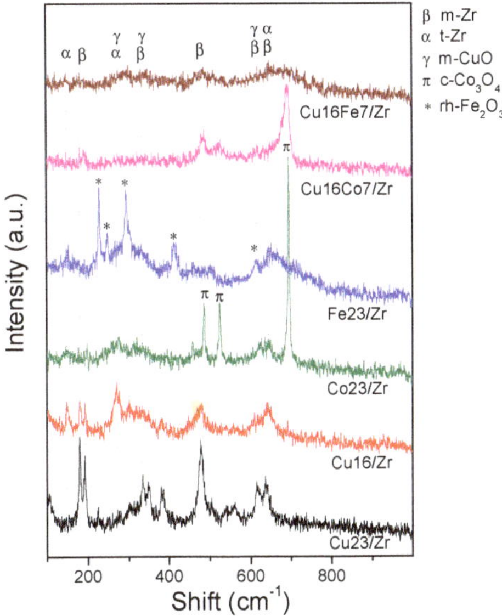

Figure 7. Laser Raman spectroscopy (LRS) spectra of the mono and bimetallic solids after the CO oxidation reaction.

In the spectra of the bimetallic catalysts, it could be observed that the zirconia signals were mainly associated with t-Zr (Figure 7). In the $Cu_{16}Fe_7/Zr$ sample, a high dispersion of iron and copper oxides was achieved given the absence of defined signals of these phases, in line with what was observed by XRD and also confirming the absence of agglomerates after the reaction. This shows that the addition of Fe to the Cu/Zr system generated highly dispersed and stable iron species, since they were kept in that situation in the solid after reaction (Figure 7). This is in contrast with the higher sintering reached in the monometallic Fe/Zr system after reaction. On the other hand, in the spectrum of sample $Cu_{16}Co_7/Zr$, the signals of a developed Co_3O_4 spinel were dominant (Figure 7), in agreement with XRD observations, showing again that the cobalt species were segregated forming big crystals at the catalyst surface.

The bimetallic sample $Cu_{16}Fe_7/Zr$ exhibited a nanoparticle system in intimate contact (Figure 8a), confirming the results of XRD and LRS discussed above, which corresponded to small domains of zirconia phases and oxides of copper and iron. The different crystalline planes of these phases in the individual crystals can be observed (Figure 8a). Figure 8b shows the analysis of the same particle in dark-field mode and its nature of aggregated nanoparticles was also highlighted. The elementary mapping performed in STEM mode showed a homogeneous distribution of the zirconium (yellow, Figure 8c), iron (green, Figure 8d), and copper (magenta, Figure 8e) phases in the nanoparticle aggregates, confirming the high dispersion and intimate contact between these nano-oxides. The characterizations performed by XRD, LRS, and TEM demonstrated the small particle size reached by the phases of these oxides in intimate contact with each other, explaining the catalytic synergy in this material, as shown in Figure 6.

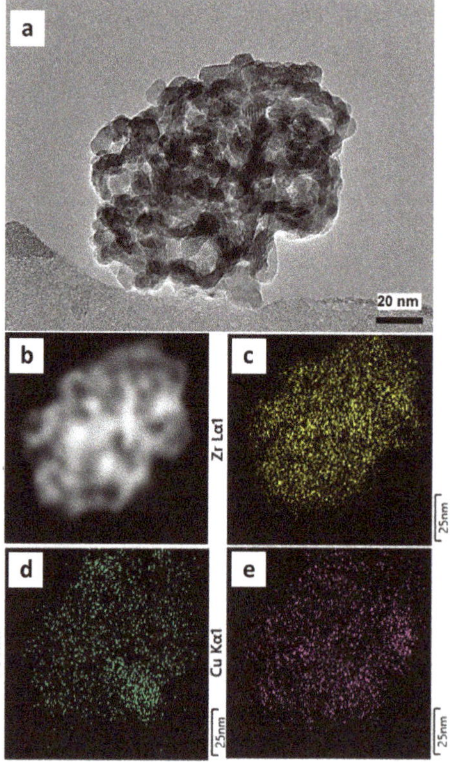

Figure 8. TEM images of $Cu_{16}Fe_7/Zr$ catalyst: (**a**) selected area for the energy-dispersive X-ray (EDX) mapping in bright field; (**b**) selected area for the EDX mapping in dark field. Elementary mapping: (**c**) zirconium (yellow); (**d**) iron (green); (**e**) copper (magenta).

3.4.2. CO Oxidation in Hydrogen-Rich Stream (COProx)

Given the high performance of the Cu_{16}/Zr, $Cu_{16}Co_7/Zr$ and $Cu_{16}Fe_7/Zr$ solids, they were analyzed in the COProx reaction. The conversion curves obtained are presented in Figure 9a and show a volcano-type shape, similar to that found for catalysts based on these types of dispersed oxides in classical supports [14,36]. Initially, the conversion increase may be due to highly dispersed CuO or superficial Cu–O–Zr type sites on the zirconia [14,32], reaching a maximum of 47% (175 °C) for the Cu_{16}/Zr sample. The fall in conversion at higher temperatures is probably due to a reduction in copper species dispersed in the hydrogen-rich atmosphere [14]. When comparing this behavior with that of the CO oxidation in air (Figure 6), a shift of the curves toward higher temperatures was observed for both the monometallic and bimetallic catalysts. This differs from that observed for dispersed cupric oxide crystals that exhibited a similar activity in both reactions [38], although this behavior was similar to that observed for other types of copper–iron mixed oxide catalysts [36]. This change in conversion levels may be due to structural differences between the active sites present and in the reaction mechanism operating under an oxidizing or reductive atmosphere [39]. Meanwhile, the selectivity was greater than 70% up to 120 °C, after which it fell sharply (Figure 9b).

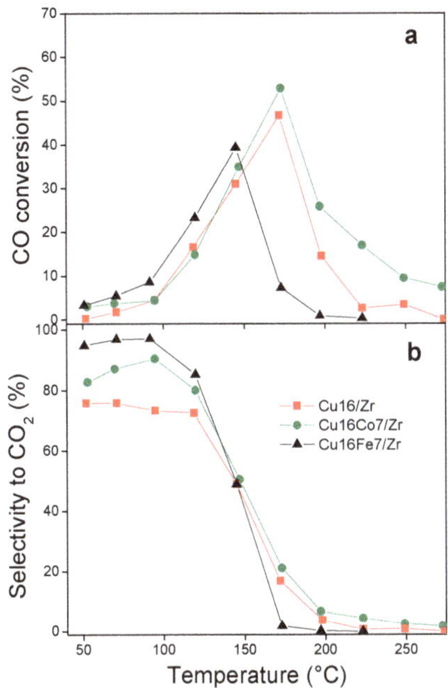

Figure 9. Catalytic evaluations in the preferential CO oxidation (COProx): (**a**) CO conversion; (**b**) selectivity toward CO$_2$.

The Cu$_{16}$Co$_7$/Zr catalyst exhibited similar characteristics to those analyzed above, with a 53% maximum conversion at the same temperature. In this sense, Co and Cu oxides would compete for the formation of M–O–Zr clusters over the zirconia support and not for the formation of Cu–Co–O–Zr species that could increase the conversion levels [14]. However, cobalt modulated the activity, given the increase in the selectivity of this system (Figure 9b). At the same time, the Cu$_{16}$Fe$_7$/Zr solid exhibited a shift of the conversion curve to lower temperature, which was compatible with the higher activity shown by this solid in COTox, previously discussed. Moreover, its selectivity was the highest of all evaluated materials (higher than 85% up to 125 °C). This behavior again accounts for the synergy between the oxide phases in this nanocatalyst.

The LRS characterization of the used catalysts after the COProx evaluation is shown in Figure 10. From the analysis of the Cu$_{16}$/Zr spectra before and after reaction, it can be inferred that the tetragonal and monoclinic phases were present in the support. Nevertheless, an increasing trend of the m-Zr strong signal at 460 cm^{-1} was observed at the expense of the t-Zr phase after reaction. This same trend can be observed in the spectra of the Co-containing materials. In the latter sample, a more acute and defined signal of Co$_3$O$_4$ can be additionally seen, from which the increase of the said particles under reducing atmosphere can be inferred. However, for the Cu$_{16}$Fe$_7$/Zr material, after being under reducing reaction conditions, t-Zr was still the main phase. Moreover, no signals of Cu or Fe oxides were observed, which highlights the high stability of these species in the said reaction atmosphere, which is also consistent with what was observed for this catalyst after CO oxidation in oxidizing atmosphere (Figure 6).

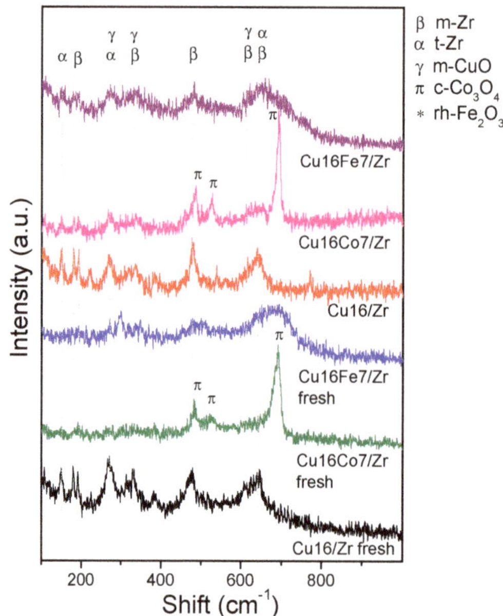

Figure 10. LRS spectra of the bimetallic nanocatalysts before and after the COProx reaction.

4. Conclusions

UiO-66 crystals obtained through a sustainable protocol were used as a dispersion matrix for copper, cobalt, and iron species, allowing the preparation of new nanostructured catalysts active in the oxidation of carbon monoxide. The MOF was modified with the said non-noble metals through simple and classic methods of incipient wetness impregnation, followed by controlled thermal treatments. It was shown that, by precisely tuning the treatment atmosphere (He), temperature (225 °C), and time (2h), the solid Cu/UiO-66 could be obtained having 10 wt.% copper species in a very high dispersion inside UiO-66 crystals, maintaining the structure of the MOF. This solid proved to be an active catalyst for the CO oxidation in air streams, representing a novel nanocatalyst not only for this reaction but also for others that demand a high dispersion of active copper species. It was also shown that, if the thermal decomposition treatments of the impregnated metal precursors were carried out in air at temperatures higher than 400 °C, the capacity of the MOF to host metallic species could be used to obtain non-noble metal-based catalysts supported on nano-zirconia derived from UiO-66. These controlled treatments in an air atmosphere of Cu, Co, or Fe-impregnated UiO-66 promoted the rapid development of a solid composed of tetragonal (t-Zr) and monoclinic (m-Zr) zirconia, supporting highly dispersed transition metal oxides. The derived bimetallic Cu–Fe/ZrO$_2$ nanocatalyst exhibited the best levels of activity and stability both in the oxidation of CO in air and in the COProx reaction, due to synergic effects of the very close contact between such oxides in the homogeneous nanomaterial.

This study shows the potential of UiO-66 as a dispersion matrix for low-cost and abundant metals such as copper, cobalt, and iron to obtain new sustainable nanocatalysts active in the CO oxidation reaction.

Supplementary Materials: The following are available online at http://www.mdpi.com/2079-4991/10/1/165/s1: S1: Incipient wet impregnation procedure; Figure S1: Diffractogram of the synthesized UiO-66 crystals and the pattern simulated from its crystallographic archive (CCDC 733458). The diffractogram of the benzenedicarboxylic acid ligand (BDC) is also included; Figure S2: SDTA of UiO-66(M), Cu/M, and copper precursor: (a) under N$_2$ atmosphere; (b) under air atmosphere; Figure S3: Thermal stability of Co/M, Fe/M solids, and their respective precursors in air: (a) TGA; (b) dTGA; (c) SDTA. Figure S4: Co/Zr and Fe/Zr nanocatalysts: (a) catalytic behavior; (b) XRD patterns after reaction.

Author Contributions: Conceptualization, M.A.U. and J.M.Z.; formal analysis, B.M.C.F., M.A.U., and J.M.Z.; funding acquisition, J.M.Z.; investigation, L.A.L., B.M.C.F., and J.M.Z.; methodology, L.A.L.; project administration, J.M.Z.; supervision, M.A.U. and J.M.Z.; writing—original draft, L.A.L., B.M.C.F., M.A.U., and J.M.Z.; writing—review and editing, M.A.U. and J.M.Z. All authors discussed the results and contributed to the manuscript. All authors read and agreed to the published version of the manuscript.

Funding: This research was funded by the Agencia Nacional de Promoción Científica y Tecnológica (ANPCyT) of Argentina under project PICT 2241 and by the Universidad Nacional del Litoral under project CAI+D 00071 LI.

Acknowledgments: We acknowledge the financial support from the Agencia Nacional de Promoción Científica y Tecnológica of Argentina (PICT 2241) and from the Universidad Nacional del Litoral (CAI+D 0071). The institutional support of CONICET is also acknowledged.

Conflicts of Interest: The authors declare no conflicts of interest.

References

1. Farrusseng, D.; Aguado, S.; Pinel, C. Metal-organic frameworks: Opportunities for catalysis. *Angew. Chem. Int. Ed.* **2009**, *48*, 7502–7513. [CrossRef]
2. Falcaro, P.; Ricco, R.; Yazdi, A.; Imaz, I.; Furukawa, S.; Maspoch, D.; Ameloot, R.; Evans, J.D.; Doonan, C.J. Application of metal and metal oxide nanoparticles@MOFs. *Coord. Chem. Rev.* **2016**, *307*, 237–254. [CrossRef]
3. Vermoortele, F.; Ameloot, R.; Alaerts, L.; Matthessen, R.; Carlier, B.; Ramós Fernandez, E.V.; Gascon, J.; Kapteijn, F.; De Vos, D.E. Tuning the catalytic performance of metal–organic frameworks in fine chemistry by active site engineering. *J. Mater. Chem.* **2012**, *22*, 10313–10321. [CrossRef]
4. Lescouet, T.; Kockrick, E.; Bergeret, G.; Pera-Titus, M.; Farrusseng, D. Engineering MIL-53(Al) flexibility by controlling amino tags. *Dalton Trans.* **2011**, *40*, 11359–11361. [CrossRef]
5. Juan-Alcañiz, J.; Gascon, J.; Kapteijn, F. Metal–organic frameworks as scaffolds for the encapsulation of active species: State of the art and future perspectives. *J. Mater. Chem.* **2012**, *22*, 10102–10118. [CrossRef]
6. Song, Y.; Li, X.; Sun, L.; Wang, L. Metal/metal oxide nanostructures derived from metal–organic frameworks. *RSC Adv.* **2015**, *5*, 7267–7279. [CrossRef]
7. Sun, J.-K.; Xu, Q. Functional materials derived from open framework templates/precursors: Synthesis and applications. *Energy Environ. Sci.* **2014**, *7*, 2071–2100. [CrossRef]
8. Zamaro, J.M.; Pérez, N.C.; Miró, E.E.; Casado, C.; Seoane, B.; Téllez, C.; Coronas, J. HKUST-1 MOF: A matrix to synthesize CuO and CuO–CeO$_2$ nanoparticle catalysts for CO oxidation. *Chem. Eng. J.* **2012**, *195–196*, 180–187. [CrossRef]
9. Green, E.; Short, S.; Shuker, L.K.; Harrison, P.T.C. Carbon Monoxide Exposure in the Home Environment and the Evaluation of Risks to Health—A UK Perspective. *Indoor Built Environ.* **1999**, *8*, 168–175. [CrossRef]
10. Xanthopoulou, G.G.; Novikov, V.A.; Knysh, Y.A.; Amosov, A.P. Nanocatalysts for Low-Temperature Oxidation of CO: Review. *Eurasian Chem. J.* **2015**, *17*, 17–32. [CrossRef]
11. Mishra, A.; Prasad, R. A Review on Preferential Oxidation of Carbon Monoxide in Hydrogen Rich Gases. *Bull. Chem. React. Eng. Catal.* **2011**, *6*, 1–14. [CrossRef]
12. Bion, N.; Epron, F.; Moreno, M.; Mariño, F.; Duprez, D. Preferential oxidation of carbon monoxide in the presence of hydrogen (PROX) over noble metal and transition metal oxides: Advantages and drawbacks. *Top. Catal.* **2008**, *51*, 76–88. [CrossRef]
13. Royer, S.; Duprez, D. Catalytic Oxidation of Carbon Monoxide over Transition Metal Oxides. *ChemCatChem* **2011**, *3*, 24–65. [CrossRef]
14. Firsova, A.A.; Khomenko, T.I.; Sil'chenkova, O.N.; Korchak, V.N. Oxidation of Carbon Monoxide in the Presence of Hydrogen on the CuO, CoO, and Fe$_2$O$_3$ oxides supported on ZrO$_2$. *Kinet. Catal.* **2010**, *51*, 299–311. [CrossRef]
15. Zhao, Z.; Yung, M.M.; Ozkan, U.S. Effect of support on the preferential oxidation of CO over cobalt catalysts. *Catal. Commun.* **2008**, *9*, 1465–1471. [CrossRef]
16. Liang, Q.; Zhao, Z.; Liu, J.; Wei, Y.C.; Jiang, G.Y.; Duan, A.J. Pd nanoparticles deposited on metal-organic framework of MIL-53(Al): An active catalyst for CO oxidation. *Acta Phys.-Chim. Sin.* **2014**, *30*, 129–134.
17. Jiang, H.; Liu, B.; Akita, T.; Haruta, M.; Sakurai, H.; Xu, Q. Au@ ZIF-8: CO oxidation over gold nanoparticles deposited to metal-organic framework. *J. Am. Chem. Soc.* **2009**, *131*, 11302–11303. [CrossRef] [PubMed]
18. Morris Bullock, R. Reaction: Earth-Abundant Metal Catalysts for Energy Conversions. *Chem* **2017**, *2*, 444–446. [CrossRef]

19. Ludwig, J.R.; Schindler, C.S. Catalyst: Sustainable Catalysis. *Chem* **2017**, *2*, 313–316. [CrossRef]
20. Cavka, J.H.; Jakobsen, S.; Olsbye, U.; Guillou, N.; Lamberti, C.; Bordiga, S.; Lillerud, K.P. A new Zirconium inorganic building brick forming metal organic frameworks with exceptional stability. *J. Am. Chem. Soc.* **2008**, *130*, 13850–13851. [CrossRef]
21. Abdel-Mageed, A.M.; Rungtaweevoranit, B.; Parlinska-Wojtan, M.; Xiaokun, P.; Yaghi, O.M.; Behm, R.J. Highly Active and Stable Single-Atom Cu Catalysts Supported by a Metal–Organic Framework. *J. Am. Chem. Soc.* **2019**, *141*, 5201–5210. [CrossRef]
22. Wang, H.L.; Yeh, H.; Chen, Y.C.; Lai, Y.C.; Lin, C.Y.; Lu, K.Y.; Ho, R.M.; Li, B.H.; Lin, C.H.; Tsai, D.H. Thermal Stability of Metal–Organic Frameworks and Encapsulation of CuO Nanocrystals for Highly Active Catalysis. *ACS Appl. Mater. Interfaces* **2018**, *10*, 9332–9341. [CrossRef] [PubMed]
23. Zhu, C.; Ding, T.; Gao, W.; Ma, K.; Tian, Y.; Li, X. CuO/CeO$_2$ catalysts synthesized from Ce-UiO-66 metal-organic framework for preferential CO oxidation. *Int. J. Hydrog. Energy* **2017**, *42*, 17457–17465. [CrossRef]
24. Yu, J.; Yu, J.; Wei, Z.; Guo, X.; Mao, H.; Mao, D. Preparation and Characterization of UiO-66-Supported Cu–Ce Bimetal Catalysts for Low-Temperature CO Oxidation. *Catal. Lett.* **2019**, *149*, 496–506. [CrossRef]
25. Lozano, L.A.; Iglesias, C.M.; Faroldi, B.M.C.; Ulla, M.A.; Zamaro, J.M. Efficient solvothermal synthesis of highly porous UiO-66 nanocrystals in dimethylformamide-free media. *J. Mater. Sci.* **2018**, *53*, 1862–1873. [CrossRef]
26. Wu, H.; Chua, Y.S.; Krungleviciute, V.; Tyagi, M.; Chen, P.; Yildirim, T.; Zhou, W. Unusual and highly tunable missing-linker defects in zirconium metal–organic framework UiO-66 and their important effects on gas adsorption. *J. Am. Chem. Soc.* **2013**, *135*, 10525–10532. [CrossRef]
27. Yan, X.; Lu, N.; Fan, B.; Bao, J.; Pan, D.; Wang, M.; Li, R. Synthesis of mesoporous and tetragonal zirconia with inherited morphology from metal–organic frameworks. *CrystEngComm* **2015**, *17*, 6426–6433. [CrossRef]
28. Gawande, M.B.; Goswami, A.; Felpin, F.X.; Asefa, T.; Huang, X.; Silva, R.; Zou, X.; Zboril, R.; Varma, R.S. Cu and Cu-Based Nanoparticles: Synthesis and Applications in Catalysis. *Chem. Rev.* **2016**, *116*, 3722–3811. [CrossRef]
29. Müller, M.; Hermes, S.; Kähler, K.; Van den Berg, M.W.E.; Muhler, M.; Fischer, R.A. Loading of MOF-5 with Cu and ZnO Nanoparticles by Gas-Phase Infiltration with Organometallic Precursors: Properties of Cu/ZnO@MOF-5 as Catalyst for Methanol Synthesis. *Chem. Mater.* **2008**, *20*, 4576–4587. [CrossRef]
30. Yang, Z.; Mao, D.; Guo, X.; Lu, G. CO oxidation over CuO catalysts supported on CeO$_2$-ZrO$_2$ prepared by microwave assisted co-precipitation: The influence of CuO content. *J. Rare Earths* **2014**, *32*, 117–123. [CrossRef]
31. Morales, F.; Viniegra, M.; Arroyo, R.; Córdoba, G.; Zepeda, T.A. CO oxidation over CuO/ZrO$_2$ catalysts: Effect of loading and incorporation procedure of CuO. *Mater. Res. Innnov.* **2010**, *14*, 183–188. [CrossRef]
32. Ribeiro, N.F.P.; Souza, M.V.M.; Schmal, M. Combustion synthesis of copper catalysts for selective CO oxidation. *J. Power Sources* **2008**, *179*, 329–334. [CrossRef]
33. Pokrovski, K.; Jung, K.T.; Bell, A.T. Investigation of CO and CO$_2$ Adsorption on Tetragonal and Monoclinic Zirconia. *Langmuir* **2001**, *17*, 4297–4303. [CrossRef]
34. Soliman, N.K. Factors affecting CO oxidation reaction over nanosized materials: A review. *J. Mater. Res. Technol.* **2019**, *8*, 2395–2407. [CrossRef]
35. Li, D.; Liu, X.; Zhang, Q.; Wang, Y.; Wan, H. Cobalt and Copper Composite Oxides as Efficient Catalysts for Preferential Oxidation of CO in H$_2$-Rich Stream. *Catal. Lett.* **2009**, *127*, 377–385. [CrossRef]
36. Yeste, M.P.; Vidal, H.; García-Cabeza, A.L.; Hernández-Garrido, J.C.; Guerra, F.M.; Cifredo, G.A.; González-Leal, J.M.; Gatica, J.M. Low temperature prepared copper-iron mixed oxides for the selective CO oxidation in the presence of hydrogen. *Appl. Catal. A Gen.* **2018**, *552*, 58–69. [CrossRef]
37. Said, A.E.A.A.; Abd El-Wahab, M.M.M.; Goda, M.N. Synthesis and characterization of pure and (Ce, Zr, Ag) doped mesoporous CuO-Fe$_2$O$_3$ as highly efficient and stable nanocatalystsfor CO oxidation at low temperature. *Appl. Surf. Sci.* **2016**, *390*, 649–665. [CrossRef]
38. Cabello, A.P.; Ulla, M.A.; Zamaro, J.M. CeO/CuOx nanostructured films for CO oxidation and CO oxidation in hydrogen-rich streams using a micro-structured reactor. *Top. Catal.* **2019**, *62*, 931–940. [CrossRef]
39. Yung, M.M.; Zhao, Z.; Woods, M.P.; Ozkan, U.S. Preferential oxidation of carbon monoxide on CoOx/ZrO$_2$. *J. Mol. Catal. A Chem.* **2008**, *279*, 1–9. [CrossRef]

 *nanomaterials*

Article

Densification-Induced Structure Changes in Basolite MOFs: Effect on Low-Pressure CH₄ Adsorption

David Ursueguía, Eva Díaz and Salvador Ordóñez *

Catalysis, Reactors and Control Research Group (CRC), Department of Chemical and Environmental Engineering, University of Oviedo, 33006-Oviedo, Spain; ursueguiadavid@uniovi.es (D.U.); diazfeva@uniovi.es (E.D.)
* Correspondence: sordonez@uniovi.es

Received: 18 April 2020; Accepted: 19 May 2020; Published: 1 June 2020

Abstract: Metal-organic frameworks' (MOFs) adsorption potential is significantly reduced by turning the original powder into pellets or granules, a mandatory step for their use at industrial scale. Pelletization is commonly performed by mechanical compression, which often induces the amorphization or pressure-induced phase transformations. The objective of this work is the rigorous study of the impact of mechanical pressure (55.9, 111.8 and 186.3 MPa) onto three commercial materials (Basolite C300, F300 and A100). Phase transformations were determined by powder X-ray diffraction analysis, whereas morphological changes were followed by nitrogen physisorption. Methane adsorption was studied in an atmospheric fixed bed. Significant crystallinity losses were observed, even at low applied pressures (up to 69.9% for Basolite C300), whereas a structural change occurred to Basolite A100 from orthorhombic to monoclinic phases, with a high cell volume reduction (13.7%). Consequently, adsorption capacities for both methane and nitrogen were largely reduced (up to 53.6% for Basolite C300), being related to morphological changes (surface area losses). Likewise, the high concentration of metallic active centers (Basolite C300), the structural breathing (Basolite A100) and the mesopore-induced formation (Basolite F300) smooth the dramatic loss of capacity of these materials.

Keywords: coordination polymers; methane storage; XRD crystallinity measurements; mechanical shaping; compaction; VAM; gas separation; MOF pelletization

---

1. Introduction

Energy demand estimations for the next decades, mainly due to the global population and industrialization process increments, boost the development of techniques and processes able to make the most of available resources [1]. What is more, the recent COVID-19 pandemic, with millions of people confined to their homes, pointed out even more our domestic reliance on electricity. In most economies that have taken strong confinement measures in response to the coronavirus, electricity demand has declined by around 15%, and the share of variable renewables like wind and solar had become higher than normal [2]. Even when electricity from wind and solar would satisfy the majority of demand, systems need to maintain flexibility in order to be able to ramp up other sources of generation quickly when the pattern of supply shifts, such as when the sun sets. That is, electricity system operators have to constantly balance demand and supply in real time to prevent blackouts, which in recent times occurred mainly during periods of low demand [2]. In this context, natural gas power plants can quickly ramp generation up or down at short notice, providing in this way flexibility, underlining the critical role of gas in the longed-for clean energy transition.

In the natural gas industry, methane purification is a major process for upgrading the streams [3]. In these streams, methane concentration is originally elevated (>90%), so satisfactory results have been reported using fixed-bed adsorption techniques [4,5]. In these cases, the usual practice is to separate the component that is in lower concentration by adsorption (typically CO_2). Adsorbents usually used for this purpose are activated carbons and zeolites, which have good CO_2 adsorption yields and their cost is relatively low [6,7]. On the other hand, these techniques present difficulties when methane is the component with the lowest concentration in the stream. Activated carbons and zeolites present low selectivity towards methane with respect to other very similar compounds in molecular size and polarity, like nitrogen [8,9]. This is the case of one of the new alternative methane sources that has begun to be studied in recent years, the recovery of methane from ventilation gases from mining exploitation (VAM). Until now, these streams, which contain typically less than 1% in methane, had been burned directly, with the need of an auxiliary fuel. VAM could be used in order to obtain energy or chemical products, as well as to prevent greenhouse gas emissions into the atmosphere [10,11]. For these operations to be profitable, it is necessary to perform a previous concentration step, whose success depends on the separation capacity of the adsorbent used [12].

Among the materials studied for this purpose, due to its amazing properties, metal-organic frameworks (MOFs) have been shown to present large adsorption and gas separation yields [13,14], these being among the most promising materials in this field. Their high specific surface area (even values up to 6255 m^2/g [15]) combined with high total pore volume (1.303 cm^3/g [16]) and great porosity (91.1% [17]) are responsible for the large adsorption capabilities, exceeding in the majority of cases other common materials [18]. The materials' structure is made up by an organic ligand, such as imidazole or pyrazine, which links different metal ions or clusters corresponding to each MOF type (copper, aluminium, etc.). These combinations form a cage-like structure that is repeated continuously, conferring on these materials a high degree of crystallinity [19]. Two of the main characteristics of the MOFs are the flexibility in the design, which means a huge variety of organic ligands and metallic ions that allow on-demand materials to be made, and the pore functionalization, presenting high interesting adsorptive and catalytic properties. The possibility of performing a large number of combinations has led to an astonishing number of works related to the synthesis of MOFs suitable for different applications, which include gas storage and separation [20]. For example, in the case of methane separation from other gases, Arami-Niya et al. [21] have tested the zeolitic imidazolate framework (ZIF-7) for the separation of methane from nitrogen, obtaining a selectivity of more than 10 for an equimolar mixture at 303 K. In addition, other authors such as Eyer et al. [22] have studied different materials capable of selectively adsorbing methane from air mixtures, obtaining promising results in the case of HKUST-1, with a selectivity methane/nitrogen of 2.8 and a large gravimetric methane adsorption capacity (171.36 mg/g) at 100 kPa and 196 K. Thus, MOFs have led to satisfactory results at the laboratory level in the case of low-concentrated methane separation from mixtures [23,24], with no experiences being performed at greater scales.

Therefore, most of the experimentation at lab scale and the properties' studies are done on the original powder form, since the most-used techniques for the MOFs synthesis are solvothermal methods, which generally produce powders [25]. Industrial-scale difficulties occur as a result of pressure drops associated with powder-filled beds, high diffusional problems and low density of the materials [26,27]. In order to reduce the pressure drop through the bed, there are techniques for increasing particle size and MOF densification: mechanical, hydraulic or hot pressing, extrusion, solid or emulsion templating, and the use of a polymeric binder [28,29]. In addition, there are also other techniques currently in development, such as the sol-gel monolithic synthesis [30]. Among them, mechanical compression is an inexpensive procedure and avoids the use of additional components like polymeric binders, which may change the physical properties of MOFs [31]. However, compression pelletization could also induce amorphization as well as phase transformations, which could influence also the adsorption capacities of the MOFs [32,33].

In this way, several studies deal with the effect of mechanical compression on hydrogen adsorption for MOF-5 [34,35] and MIL-101 [36] MOFs; as well as on CO_2 adsorption [37]. By contrast, there are fewer works related to the influence of MOFs' densification on the methane adsorption. For example, Yuan et al. have studied the behavior of PCN-250 on the methane and nitrogen adsorption at densification pressures up to 300 MPa [38]. General pressure-effects are, added to the loss of gravimetric performance, an increase in the volumetric adsorption capacity, in addition to higher stability in a humid ambient. Typically, these adsorption studies are done at elevated gas pressures since the main objective is to increase material density and volumetric adsorption capacity for meeting gas storage challenges. In this work, the aim is the separation of methane from low-concentration streams, so adsorption studies have been conducted at low pressure (0.1 MPa).

Therefore, the aim of this work is to study, firstly, the pressure-induced changes on the morphology and structure of three of the most common (and commercially available) MOFs, Basolite C300, Basolite F300 and Basolite A100; and, secondly, on the methane and nitrogen adsorption capacity at low pressure (0.1 MPa). The study of the adsorption capacity for methane (component to be recovered in VAM) and nitrogen (majority component in VAM) establishes a benchmark for the use of these commercial materials at industrial scale for obtaining profiting lean emissions as a novel energy source.

2. Materials and Methods

Basolite C300 [$Cu_3(C_9H_3O_6)_2$], Basolite F300 ($C_9H_3FeO_6$) and Basolite A100 ($C_8H_5AlO_5$) were manufactured by BASF and supplied by Aldrich (96% mass basis purity, Steinheim, Germany). All three materials were stored in a desiccator in order to avoid its contact with the ambient air. Particles were used in powder form, being the commercial size: Basolite C300 (16 μm, D50), Basolite F300 (5 μm) and Basolite A100 (32 μm, D50). Methane (CH_4), nitrogen (N_2) and helium (He), with a purity >99.995% mol, were supplied by Air Liquide (Madrid, Spain).

The pelletization method was performed using a hydraulic press (Graseby SPECAC 15.011, Orpington, UK) at compression pressures of 55.9, 111.8 and 186.3 MPa, for 30 s. Starting pressure was selected considering two considerations: ensuring the pelletization of the material to work at the actual conditions, and the lower operating limit of the hydraulic press used for this purpose. The resulting pellets were crushed and sieved in order to obtain powder (<50 μm) to perform all the successive analysis.

Breakthrough adsorption curves were obtained by flowing either CH_4 or N_2 (60%) diluted in He with a total flowrate of 50 mL/min, 298 K and 0.1 MPa of total pressure in a Micromeritics AutoChem II 2920 apparatus (Norcross, GA, USA) through a fixed bed of each sample (30 mg). The evolution of CH_4, N_2 and He signals were followed in a Pfeiffer vacuum Omnistar Prisma mass spectrometer (Pfeiffer Vacuum, Asslar, Germany). Adsorption gravimetric capacity was obtained from desorption experiments that were performed in the same apparatus flowing a He stream (20 mL/min and 0.1 MPa) with a temperature ramp of 5 K/min from 298 K to 463 K, recording also the outlet with the mass spectrometer.

The textural characteristics of specific surface area and pore volume were estimated by N_2 physisorption at 77 K in a Micromeritics ASAP 2020 surface area and porosity analyzer (Norcross, GA, USA). Physisorption data was processed using Brunauer–Emmett–Teller (BET), Barrett–Joyner–Halenda (BJH) and t-plot approaches for determining surface area, total mesopore volume and total micropore volume, respectively. Variations in average pore size were calculated by assuming pore cylindrical geometry. Scanning electron microscopy (SEM) images were obtained by using a JEOL 6610LV scanning electron microscope (JEOL, Yvelines, France). The samples were coated with gold prior to observation.

The crystallographic structures of the materials were determined by powder X-ray diffraction (PXRD) using a Philips PW 1710 diffractometer (Koninklijke Philips, Amsterdam, The Netherlands), working with the Cu-K$_\alpha$ line (λ = 0.154 nm) in the 2θ range of 5–85° at a scanning rate of 2°/min. Variations in the materials cell structure were verified by the Bragg law. Consequently, variations in lattice parameters of the structures were obtained through the standard equations for cubic, orthorhombic and monoclinic structures.

3. Results

3.1. Materials Characterization

Figures 1–3 show the SEM images of powder in the commercial form, as well as the sieved powder after the three pressure treatments. As can be observed, materials in the original form show well-defined particulate shapes, polyedric form in case of Basolite C300, and rounded shape in the case of Basolite F300 and Basolite A100. Size distribution seems to be wide for all of them, being the original size order: Basolite A100 > Basolite C300 > Basolite F300. Pressure increments lead to particle fragmentation, with the subsequent formation of irregular particle agglomerates. At the highest pressure (186.3 MPa) individual particles are practically indistinguishable, which become part of a large individual no-shaped bulk, especially for Basolite C300 and A100.

Figure 1. Scanning electron microscope (SEM) images of Basolite C300 (zoom in 50 μm) ((**A**): original, (**B**): 55.9 MPa, (**C**): 111.8 MPa, (**D**): 186.3 MPa).

Figure 2. *Cont.*

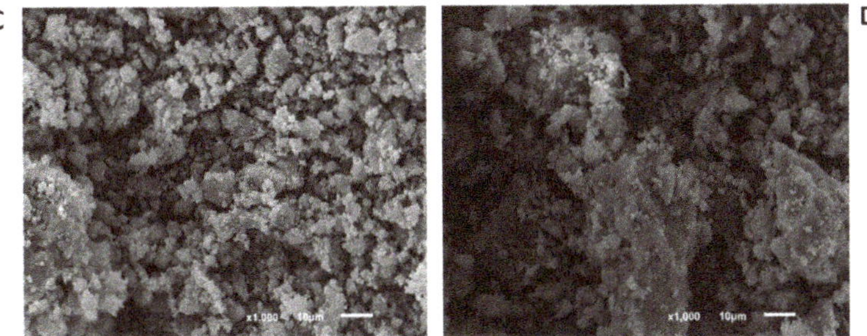

Figure 2. SEM images of Basolite F300 (zoom in 10 μm) ((**A**): original, (**B**): 55.9 MPa, (**C**): 111.8 MPa, (**D**): 186.3 MPa).

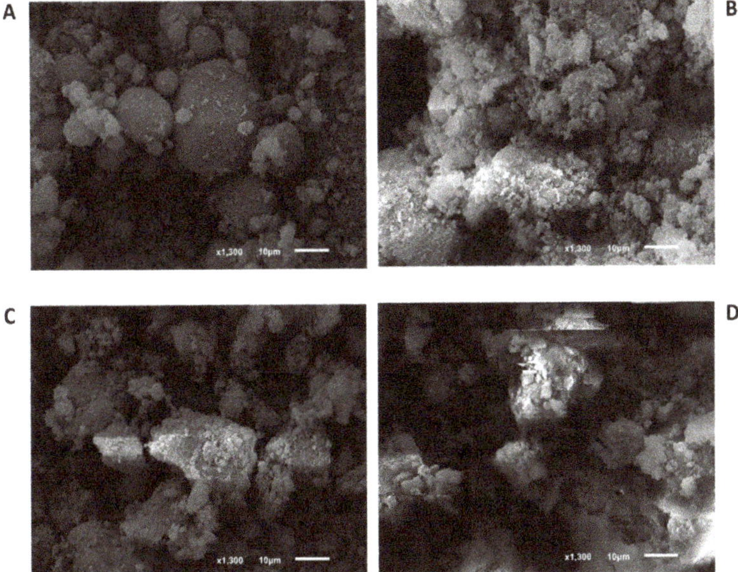

Figure 3. SEM images of Basolite A100 (zoom in 10 μm) ((**A**): original, (**B**): 55.9 MPa, (**C**): 111.8 MPa, (**D**): 186.3 MPa).

Figure 4 shows the adsorption-desorption isotherms determined by N_2 physisorption analysis at 77 K. As shown in the figure, pristine samples exhibit a combination of type I (b) and type II isotherms, according to the International Union of Pure and Applied Chemistry IUPAC. The first zone (up to $P/P_0 = 0.8$) resembles a type I (b) isotherm, with a steep elevation of the adsorbed quantity at very low pressure, and a subsequent maintenance. It is characteristic of microporous materials with wide micropores and possibly narrow mesopores [39]. The second area, up to a $P/P_0 = 1$, shows a more pronounced increase of adsorbate retained, which resembles the final part of a type II isotherm. This indicates the adsorption onto macroporous or non-porous materials in multilayer disposition, which corresponds to the external phase of MOFs [40]. A combination of these two isotherms usually results in a type IV isotherm, but in this case no characteristic hysteresis is observed, and the end of the isotherms is not a plateau [41]. As the densification pressure increases, the isotherms are closer to type I (b), due to the material agglomeration and the consequent loss of external surface availability. In addition, in all the materials a marked reduction is observed in the quantity adsorbed at low P/P_0 after pressure compression, indicative of a reduction in the total micropore volume, as it can be seen in the expanded graph (Figure 4). Micropores are clogged when particles are agglomerated with each other, in agreement with SEM images (Figure 1). The results show a significant effect of pelletization pressure on the morphology of the three MOFs (Table 1). Basolite C300 exhibits the highest BET surface loss (95.4%) at the highest pressure, although even at 55.9 MPa, the BET surface decrease reaches a value of 54.2%, in addition to 69.4% for total pore volume, which rules out the appearance of mesopores in the structure. In agreement, Casco et al. [42] have observed a great structural collapse by applying mechanical pressure (1.5 tons) to this material.

Basolite F300 presents high decreases in specific surface area (up to 93.3%) and micropore volume (96.3%), but lower in mesopore volume (up to 56.8%). The sharp BET decrease at 55.9 MPa shows the ease with which micropores collapse. However, the scarce total mesopores volume variation in the whole pressure range indicates the appearance of narrow mesopores in the structure (Table 1), as it is confirmed by the presence of some hysteresis (H4 type, according to IUPAC) at high P/P_0 values, marked in case of 55.9 and 111.8 MPa. For this material, the appearance of two leaps in total mesopore volume value is also remarkable, one between original material and 55.9 MPa and the other between 111.8 and 186.3 MPa. This indicates that the appearance of mesopores is higher at 111.8 MPa, increasing the total mesopore volume even above of the previous applied pressure (55.9 MPa). Despite that, the total pore volume is reduced (0.15 to 0.13 cm^3/g) in that pressure increment. This could be attributed to the formation from the voids of interparticular pore volume, as a result of the compaction.

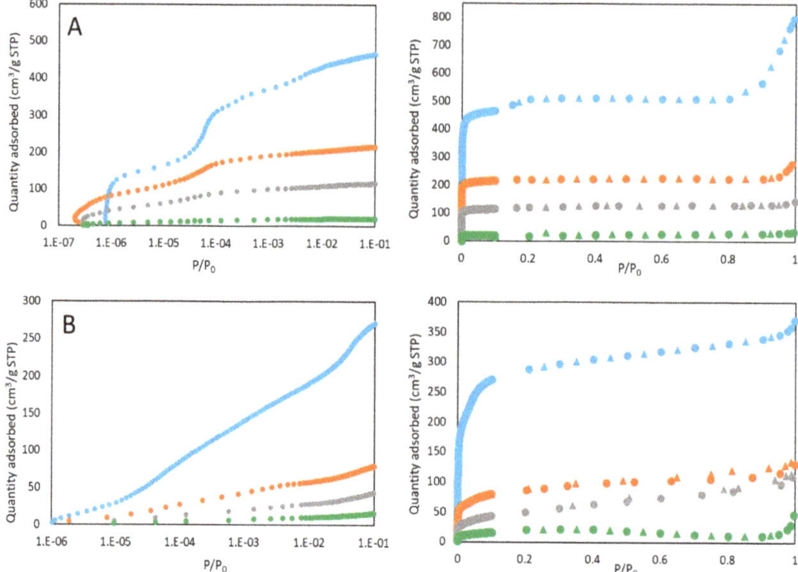

Figure 4. *Cont.*

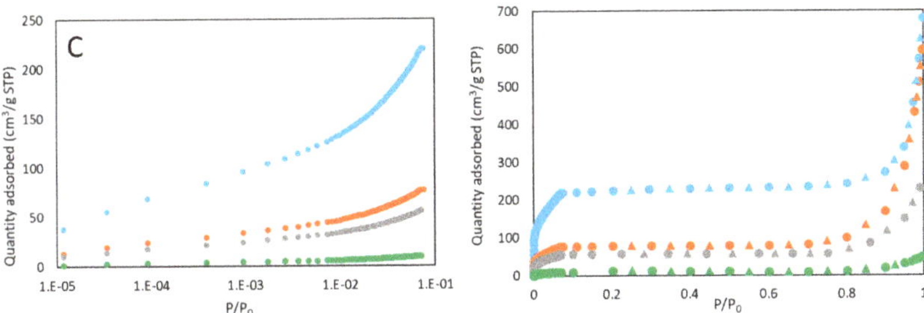

Figure 4. Adsorption (●) and desorption (▲) N_2 isotherms (77 K). Basolite C300 (**A**), Basolite F300 (**B**) and Basolite A100 (**C**). Original sample (Blue), 55.9 MPa (Orange), 111.8 MPa (Grey) and 186.3 MPa (Green). The graphs on the left are zoom of the low pressure zone (up to $P/P_0 = 0.1$) on logarithmic scale.

Table 1. Variations of Brunauer–Emmett–Teller (BET) surface area, Barrett–Joyner–Halenda (BJH) total mesopore volume, *t*-plot total micropore volume and average pore size with applied pressure.

Material	Applied Pressure (MPa)	BET Specific Surface Area (m^2/g)	BJH Mesopore Volume (cm^3/g)	*t*-Plot Micropore Volume (cm^3/g)	Average Pore Size (Å)
C300	0	1466.8	0.53	0.71	33.7
	55.9	671.2	0.09	0.29	22.9
	111.8	364.8	0.07	0.14	23.2
	186.3	67.6	0.06	0.02	47.3
F300	0	1015.4	0.15	0.27	16.5
	55.9	276.2	0.09	0.06	22.3
	111.8	173.4	0.11	0.02	28.8
	186.3	67.8	0.06	0.01	41.2
A100	0	655.9	0.77	0.28	64.0
	55.9	380.2	0.57	0.06	65.8
	111.8	362.1	0.54	0.05	65.1
	186.3	35.4	0.06	0.01	70.1

Finally, Basolite A100 presents also high specific surface and total pore volume losses, but with a different trend than the others. The first applied pressure (55.9 MPa) provokes the highest BET surface decrease (42.1%). However, the following pressure does not affect greatly either the BET surface or the pore volume (Table 1). In agreement, Ribeiro et al. [43], after application of 62 and 125 MPa to the material, observed null relation between applied pressure and the morphological parameters, obtaining really similar results for both pressures. Finally, at the maximum pressure, a BET surface and total pore volume decrease of 94.6% and 93.3% was reached, respectively.

Figure 5 illustrates the effect of the applied pressure on the crystallinity of both pristine and pressure-modified MOFs. The relative crystallinity is obtained by comparison of the main peak among the series of each material, assuming 100% of crystallinity for the commercial material (Table 2) [44]. In addition, peaks' displacement along the x-axis and the appearance of new ones may mean changes in the material structure (Table 3).

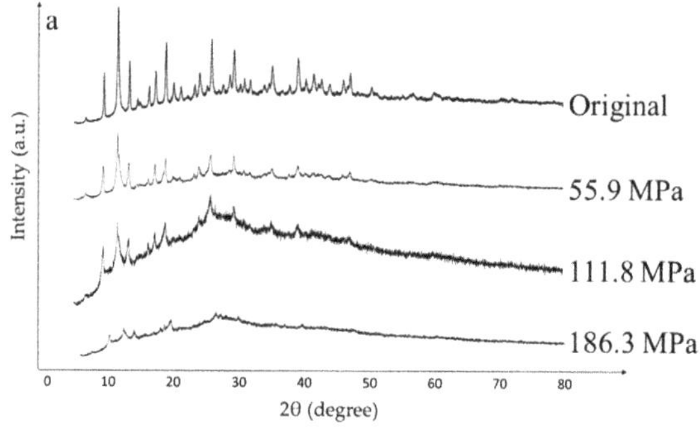

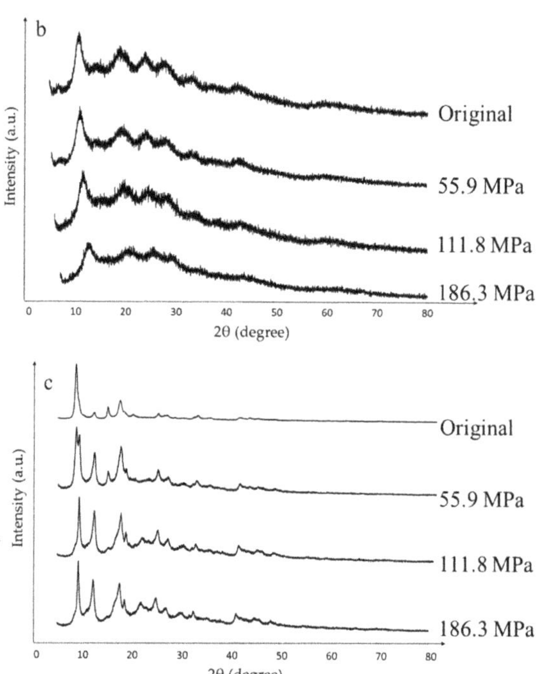

Figure 5. Powder X-ray diffraction (PXRD) patterns of the three materials at different applied pressures ((**a**): Basolite C300, (**b**): Basolite F300 and (**c**): Basolite A100). Applied pressures are ordered from top to bottom in increasing order (0, 55.9, 111.8 and 186.3 MPa).

Table 2. Relative crystallinity losses associated with applied pressure (referred to the original material).

Material	Pressure (MPa)	Crystallinity Loss (%)
C300	0	0
	55.9	19.51
	111.8	62.07
	186.3	69.93
F300	0	0
	55.9	0
	111.8	4.31
	186.3	14.08
A100	0	0
	55.9	69.35
	111.8	68.52
	186.3	68.38

Table 3. Cell total volume and lattice parameters for each structure depending on applied pressure.

Material	Pressure (MPa)	a (Å)	b (Å)	c (Å)	α (°)	β (°)	γ (°)	Volume (Å^3)
C300	0	25.9	25.9	25.9	90	90	90	17,452.2
	55.9	26.2	26.2	26.2	90	90	90	17,992.8
	111.8	26.2	26.2	26.2	90	90	90	17,992.8
	186.3	26.1	26.1	26.1	90	90	90	17,901.2
A100	0	16.1	6.56	13.2	90	90	90	1397.4
	55.9	6.56	14.3	14.8	90	105	90	1351.7
	111.8	6.43	12.9	16.2	90	108	90	1280.5
	186.3	5.83	13.7	16.1	90	110	90	1205.4

The pristine Basolite C300 powder X-ray diffraction (PXRD) pattern shows the typical peaks reported for this material at 2θ = 6.7°, 9.5°, 11.65°, 13.5°, 19.3° and 26°, in addition to three little peaks at 35.5°, 38.7° and 36.43°, which indicate some CuO and Cu_2O impurities [45,46]. After pressure is applied, the intensity of the peaks decreases progressively, indicative of crystallinity loss (Table 2). As its PXRD pattern is practically coincident with HKUST-1, a face-centered cubic structure is assumed [47], consisting of 16 copper atoms, 8 at the corners, as well as 6 at the center of the cube faces. Low-angle peaks (9.5°, 11.65° and 13.5°) present (220), (222) and (400) as Miller indices [48]. The net parameter (a) is obtained from the lattice plane of (222), Table 3. As shown, the cell volume remains practically unalterable (maximum variation of 1%), due to the structure rigidity [49]. In agreement, McKellar et al. reported variations of 2.6% for densification pressures of 3.9 GPa [50]. Likewise, the non-appearance of new crystalline peaks indicates that the cubic structure is maintained [51,52]. Therefore, the pressure effect on Basolite C300 consists of crystallinity destruction, in agreement with the BET surface area and pore volume reduction with the pressure, but remaining unaltered the cubic structure of unaltered cells. In agreement, Peng et al. [53] have studied the effect of mechanical pressure (up to 5 tons) onto HKUST-1, indicating a great micropore volume loss (N_2 physisorption analysis), in addition to a total collapse of the crystalline structure (PXRD analysis).

For pristine Basolite F300, a characteristic peak at 2θ = 11° is observed, despite the low resolution of the pattern as a consequence of the semiamorphous nature of the material and the elevated background values due to the iron fluorescence [54]. In fact, Basolite F300 is a distorted form of crystalline MIL-100(Fe) [55], and possesses a zeolite MTN topology [56]. In this case, the semiamorphous nature of the material just allows observing an increase of the amorphous matter with the pressure (Table 2). As it is a semiamorphous material, crystallinity is slightly reduced in relative terms [57], not reflected in the BET surface, which does not depend on crystallinity and is severely affected by increased pressure. Particle agglomeration causes the collapse of micropores, as it was demonstrated in Figure 4.

Finally, in the case of Basolite A100, this shows a structure practically coincident with MIL-53(Al) MOF, with characteristic peaks at $2\theta = 8.8°$, $15.25°$ and $17.75°$ [58]. The original pattern obtained is close to that of the large-pore (lp) phase of MIL-53(Al), which is coincident with an orthorhombic structure [59]. For this result, three different net parameters make up the structure and all the angles are right. Diffracting planes that match the characteristic peaks are (101), (011) and (210) [60,61]. As the applied pressure progresses, the appearance of new peaks around $2\theta = 20°$ indicates a movement to the narrow-pore (np) phase [62,63]. In this case, structural changes are high, due to the phase transition, reaching differences up to 13.7% for total cell volume (Table 3). In fact, according to Ghoufi et al. [64], the cell shows a monoclinic structure [65] from, approximately, 53 MPa onwards. As observed, transition to the np structure has an associated reduction of the total cell volume, as well as a decrease of the a parameter, in conjunction with increasing trends in the rest of the parameters, including the β angle. This increase in the β angle denotes a flattening on one of its axes [62,66], being these phase changes reversible [59]. Thus, this structure is characterized by its great flexibility. Regarding crystallinity, after the initial loss at the lowest pressure, it remains practically unchanged. The first applied pressure changes the material structure to np phase, which is known for its high resistance to external pressure and flexibility [63], thus maintaining crystallinity for successive applied pressures. The same occurs at 55.9 and 111.8 MPa in the case of BET available surface and total pore volume, which are practically maintained after an abrupt decrease despite the increase of applied pressure.

3.2. Performance Analysis

The gravimetric adsorption capacity of the samples was calculated from desorption analyses. Figure 6 plots adsorption capacity at different applied pressures as well as the relationship between adsorption capacity and BET specific surface area for each material. Basolite C300 shows a dramatic total decrease of its adsorption capacity with applied pressure, following a progressive trend as in the case of crystallinity and BET surface area. After the first pressure applied, some microporosity is still available, observing decreases of the adsorption capacity of 10.8% for nitrogen and 6.25% for methane. A further pressure increase will led to the total loss of adsorption capacity, BET surface and crystallinity. Additionally, the adsorption capacity/BET surface area ratio is practically linear at low applied pressures, showing certain dependence on BET surface. At the highest pressure, a sharp increase is observed, probably due to the increased role of active metal centres in the adsorption, once the crystalline structure was collapsed.

In the case of Basolite F300, a decreasing trend of the capacity of adsorption with the applied pressure is observed, the downward trend being more pronounced at the highest pressure (loss of 41.3% for N_2 and 36.5% for CH_4), Figure 6B. Adsorption capacity follows a similar trend to BJH total mesopore volume (Table 1), which could be related to its originally semi-amorphous properties, in which the adsorption capacity is not drastically reduced until a certain pressure limit. The accessibility to metal adsorption sites is maintained due to the appearance of mesopores and, thus, the intracrystalline diffusivity increases. This increase in accessibility is closely related to the smooth downward trend in adsorption capacity, showing an almost linear relationship with the specific available surface.

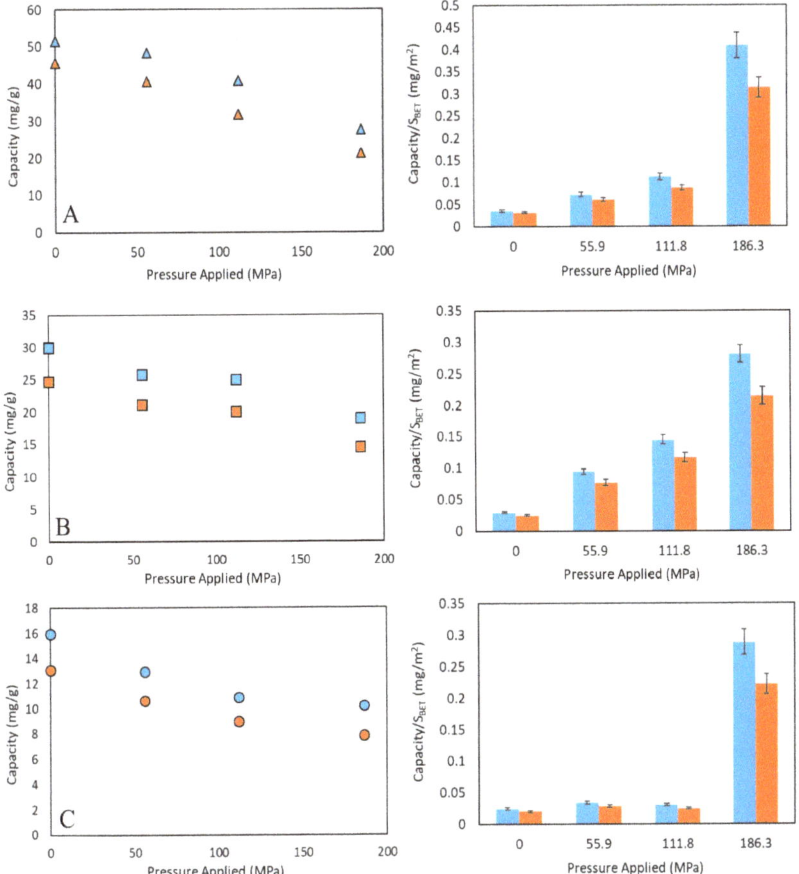

Figure 6. Adsorption capacity for pure methane (blue) and nitrogen (orange) for different applied pressures at 298 K and 0.1 MPa of total pressure (figures on the left), and its relation with BET specific surface area (figures on the right). Basolite C300 ((**A**), ▲), Basolite F300 ((**B**), ■) and Basolite A100 ((**C**), ●).

For Basolite A100, a sharp decrease is observed after the first applied pressure, coincident with the asymptotic trend of BET surface area to the last applied pressures (Figure 6). This may be due to the presence of pure CH_4 and N_2, which provokes the transition to the lp phase at ambient conditions, thus increasing the adsorption capacity by increasing the accessibility to metallic adsorption centers [67]. In fact, from adsorption capacity/BET surface ratio, a constant behavior is observed at the lowest pressures, and a sudden increase at the highest one, due to the drastic reduction of specific surface area after the transition to the lp phase which allows the metallic adsorption centers to have great relevance in the adsorption. Comparing this with other techniques, Finsy et al. [68] have studied the effect of making pellets of MIL-53(Al) using polyvinyl alcohol as a binder. They indicated a reduction of 32% in micropore volume with a pore accessibility reduction of 19% in the best of the cases, which hinders adsorption processes. In fact, it must be pointed out that the presence of a binder can affect the adsorption behavior of the material [69].

From Figure 6 it is observed that the relative adsorption capacity decreases are higher for N_2 than for CH_4, and it may be related to metal adsorption centers being available, and more selective towards CH_4 than N_2 [70]. Thus, after surface area and total pore volume reduction, the available active metallic centers play a more relevant role in the selective gas adsorption, especially in Basolite C300 and A100 cases. The influence of the applied pressure in the CH_4/N_2 selectivity (mass basis) is shown in Figure 7. The increasing slope for Basolite C300 is markedly higher than for the other materials, due to the presence of a higher percentage of metal in its structure (31.5% of copper, vs. 21.2% and 12.9% of iron and aluminum for Basolite F300 and A100, respectively). Therefore, the higher metal content in the structure, the greater the selectivity-increasing trend with applied pressure.

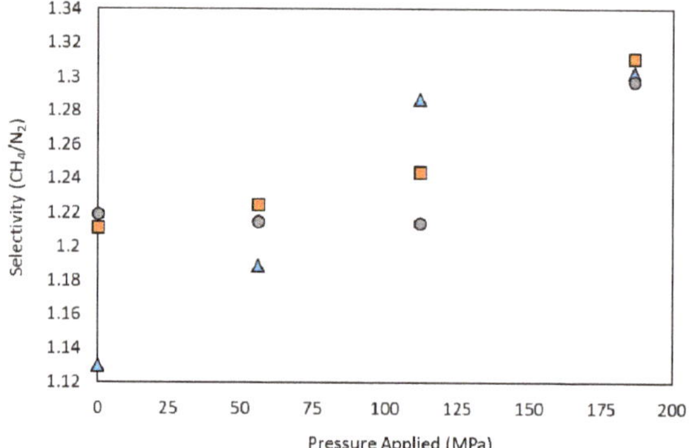

Figure 7. Adsorption CH_4/N_2 selectivity (mass basis) for each material at different applied pressures, at 298 K and 0.1 MPa of total pressure. Basolite C300 (▲), Basolite F300 (■) and Basolite A100 (●).

Breakthrough adsorption curves for CH_4 and N_2 in a fixed bed are shown in Figure 8. In general, for all the samples, breakthrough times (hence, adsorption capacity) are higher for CH_4 than for N_2, being attributed to the presence of metallic active adsorption sites and the difference in polarizability of both molecules [70].

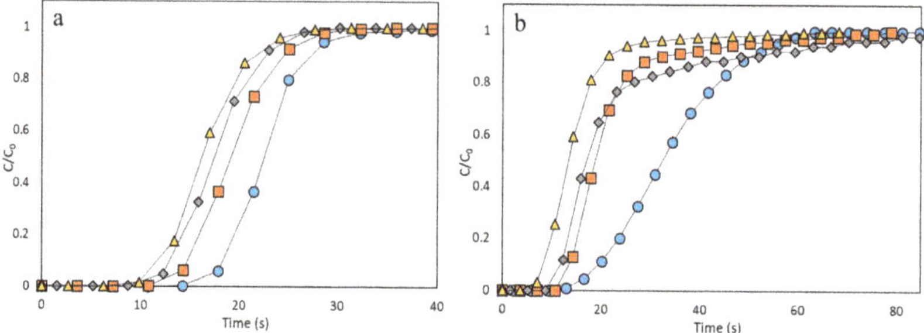

Figure 8. *Cont.*

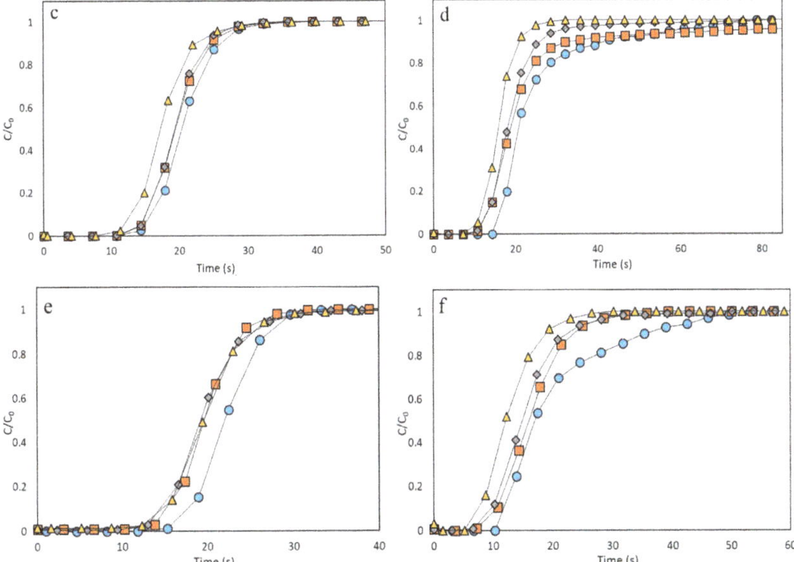

Figure 8. Adsorption breakthrough curves for CH_4 and N_2 onto the three MOFs at different applied pressures (Original: blue ●, 55.9 MPa: orange ■, 111.8 MPa: grey ♦, 186.3 MPa: yellow ▲). Basolite C300 ((**a**): methane, (**b**): nitrogen), Basolite F300 ((**c**): methane, (**d**): nitrogen) and Basolite A100 ((**e**): methane, (**f**): nitrogen). Black lines are used to guide the view.

In the case of Basolite C300, there is a slight difference in the slope between the original material and the others, most obvious in N_2 case. As N_2 molecular size is lower than CH_4 (3.65 and 3.82 Å, respectively), this molecule may penetrate in narrower pores than CH_4. Likewise, a decrease in the Knudsen diffusion coefficient led to more inclined curves [71,72]. The Knudsen diffusion coefficient (D_K) depends on the pore diameter (d_p), since the other parameters are constant for all the experiments. Variations in the Knudsen diffusion coefficient affects directly the breakthrough curve, since it influences adsorbate mass transfer kinetics within the microporous adsorbent.

The reduction in total available specific surface, especially in the micropores zone (low P/P_0), indicates that these narrower pores have been totally collapsed by compression (Figure 4). This collapse is common in MOFs when pressure is applied, due to their extraordinary initial porosity [73]. This provokes the following applied pressures to present less-inclined breakthrough curve slopes, but also having less adsorptive capacity, as evidenced by the x-axis order of their breakthrough times (Figure 8). Breakthrough times follow, approximately, the same trend as BET surface area.

In the case of Basolite F300, all the samples, except the original one, show the same slope for breakthrough curves, but in this case the difference is lower than in C300 case. The original sample presents a more inclined breakthrough curve for both adsorbates, which indicates a lower Knudsen diffusion coefficient. Applied pressure modified the pore structure, plugging the micropores, but without reducing greatly the total pore volume by the appearance of mesopores that facilitate the penetration, so the differences in accessibility are softer (Figure 4). The high resemblance between the 55.9 and 111.8 MPa curves (Figure 8) is remarkable and can be related to the no-clear total mesopore volume dependence on pressure (Table 1). The appearance of mesopores in the structure enhance the intracrystalline diffusivity [74], which may be the dominant factor in this case, since the crystallinity is not great affected by mechanical pressure. Dhakshinamoorthy et al. have studied the high relevance of the intracrystalline diffusivity in this material, applied to the case of an oxidation reaction [75].

Finally, in the case of Basolite A100, despite the change from orthorhombic to monoclinic structure and the total cell volume reduction, the presence of pure CH_4 and N_2 causes the return to the lp phase at ambient conditions, for which the penetration is easier, obtaining a steep curve for all cases due to the structure flexibility [76]. As is also observed, the breakthrough curves of the original material present more resistance than the others, especially for N_2. Despite the return to lp phase, the agglomeration provoked a certain irreversible reduction of micropore volume, which increases the Knudsen diffusion coefficient since the average available pore size is higher (Table 1). It is remarkable that differences in CH_4 breakthrough times follow almost the same trend as crystallinity, whereas in the N_2 case, the trend is similar to specific surface or total pore volume.

4. Conclusions

Structural and morphological transformations of three MOFs (Basolite C300, Basolite F300 and Basolite A100), as well as CH_4 and N_2 uptakes variation, were studied after appliance of mechanical pressure to the materials. Basolite C300, a rigid crystalline material, experimented a dramatic and progressive loss of crystallinity, as well as surface area and pore volume, which implies lower adsorption capacity due to its characteristic pores collapse. In the case of Basolite F300, a semiamorphous material, this experienced also a high decrease of surface area and micropore collapse due to agglomeration, but keeping total pore volume due to the appearance of mesopores in the structure. This transformation implies an increase of intracrystalline diffusivity and, then, lower adsorption capacity losses. For Basolite A100, a flexible crystalline MOF, a transformation is observed from orthorhombic disposition to monoclinic structure from 55.9 MPa onwards, in addition to high permanent losses of microporosity due to agglomeration. This structure change is reversible, returning to the lp phase in presence of CH_4 and N_2 at ambient conditions. This fact increases the accessibility to metallic active centers and an asymptotic decrease of the adsorption capacity is observed. Additionally, the key role of metal active sites in the CH_4/N_2 selectivity was pointed out. In fact, an increased selectivity for the three MOFs was observed with the applied pressure, decreasing this positive effect in the order: Basolite C300 (Cu, 31.5%) > Basolite F300 (Fe, 21.2%) > Basolite A100 (Al, 12.9%). However, the total gravimetric adsorption capacity has experienced high losses for all of them. Despite that, Basolite C300 stands out above the other two. It has greater adsorption capacity and also a higher metallic content in its structure. In addition, it is able to retain 94% of its adsorption capacity when applying a pressure of 55.9 MPa, enough to increase its particle size and be able to operate in real adsorption stages.

Author Contributions: Conceptualization, S.O.; methodology, E.D.; formal analysis, D.U.; investigation, D.U.; data curation, E.D.; writing—original draft preparation, D.U.; writing—review and editing, E.D. and S.O.; supervision, S.O.; All authors have read and agreed to the published version of the manuscript.

Funding: This research was funded by the Research Fund for Coal and Steel of the European Union, contract 754077 METHENERGY PLUS.

Acknowledgments: D. Ursueguía acknowledges the Spanish Government for the FPU fellowship (FPU18/01448). The authors would like to acknowledge the technical support provided by *Servicios Científico-Técnicos* de la Universidad de Oviedo.

Conflicts of Interest: The authors declare that they have no known competing financial interests or personal relationships that could have appeared to influence the work reported in this paper.

References

1. Newell, R.; Raimi, D.; Aldana, G. *Global Energy Outlook 2019: The Next Generation of Energy*; Report 8–19; Resources for the Future: Washington, DC, USA, 2019.
2. Coronavirus Has Reminded Us How Much We Depend on Electricity. Available online: www.weforum.org/agenda/2020/03/coronavirus-crisis-future-of-energy (accessed on 29 March 2020).
3. Ferreira, A.; Ribeiro, A.; Kulaç, S.; Rodrigues, A. Methane purification by adsorptive processes on MIL-53(Al). *Chem. Eng. Sci.* **2015**, *124*, 79–95. [CrossRef]
4. Cavenati, S.; Grande, C.; Rodrigues, A. Separation of $CH_4/CO_2/N_2$ mixtures by layered pressure swing adsorption for upgrade of natural gas. *Chem. Eng. Sci.* **2006**, *61*, 3893–3906. [CrossRef]

5. Moreira, M.; Ribeiro, A.; Ferreira, A.; Rodrigues, A. Cryogenic pressure temperature swing adsorption process for natural gas upgrade. *Sep. Purif. Technol.* **2017**, *173*, 339–356. [CrossRef]

6. Campo, M.; Ribeiro, A.; Ferreira, A.; Santos, J.; Lutz, C.; Loureiro, J.; Rodrigues, A. Carbon dioxide removal for methane upgrade by a VSA process using an improved 13X zeolite. *Fuel Process. Technol.* **2016**, *143*, 185–194. [CrossRef]

7. Alonso-Vicario, A.; Ochoa-Gómez, J.; Gil-Río, S.; Gómez-Jiménez, O.; Ramírez-López, C.; Torrecilla-Soria, J.; Domínguez, A. Purification and upgrading of biogas by pressure swing adsorption on synthetic and natural zeolites. *Microporous Mesoporous Mater.* **2010**, *134*, 100–107. [CrossRef]

8. Li, P.; Tezel, H. Adsorption separation of N_2, O_2, CO_2 and CH_4 gases by β-zeolite. *Microporous Mesoporous Mater.* **2007**, *98*, 94–101. [CrossRef]

9. Yi, H.; Li, F.; Ning, P.; Tang, X.; Peng, J.; Li, Y.; Deng, H. Adsorption separation of CO_2, CH_4 and N_2 on microwave activated carbon. *Chem. Eng. J.* **2013**, *215*, 635–642. [CrossRef]

10. Singh, A.; Kumar, J. Fugitive methane emissions from Indian coal mining and handling activities: Estimates, mitigation and opportunities for its utilization to generate clean energy. *Energy Procedia* **2016**, *90*, 336–348. [CrossRef]

11. Cluff, D.; Kennedy, G.; Bennett, J.; Foster, P. Capturing energy from ventilation air methane a preliminary design for a new approach. *Appl. Therm. Eng.* **2015**, *90*, 1151–1163. [CrossRef]

12. Karakurt, I.; Aydin, G.; Aydiner, K. Mine ventilation air methane as a sustainable energy source. *Renew. Sustain. Energy Rev.* **2011**, *15*, 1042–1049. [CrossRef]

13. Li, Q.; Ruan, M.; Zheng, Y.; Mei, X.; Lin, B. Investigation on the selective adsorption and separation properties of coal mine methane in ZIF-68 by molecular simulations. *Adsorption* **2017**, *23*, 163–174. [CrossRef]

14. Norouzi, A. Modeling of adsorption in a packed bed tower, the case study of methane removal and parametric calculation. *J. Environ. Treat. Tech.* **2019**, *7*, 324–333.

15. Grünker, R.; Bon, V.; Müller, P.; Stoeck, U.; Krause, S.; Mueller, U.; Senkovska, I.; Kaskel, S. A new metal-organic framework with ultra-high surface area. *Chem. Commun.* **2014**, *50*, 3450–3452. [CrossRef] [PubMed]

16. Senkovska, I.; Kaskel, S. High pressure methane adsorption in the metal-organic frameworks $Cu_3(btc)_2$, $Zn_2(bdc)_2dabco$, and $Cr_3F(H2O)_2O(bdc)_3$. *Microporous Mesoporous Mater.* **2008**, *112*, 108–115. [CrossRef]

17. Rowsell, J.; Yaghi, O. Metal-organic frameworks: A new class of porous materials. *Microporous Mesoporous Mater.* **2004**, *73*, 3–14. [CrossRef]

18. Liang, Z.; Marshall, M.; Chaffee, A. CO_2 adsorption-based separation by metal organic framework (Cu-BTC) versus zeolite (13X). *Energy Fuels* **2009**, *23*, 2785–2789. [CrossRef]

19. Blanco-Brieva, G.; Campos-Martin, J.M.; Al-Zahrani, S.M.; Fierro, J.L.G. Effectiveness of metal-organic frameworks for removal of refractory organo-sulfur compound present in liquid fuels. *Fuel* **2011**, *90*, 190–197. [CrossRef]

20. Jiao, L.; Seow, J.; Skinner, W.; Wang, Z.; Jiang, H. Metal-organic frameworks: Structures and functional applications. *Mater. Today* **2019**, *27*, 43–68. [CrossRef]

21. Arami-Niya, A.; Birkett, G.; Zhu, Z.; Rufford, T. Gate opening effect of zeolitic imidazolate framework ZIF-7 for adsorption of CH_4 and CO_2 from N_2. *J. Mater. Chem. A* **2017**, *5*, 21389–21399. [CrossRef]

22. Eyer, S.; Stadie, N.; Borgschulte, A.; Emmenegger, L.; Mohn, J. Methane preconcentration by adsorption: A methodology for materials and conditions selection. *Adsorption* **2014**, *20*, 657–666. [CrossRef]

23. Bastin, L.; Bárcia, P.; Hurtado, E.; Silva, J.; Rodrigues, A.; Chen, B. A microporous metal-organic framework for separation of CO_2/N_2 and CO_2/CH_4 by fixed-bed adsorption. *J. Phys. Chem. C* **2008**, *112*, 1575–1581. [CrossRef]

24. Bloch, E.; Queen, W.; Krishna, R.; Zadrozny, J.; Brown, C.; Long, J. Hydrocarbon separations in a metal-organic framework with open iron(II) coordination sites. *Science* **2012**, *335*, 1606–1610. [CrossRef] [PubMed]

25. Evans, J.; Garai, B.; Reinsch, H.; Li, W.; Dissegna, S.; Bon, V.; Senkovska, I.; Fischer, R.; Kaskel, S.; Janiak, C.; et al. Metal-organic frameworks in Germany: From synthesis to function. *Coord. Chem. Rev.* **2019**, *380*, 378–418. [CrossRef]

26. Koekemoer, A.; Luckos, A. Effect of material type and particle size distribution on pressure drop in packed beds of large particles: Extending the Ergun equation. *Fuel* **2015**, *158*, 232–238. [CrossRef]

27. Malkoc, E.; Nuhoglu, Y.; Abali, Y. Cr(VI) adsorption by waste acorn of *Quercus* Ithaburensis in fixed beds: Prediction of breakthrough curves. *Chem. Eng. J.* **2006**, *119*, 61–68. [CrossRef]

28. Edubilli, S.; Gumma, S. A systematic evaluation of UiO-66 metal organic framework for CO_2/N_2 separation. *Sep. Purif. Technol.* **2019**, *224*, 85–94. [CrossRef]

29. Hou, P.; Orikasa, H.; Itoi, H.; Nishihara, H.; Kyotani, T. Densification of ordered microporous carbons and controlling their micropore size by hot-pressing. *Carbon* **2007**, *45*, 2011–2016. [CrossRef]

30. Tian, T.; Zeng, Z.; Vulpe, D.; Casco, M.; Divitini, G.; Midgley, P.; Silvestre-Albero, J.; Tan, J.; Moghadam, P.; Fairen-Jimenez, D. A sol-gel monolithic metal-organic framework with enhanced methane uptake. *Nat. Mater.* **2018**, *17*, 174–179. [CrossRef]

31. Hu, Z.; Wang, Y.; Shah, B.; Zhao, D. CO_2 capture in metal-organic framework adsorbents: An engineering perspective. *Adv. Sustain. Syst.* **2018**, *3*, 1800080. [CrossRef]

32. Beurroies, I.; Boulhout, M.; Llewellyn, P.; Kuchta, B.; Férey, G.; Serre, C.; Denoyel, R. Using pressure to provoke the structural transition of metal-organic frameworks. *Angew. Chem. Int. Ed.* **2010**, *49*, 7526–7529. [CrossRef]

33. Manos, G.; Dunne, L. Predicting the features of methane adsorption in large pore metal-organic frameworks for energy storage. *Nanomaterials* **2018**, *8*, 818. [CrossRef]

34. Purewall, J.J.; Liu, D.; Yang, J.; Sudik, A.; Siegel, D.J.; Maurer, S.; Müller, U. Increased volumetric hydrogen uptake of MOF-5 by powder densification. *Int. J. Hydrogen Energy* **2012**, *37*, 2723–2727. [CrossRef]

35. Nandasiri, M.; Jambovane, S.; McGrail, B.; Schaef, H.; Nune, S. Adsorption, separation, and catalytic properties of densified metal-organic frameworks. *Coord. Chem. Rev.* **2016**, *311*, 38–52. [CrossRef]

36. Anderlean, O.; Blanita, G.; Borodi, G.; Lazar, M.; Misan, I.; Coldea, I.; Lupu, D. Volumetric hydrogen adsorption capacity of densified MIL-101 monoliths. *Int. J. Hydrogen Energy* **2013**, *38*, 7046–7055.

37. Majchrzak-Kuçeba, I.; Sciubidlo, A. Shaping metal-organic framework (MOF) powder materials for CO_2 capture applications—A thermogravimetric study. *J. Therm. Anal. Calorim.* **2019**, *138*, 4139–4144. [CrossRef]

38. Yuan, S.; Sun, X.; Pang, J.; Sun, D.; Liu, D.; Zhou, H. PCN-250 under pressure: Sequential phase transformation and the implications for MOF densification. *Joule* **2017**, *1*, 806–815. [CrossRef]

39. Thommes, M.; Kaneko, K.; Neimark, A.; Olivier, J.; Rodríguez-Reinoso, F.; Rouquerol, J.; Sing, K. Physisorption of gases, with special reference to the evaluation of surface and pore size distribution (IUPAC technical report). *Pure Appl. Chem.* **2015**, *87*, 1051–1069. [CrossRef]

40. Sing, K.; Everett, D.; Haul, R.; Moscou, L.; Pierotti, R.; Rouquérol, J.; Siemieniewska, T. Reporting physisorption data for gas/solid systems with special reference to the determination of surface area and porosity. *Pure Appl. Chem.* **1985**, *57*, 603–619. [CrossRef]

41. Muttakin, M.; Mitra, S.; Thu, K.; Ito, K.; Saha, B. Theoretical framework to evaluate minimum desorption temperature for IUPAC classified adsorption isotherms. *Int. J. Heat Mass Trans.* **2018**, *122*, 795–805. [CrossRef]

42. Casco, M.; Fernández-Catalá, J.; Martínez-Escandell, M.; Rodríguez-Reinoso, F.; Ramos-Férnandez, E.; Silvestre-Albero, J. Improved mechanical stability of HKUST-1 in confined nanospace. *Chem. Commun.* **2015**, *51*, 14191–14194. [CrossRef]

43. Ribeiro, R.; Antunes, C.; Garate, A.; Portela, A.; Plaza, M.; Mota, J.; Esteves, A. Binderless shaped metal-organic framework particles: Impact on carbon dioxide adsorption. *Microporous Mesoporous Mater.* **2019**, *275*, 111–121. [CrossRef]

44. Wang, J.; Liu, B.; Nakata, K. Effects of crystallinity, {001}/{101} ratio, and Au decoration on the photocatalytic activity of anatase TiO_2 crystals. *Chin. J. Catal.* **2019**, *40*, 403–412. [CrossRef]

45. Nobar, S. Cu-BTC synthesis, characterization and preparation for adsorption studies. *Mater. Chem. Phys.* **2018**, *213*, 343–351. [CrossRef]

46. Schlichte, K.; Kratzke, T.; Kaskel, S. Improved synthesis, thermal stability and catalytic properties of the metal-organic framework compound $Cu_3(BTC)_2$. *Microporous Mesoporous Mater.* **2004**, *73*, 81–88. [CrossRef]

47. Prestipino, C.; Regli, L.; Vitillo, J.G.; Bonino, F.; Damin, A.; Lamberti, C.; Zecchina, A.; Solari, P.; Kongshaug, K.; Bordiga, S. Local structure of framework Cu(II) in HKUST-1 metallorganic framework: Spectroscopic characterization upon activation and interaction with adsorbates. *Chem. Mater.* **2006**, *18*, 1337–1346. [CrossRef]

48. Yang, A.; Li, P.; Zhong, J. Facile preparation of low-cost HKUST-1 with lattice vacancies and high-efficiency adsorption for uranium. *RCS Adv.* **2019**, *9*, 10320–10325. [CrossRef]

49. Yang, K.; Zhou, G.; Xu, Q. The elasticity of MOFs under mechanical pressure. *RCS Adv.* **2016**, *44*, 37506–37514. [CrossRef]

50. McKellar, S.; Moggach, S. Structural studies of metal-organic frameworks under high pressure. *Acta Cryst.* **2015**, *B71*, 587–607. [CrossRef]

51. Terracina, A.; Todaro, M.; Mazaj, M.; Agnello, S.; Gelardi, F.; Buscarino, G. Unveiled the source of the structural instability of HKUST-1 powders upon mechanical compaction: Definition of a fully preserving tableting method. *J. Phys. Chem. C* **2018**, *123*, 1730–1741. [CrossRef]

52. Wu, H.; Yildirim, T.; Zhou, W. Exceptional mechanical stability of highly porous zirconium metal-organic framework UiO-66 and its important implications. *J. Phys. Chem. Lett.* **2013**, *4*, 925–930. [CrossRef]

53. Peng, Y.; Krungleviciute, V.; Eryazici, I.; Hupp, J.; Farha, O.; Yildirim, T. Methane storage in metal-organic frameworks: Current records, surprise findings, and challenges. *J. Am. Chem. Soc.* **2013**, *135*, 11887–11894. [CrossRef] [PubMed]

54. Sánchez-Sánchez, M.; Asua, I.; Ruano, D.; Díaz, K. Direct synthesis, structural features, and enhanced catalytic activity of the Basolite F300-like semiamorphous Fe-BTC framework. *Cryst. Growth Des.* **2015**, *15*, 4498–4506. [CrossRef]

55. Dhakshinamoorthy, A.; Alvaro, M.; Horcajada, P.; Gibson, E.; Vishnuvarthan, M.; Vimont, A.; Greneche, J.; Serre, C.; Daturi, M.; Garcia, H. Comparison of porous iron trimesates Basolite F300 and MIL-100(Fe) as heterogeneous catalysts for Lewis acid and oxidation reactions: Roles of structural defects and stability. *ACS Catal.* **2012**, *2*, 2060–2065. [CrossRef]

56. Seo, Y.; Yoon, J.; Lee, J.; Lee, U.; Hwang, Y.; Jun, C.; Horcajada, P.; Serre, C.; Chang, J. Large scale fluorine-free synthesis of hierarchically porous iron(III) trimesate MIL-100(Fe) with a zeolite MTN topology. *Microporous Mesoporous Mater.* **2012**, *157*, 137–145. [CrossRef]

57. Bennet, T.; Cheetham, A. Amorphous Metal-Organic Frameworks. *Acc. Chem. Res.* **2014**, *47*, 1555–1562. [CrossRef] [PubMed]

58. Chowdhury, T.; Zhang, L.; Zhang, J.; Aggarwal, S. Removal of arsenic(III) from aqueous solution using metal organic framework-graphene oxide nanocomposite. *Nanomaterials* **2018**, *8*, 1062. [CrossRef]

59. Mishra, P.; Uppara, H.; Mandal, B.; Gumma, S. Adsorption and separation of carbon dioxide using MIL-53(Al) metal-organic framework. *Ind. Eng. Chem. Res.* **2014**, *53*, 19747–19753. [CrossRef]

60. Llewellyn, P.; Horcajada, P.; Maurin, G.; Devic, T.; Rosenbach, N.; Bourrelly, S.; Serre, C.; Vincent, D.; Loera-Serna, S.; Filinchuk, Y.; et al. Complex adsorption of short linear alkanes in the flexible metal-organic-framework MIL-53(Fe). *J. Am. Chem. Soc.* **2009**, *131*, 13002–13008. [CrossRef]

61. Alaerts, L.; Kirschhock, C.; Maes, M.; Veen, M.; Finsy, V.; Depla, A.; Martens, J.; Baron, G.; Jacobs, P.; Denayer, J.; et al. Selective adsorption and separation of xylene isomers and ethylbenzene with the microporous vanadium(IV) terephthalate MIL-47. *Angew. Chem.* **2007**, *119*, 4371–4375. [CrossRef]

62. Neimark, A.; Coudert, F.; Triguero, C.; Boutin, A.; Fuchs, A.; Beurroies, I.; Denoye, R. Structural transitions in MIL-53(Cr): View from outside and inside. *Langmuir* **2011**, *27*, 4734–4741. [CrossRef]

63. Serra-Crespo, P.; Dikhtiarenko, A.; Stavitski, E.; Juan-Alcañiz, J.; Kapteijn, F.; Coudert, F.; Gascon, J. Experimental evidence of negative linear compressibility in the MIL-53 metal-organic framework family. *CrystEngComm* **2015**, *17*, 276–280. [CrossRef] [PubMed]

64. Ghoufi, A.; Subercaze, A.; Ma, Q.; Yot, P.; Ke, Y.; Puente-Orench, I.; Devic, T.; Guillerm, V.; Zhong, C.; Serre, C.; et al. Comparative guest, thermal, and mechanical breathing of the porous metal organic framework MIL-53(Cr): A computational exploration supported by experiments. *J. Phys. Chem. C* **2012**, *116*, 13289–13295. [CrossRef]

65. Reinsch, H.; Pillai, R.; Siegel, R.; Senker, J.; Lieb, A.; Maurin, G.; Stock, N. Structure and properties of Al-MIL-53-ADP, a breathing MOF based on the aliphatic linker molecule adipic acid. *Dalton Trans.* **2016**, *45*, 4179–4186. [CrossRef] [PubMed]

66. Ghysels, A.; Vanduyfhys, L.; Vandichel, M.; Waroquier, M.; Speybroeck, V.; Smit, B. On the thermodynamics of framework breathing: A free energy model for gas adsorption in MIL-53. *J. Phys. Chem. C* **2013**, *117*, 11540–11554. [CrossRef]

67. Boutin, A.; Couck, S.; Coudert, F.; Serra-Crespo, P.; Gascon, J.; Kapteijn, F.; Fuchs, A.; Denayer, J. Thermodynamic analysis of the breathing of amino-functionalized MIL-53(Al) upon CO_2 adsorption. *Microporous Mesoporous Mater.* **2011**, *140*, 108–113. [CrossRef]

68. Finsy, V.; Ma, L.; Alaerts, L.; Vos, D.; Baron, G.; Denayer, J. Separation of CO_2/CH_4 mixtures with the MIL-53(Al) metal-organic framework. *Microporous Mesoporous Mater.* **2009**, *120*, 221–227. [CrossRef]

69. Valekar, A.; Cho, K.; Lee, U.; Lee, J.; Yoon, J.; Hwang, Y.; Lee, S.; Cho, S.; Chang, J. Shaping of porous metal-organic framework granules using mesoporous p-alumina as a binder. *RSC Adv.* **2017**, *7*, 55767–55777. [CrossRef]

70. Ursueguía, D.; Díaz, E.; Ordóñez, S. Adsorption of methane and nitrogen on Basolite MOFs: Equilibrium and kinetic studies. *Microporous Mesoporous Mater.* **2020**, *298*, 110048. [CrossRef]

71. Kosuge, K.; Kubo, S.; Kikukawa, N.; Takemori, M. Effect of pore structure in mesoporous silicas on VOC dynamic adsorption/desorption performance. *Langmuir* **2007**, *23*, 3095–3102. [CrossRef]

72. Murillo, R.; García, T.; Aylón, E.; Callén, M.; Navarro, M.; López, J.; Mastral, A. Adsorption of phenanthrene on activated carbons: Breakthrough curve modeling. *Carbon* **2004**, *42*, 2009–2017. [CrossRef]

73. Howarth, A.; Liu, Y.; Li, P.; Li, Z.; Wang, T.; Hupp, J.; Farha, O. Chemical, thermal and mechanical stabilities of metal-organic frameworks. *Nat. Rev. Mater.* **2016**, *1*, 1–15. [CrossRef]

74. Mehlhorn, D.; Valiullin, R.; Kärger, J.; Cho, K.; Ryoo, R. Intracrystalline diffusion in mesoporous zeolites. *ChemPhysChem* **2012**, *13*, 1495–1499. [CrossRef] [PubMed]

75. Dhakshinamoorthy, A.; Alvaro, M.; Hwang, Y.; Seo, Y.; Corma, A.; García, H. Intracrystalline diffusion in Metal Organic Framework during heterogeneous catalysis: Influence of particle size on the activity of MIL-100 (Fe) for oxidation reactions. *Dalton Trans.* **2011**, *40*, 10719–10724. [CrossRef]

76. Lyubchyk, A.; Esteves, I.; Cruz, F.; Mota, J. Experimental and theoretical studies of supercritical methane adsorption in the MIL-53(Al) metal organic framework. *J. Phys. Chem. C* **2011**, *115*, 20628–20638. [CrossRef]

 *nanomaterials*

Article

BaFe$_{1-x}$Cu$_x$O$_3$ Perovskites as Active Phase for Diesel (DPF) and Gasoline Particle Filters (GPF)

Verónica Torregrosa-Rivero, Carla Moreno-Marcos, Vicente Albaladejo-Fuentes, María-Salvadora Sánchez-Adsuar and María-José Illán-Gómez *

Carbon Materials and Environment Research Group, Department of Inorganic Chemistry, Faculty of Science, University of Alicante, Av. Alicante s/n, San Vicente del Raspeig, 03690 Alicante, Spain; vero.torregrosa@ua.es (V.T.-R.); carlamorenomarcos1@gmail.com (C.M.-M.); vicentealbaladejo@gmail.com (V.A.-F.); dori@ua.es (M.-S.S.-A.)
* Correspondence: illan@ua.es; Tel.: +34-965-903-975

Received: 29 July 2019; Accepted: 29 October 2019; Published: 31 October 2019

Abstract: BaFe$_{1-x}$Cu$_x$O$_3$ perovskites (x = 0, 0.1, 0.3 and 0.4) have been synthetized, characterized and tested for soot oxidation in both Diesel and Gasoline Direct Injection (GDI) exhaust conditions. The catalysts have been characterized by BET, ICP-OES, SEM-EDX, XRD, XPS, H$_2$-TPR and O$_2$-TPD and the results indicate the incorporation of copper in the perovskite lattice which leads to: (i) the deformation of the initial hexagonal perovskite structure for the catalyst with the lowest copper content (BFC1), (ii) the modification to cubic from hexagonal structure for the high copper content catalysts (BFC3 and BFC4), (iii) the creation of a minority segregated phase, BaO$_x$-CuO$_x$, in the highest copper content catalyst (BFC4), (iv) the rise in the quantity of oxygen vacancies/defects for the catalysts BFC3 and BFC4, and (v) the reduction in the amount of O$_2$ released in the course of the O$_2$-TPD tests as the copper content increases. The BaFe$_{1-x}$Cu$_x$O$_3$ perovskites catalyze both the NO$_2$-assisted diesel soot oxidation (500 ppm NO, 5% O$_2$) and, to a lesser extent, the soot oxidation under fuel cuts GDI operation conditions (1% O$_2$). BFC0 is the most active catalysts as the activity seems to be mainly related with the amount of O$_2$ evolved during an. O$_2$-TPD, which decreases with copper content.

Keywords: Iron-based perovskites; copper; NO oxidation to NO$_2$; NO$_2$-assisted diesel soot oxidation; soot oxidation under GDI exhaust conditions

---

1. Introduction

The high toxicity of particulate matter (PM) or soot, mainly produced by internal combustion engines, is well established. As in Europe the transport sector generates a 14% of PM2.5 (particulates with a size lesser than 2.5 m, the most hazardous portion), the actual European emissions legislation (Euro 6c) for new passengers vehicles meet or decreases the Particulates Numbers (PN) generated by Gasoline Direct Injection (GDI) to the level corresponding to Diesel engines [1]. GDI engines are considered more effective than diesel engines due to the substantial decrease of fuel intake and CO$_2$ emissions [2]. Consequently, a growth in the US and European market of GDI cars is being observed. To attend the actual European emission legislation, the use of Gasoline Particulate Filter (GPF) is necessary for GDI vehicles, as the Diesel Particulate Filter (DPF) was for Diesel vehicles. In both filters, periodic regeneration is demanded to avoid soot accumulation in the channels of the filter [3–5].

In Diesel engine, as NO$_2$ promotes soot oxidation, a catalyst able to oxidize NO to NO$_2$ is incorporated into the DPF to carry out the NO$_2$-assisted soot oxidation. In fact, several systems (most of them containing Platinum Group Metals, PGM) were developed and implemented in diesel cars to oxidize soot. However, recently, the EU [6] has highlighted that the use of critical raw materials (such as PGM) must be optimized.

Based on the success of DPF in diesel engines, GPF is proposed as a solution for GDI engines. The operating requirements of GPF differ largely from those of the DPF, as NO_2 is not present and a very low amount of O_2 is available in the GDI exhaust downstream the TWC [7–9]. Thus, active catalysts to oxidize soot in poor (or even null) oxygen conditions must be developed. However, even though it is a challenging issue, the soot oxidation reaction in the severe GDI exhaust requirements (i.e., <10,000 ppm of O_2) has been scarcely studied [8–10].

Among the catalysts suggested for O_2-soot oxidation [8–18], one of the most interesting are mixed oxides with perovskite structure (ABO_3), as their properties can be tailored by selecting the nature of the A and B ions according to the specific needs of the oxidation reaction [19]. Certainly, perovskites are an option with future potential as soot oxidation catalysts in DPF conditions [16–26], as well as for other oxidation reactions such as CO, hydrocarbons and volatile organic compounds [27–33]. In previous papers [26,34], the beneficial result of the incorporation of copper into the structure of $BaMnO_3$ and $BaTiO_3$ perovskites for NO_2-assisted diesel soot oxidation was explored. Lately, Hernández et al. [8,9] stated that iron-based perovskites are also appealing as soot oxidation catalysts in GDI exhaust requirements.

Considering this background, and taking into account the promising performance previously featured by a $BaFe_{1-x}Cu_xO_3$ catalysts series for soot oxidation in the most severe GDI exhaust requirements (regular stoichiometric GDI operation, i.e., 0% O_2) [35], the objective of this research is to further study the influence of the partial replacement of iron by copper in the properties of a $BaFeO_3$ perovskite which will define its catalytic performance for soot oxidation. Therefore, $BaFe_{1-x}Cu_xO_3$ catalysts ($x = 0, 0.1, 0.3$ and 0.4) were synthetized, characterized and tested for soot oxidation in both diesel and "fuel cuts" GDI exhaust conditions (i.e., 1% O_2).

2. Materials and Methods

2.1. Catalyst Preparation

$BaFe_{1-x}Cu_xO_3$ catalysts ($x = 0, 0.1, 0.3, 0.4$) have been obtained using a citrate sol-gel method [26]. $Ba(CH_3COOH)_2$ (Sigma-Aldrich, 99%), $Fe(NO_3)_2·9H_2O$ (Sigma-Aldrich, 97%) and $Cu(NO_3)_2·3H_2O$ (Panreac, 99%) have been employed as metal precursors. Briefly, a 1M citric acid solution, with a 1:2 molar ratio with respect to barium has been heated to 60 °C. The solution pH has been raised to 8.5 with ammonia solution. Subsequently, the corresponding amounts of barium, iron, and copper precursors have been incorporated, and the pH value has been readjusted to 8.5 with ammonia solution. The solution was hold at 65 °C during 5 h and later dried at 90 °C for 48 h. The dried gel has been calcined at 150 °C for 1 h and then, at 850 °C 6 h [26]. Table 1 includes the catalysts nomenclature.

Table 1. Molecular composition, specific surface area, copper content, and Goldschmidt tolerance factor (*t*) * values for $BaFe_{1-x}Cu_xO_3$ catalysts.

Catalyst	Molecular Composition	BET Specific Surface Area (m^2/g) *	Nominal Cu (wt %)	ICP Cu (wt %)	*T* (Fe^{3+}) **	*t* (Fe^{4+}) **
BFC0	$BaFeO_3$	4	–	–	1.09	1.12
BFC1	$BaFe_{0.9}Cu_{0.1}O_3$	1	2.5	2.5	1.09	1.12
BFC3	$BaFe_{0.7}Cu_{0.3}O_3$	1	7.7	7.0	1.08	1.10
BFC4	$BaFe_{0.6}Cu_{0.4}O_3$	3	9.1	9.1	1.07	1.09

* In the range of experimental detection limit. ** Calculated as: $t = \dfrac{R_{Ba}+R_O}{\sqrt{2}·(((1-x)·R_{Fe}+xR_{Cu})+R_O)}$.

2.2. Characterization

To measure the metal content in the samples by Inductively Coupled Plasma Atomic Emission Spectroscopy (ICP-OES), a Perkin-Elmer equipment (Optima 4300 DV) has been used. For the analysis, copper was extracted dissolving the samples with magnetic stirring in 8M HCl solution by reflux heating.

An Autosorb-6B instrument (Quantachrome Instruments, Boynton Beach, FL, USA) was used to determine, by N_2 adsorption at -196 °C, the Brunauer Emmet Teller (BET) surface area of the samples, which were previously degasified at 250 °C for 4 h.

X-ray diffraction (XRD) tests were performed between 20–80° 2θ angles with a step rate of 1.5°/2 min and using CuKα (0.15418 nm) radiation in a Bruker D8-Advance device. The Rietveld analysis of XRD data was developed with the Automatic Rietveld Refinement (HIGHScore Plus from PANalytical program).

A ZEISS Merlin VP Compact Field Emission Scanning Electron Microscopy (FESEM) equipment (Quantax 400 from Bruker, Berlin, Germany) was employed to analyze the morphology of the catalysts and to determine the elemental composition of the catalysts (by Energy Dispersive X-Ray analysis, EDX).

X-Ray Photoelectron Spectroscopy (XPS) was used to obtain the surface composition. To register the XPS spectra, a K-Alpha photoelectron spectrometer by Thermo-Scientific, with an Al Kα (1486.6 eV) radiation source, was used in the following conditions: 5×10^{-10} mbar pressure in the chamber and setting the C1s transition at 284.6 eV. The binding energy (BE) and kinetic energy (KE) values were then determined with the peak-fit software of the spectrophotometer, to regulate the BE and KE scales.

Reducibility of the catalysts was evaluated by Temperature Programmed Reduction with H_2 (H_2-TPR). The experiments were developed in a Pulse Chemisorb 2705 device from Micromeritics fitted with a Thermal Conductivity Detector (TCD to find out the outlet gas composition changes. 20 mg of the sample was heated at 10 °C/min from room temperature to 1000 °C in 5% H_2/Ar atmosphere (40 mL/min, $P_t = 1$ atm). The H_2 consumption amount was determined using a CuO sample supplied by Micromeritics.

2.3. Activity Tests

The catalytic activity for NO to NO_2 oxidation and NO_2-assisted diesel soot oxidation was established by Temperature Programmed Reaction (NOx-TPR) using of a gas mixture composed of 500 ppm NOx and 5% O_2, balanced with N_2 (500 mL/min gas flow). For NO oxidation experiments, 80 mg of the catalyst were mixed with SiC, in a 1:4 mass ratio, and warmed from 25 to 800 °C, at 10 °C/min, in a quartz fixed-bed reactor. The activity for diesel soot oxidation was evaluated adding 20 mg of Printex U from Degussa (employed as surrogated soot, which represents the least reactive fraction of particulate matter [8–10,16,26,34–38]), in loose contact with the catalyst. For the catalyst with the highest activity, isothermal soot oxidation reactions at 450 °C were also performed. The gas composition was monitored by specific Non-dispersive Infrared Ultraviolet NDIR-UV gas analyzers for NO, NO_2, CO, CO_2, and O_2 (Rosemount Analytical Model BINOS 1001, 1004 and 100). The NO_2 generation, soot conversion and CO_2 selectivity percentages were calculated using Equations (1), (2), and (3), respectively:

$$NO_2(\%) = \frac{NO_{2,\,out}}{NO_{x,out}} \times 100 \tag{1}$$

$$\text{Soot conversion}(\%) = \frac{\sum_0^t CO_2 + CO}{(CO_2 + CO)_{total}} \times 100 \tag{2}$$

$$CO_2 \text{ selectivity } (\%) = \frac{CO_{2,\,total}}{(CO_2 + CO)_{total}} \times 100 \tag{3}$$

where $NO_{2,out}$ and $NO_{x,out}$ are the NO_2 and NOx concentrations determined at the reactor exit, $\sum_0^t CO_2 + CO$ is the quantity of CO_2 and CO evolved at a time t, and and $(CO_2 + CO)_{total}$ are the CO and CO_2 + CO evolved during all the experiment time.

To determine the catalysts performance for soot oxidation in GDI exhaust conditions, a gas mixture with 1% O_2 in He was used as it is the typical O_2 concentration at the turbine-GDI engine exit, i.e., upstream the TWC [8], but also because it simulates the fuel cut GDI operation conditions (<20% O_2) [7]. These experiments were developed as Temperature Programmed Reactions (6 °C/min from

150 °C temperature till 900 °C, 500 mL/min gas flow) in a quartz fixed-bed reactor using 80 mg of the catalysts and 20 mg of Printex-U (1:4 soot/catalyst ratio in loose contact) mixed with SiC. A Gas Chromatograph (Hewlet Packard 8690) with two packed columns (Porapak Q and Molecular Sieve 5a) connected to a Thermal Conductivity Detector (TCD) was used for the measure of the gas composition. Previous to the soot oxidation reaction, the catalysts were preheated in the reaction mixture (1% O_2 in He) at 150 °C during 1 hour. The soot conversion and CO_2 selectivity percentages were calculated using Equations (2) and (3), respectively.

3. Results and Discussion

3.1. Characterization of the Fresh Catalysts

3.1.1. Chemical, Morphological, and Structural Properties

Table 1 features the real copper content and BET surface area of the $BaFe_{1-x}Cu_xO_3$ (x = 0, 0.1, 0.3, 0.4) perovskites obtained by ICP-OES and N_2 adsorption, respectively. All the $BaFe_{1-x}Cu_xO_3$ catalysts present a very low surface area, as it corresponds to mixed oxides with perovskite structure [19]. The data of the real copper content (very close to the nominal corresponding to the stoichiometric formula) reveal that nearly all the copper used in the synthesis appears in the catalysts. Concerning morphology, FESEM images (Figure S1 in Supplementary Information) show that catalysts are formed by highly agglomerated irregular grains with a size in the range of micrometer. The presence of a low amount of copper (BFC1 and BFC3) does not significantly change the morphology of the bare perovskite; however, for the catalyst with highest copper content (BFC4), a different type of grains is detected which could correspond to a new phase. The EDX data (see Table S1 in Supplementary Information) reveal an identical atomic percentage of Ba and Fe for BFC0, as expected according to perovskite composition ($BaFeO_3$). However, for BFC4, in addition to the presence of Cu, larger atomic percentages of Ba and O are detected. This fact supports the existence of a new phase composed by barium, oxygen, and copper in the surface of this catalyst.

Figure 1 features the XRD profiles, showing (as expected according to the calculated t values shown in Table 1) a perovskite structure as an almost unique crystalline phase for all catalysts. Additionally, a Fe(III) and Fe(IV) mixed-oxide with triclinic structure is identified as a minority phase for BFC0 and BFC1, while, for BFC4, a BaO_x-CuO_x phase (with a suggested stoichiometry of $BaCuO_2$) appears. This oxide could correspond to the different type of grains observed by FESEM for BFC4 catalyst (Figure S1d in Supplementary Information) and justifies the EDX data (Table S1 in Supplementary Information) for this catalyst.

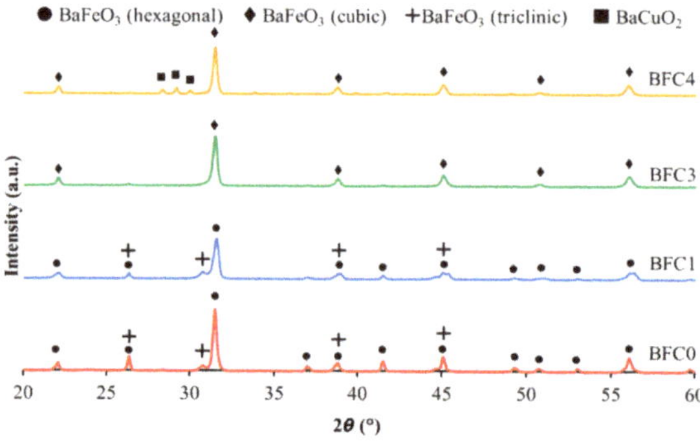

Figure 1. XRD patterns for fresh $BaFe_{1-x}Cu_xO_3$ catalysts.

For BFC0 and BFC1 catalysts, the diffraction peaks are assigned to a hexagonal perovskite structure; however, for BFC3 and BFC4, the peaks are consistent with a cubic structure. These results agree with the decrease in the *t* parameter values (Table 1), which becoming closer to 1 (corresponding to an ideal cubic structure) as the copper content increases. This structural modification (which has been verified by the Rietveld analysis presented in Figure 2 was previously noticed for other barium-based perovskites [26,34,38] and also for Sn- doped $BaFeO_3$ perovskites [39], and supports that Cu has been introduced into the perovskite lattice. Concerning BFC1, the reduction in the intensity of the main perovskite peak (at. 31.5°) evidences that copper has been inserted into the perovskite structure [26,34,38]. Moreover, except for the catalyst with a highest copper content (BFC4), peaks corresponding to a copper segregated phase are not clearly identified, revealing that copper species are not segregated or, if they are, they would present a size under the detection limit of XRD. Finally, for BFC4, the presence of the BaO_x-CuO_x phase as minority segregated phase shows a limit in the amount of copper introduced into the perovskite framework [26,34,35,38].

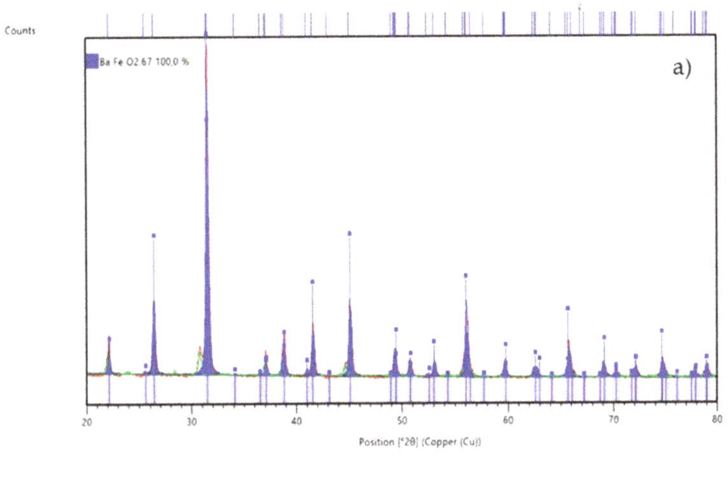

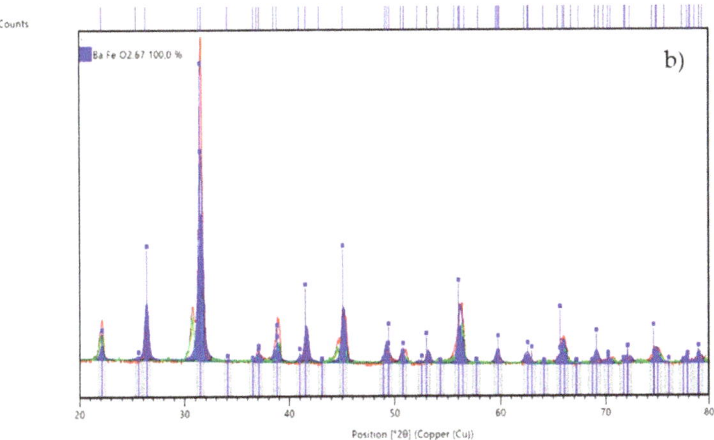

Figure 2. *Cont.*

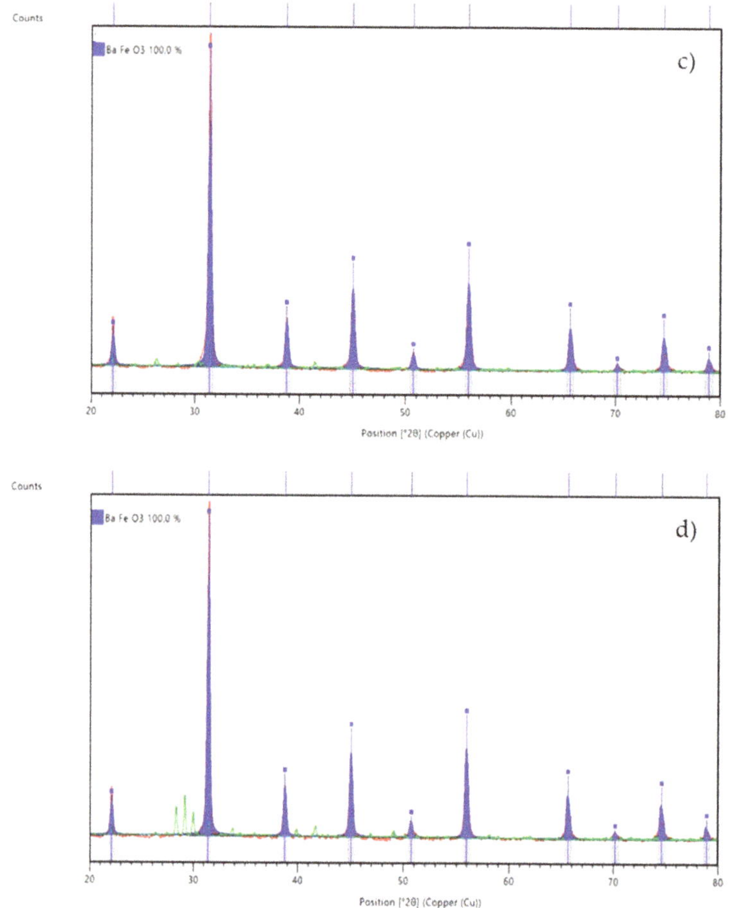

Figure 2. Rietveld analysis for (minority phase not included in the analysis): (**a**) BFC0: in red the original XRD pattern, in blue the Rietveld simulation corresponding to hexagonal perovskite structure and in green the residual data corresponding to triclinic structure; (**b**) BFC1: in red the original XRD pattern, in blue the Rietveld simulation corresponding to hexagonal perovskite structure and in green the residual data corresponding to triclinic structure; (**c**) BFC3: in red the original XRD pattern, in blue the Rietveld simulation corresponding to cubic perovskite structure and in green the residual data corresponding to $BaCuO_2$; (**d**) BFC4: in red the original XRD pattern, in blue the Rietveld simulation corresponding to cubic perovskite structure and in green the residual data corresponding to $BaCuO_2$.

The average crystal size for the catalyst has been determined from the Full Width at Half Maximum (FWHM) of the main perovskite XRD peak (in hexagonal or cubic structure) applying the Scherrer equation [40]; data are included in Table 2. The average crystal size is smaller for the catalyst containing copper with hexagonal structure (BFC1) than for the bare perovskite (BFC0). On the contrary, for catalysts with cubic structure (BFC3 and BFC4), the average crystal size increases with the copper content. The lattice parameter for hexagonal (*a* and *c*) and cubic (*a*) perovskites have also been estimated from XRD data (Table 2). As the average crystal size, the decrease in *a* and *c* values is observed in the presence of copper for the catalyst with hexagonal structure (BFC1), confirming that copper has been inserted into the lattice. However, as the ionic radii of copper (as Cu^{2+}, 0.73 Å) is larger than the Fe^{3+} ionic radii (0.65 Å) or Fe^{4+} (0.59 Å), an increase in the lattice parameters would be expected if this was the unique factor affecting the values. Nevertheless, it has been reported that a modification in the amount of the oxygen vacancies affects the lattice parameter [41], thus, it seems that the amount of oxygen vacancies is also being affected by copper incorporation into the $BaFeO_3$ perovskite framework. For cubic perovskites (BFC3 and BFC4), the lattice parameter is almost constant but larger than the corresponding to a reference $BaFeO_3$ with cubic structure (4.012 Å) [39], again supporting that copper has been inserted into the lattice.

Table 2. XRD characterization data of $BaFe_{1-x}Cu_xO_3$ catalysts.

Catalyst	XRD Phase Identification	Average Crystal Size (nm) *	a (Å) *	c (Å) *
BFC0	$BaFeO_{2.67}$, hexagonal	17.0	5.684	13.925
BFC1	$BaFeO_{2.67}$, hexagonal	12.4	5.667	13.908
BFC3	$BaFeO_3$, cubic	14.3	4.019	-
BFC4	$BaFeO_3$, cubic, $Ba_{0.9}Cu_{1.06}O_{2.43}$	16.9	4.018	-

* Calculated using the main XRD perovskite peak.

Summarizing, from XRD data, it can be concluded that copper is inserted into the perovskite structure causing: (i) the distortion of the original hexagonal perovskite structure for the catalyst with the lowest copper content (BFC1), (ii) the modification from hexagonal to cubic structure for the catalysts with high copper content (BFC3 and BFC4), (iii) the formation of a BaOx-CuOx oxide as minority segregated phase for BFC4 catalyst, and iv) a possible increase in the amount of oxygen vacancies/defects.

3.1.2. Surface Properties

XPS analysis provides data about the surface composition of the $BaFe_{1-x}Cu_xO_3$ perovskite catalysts. The XPS profiles corresponding to the Cu $2p^{3/2}$ transition are presented in Figure 3. Reduced copper species, such as metallic copper or Cu_2O, usually appear at a binding energy (BE) close to 933 eV, while, for Cu(II) species, the Cu $2p^{3/2}$ transition appears above 933 eV [36,42–44]. In Figure 3, the BE maximum of the main XPS band appears slightly above 933 eV in the three catalysts containing copper, suggesting the presence of Cu(II) species. Moreover, Cu(I) and Cu(II) species can be distinguished by the presence of a satellite peak at 942–945 eV, due to an electron transfer from Cu $2p^{3/2}$ to 3d free level in Cu(II) [45]. The existence of the satellite peak for the three copper-content catalysts, that reveals the presence of Cu(II) species [45], confirms that copper is present as Cu(II) species. Additionally, based on the use of Auger data (Cu LMM) [45], the existence of Cu(II) species has been verified as reveals the Wagner (chemical state) plot shown in Figure S2 (Supplementary information). The deconvolution of the normalized Cu $2p^{3/2}$ bands reveals two contributions with maxima at around 933 eV and 935 eV, which seem to correspond to two different Cu(II) species [42–44]: (i) the band at lower BE, assignable to copper species with a weak electronic interaction with perovskite, that is, to CuO species located on the surface (Cu_S) and (ii) the band at higher BE, corresponding to copper species with a strong electronic interaction with perovskite (Cu_L), i.e., copper inserted in the lattice, near the surface. As the percentage of the area for the XPS band at 935 eV (Cu_L band) increases with the copper content from 26% to 33%, it seems that the presence of copper with a strong electronic interaction with perovskite is favored as copper content increases. However, a slight decrease of this value is observed for the BFC4 catalyst with respect to BFC3 (33% versus 35%), confirming that a limit for the copper insertion has been achieved. In fact, a comparison between the Cu/Cu+Fe+Ba ratio calculated by XPS and the corresponding nominal ratio (both data included in Table 3) confirms that copper has been inserted into the perovskite structure, as the XPS ratio are lower (for BFC1 and BFC3) or similar than (for BCF4) the nominal ratio [25,34–38]. It is remarkable that the smallest difference between these two values is presented by BFC4 catalyst, supporting, again, the limit in the copper insertion. Therefore, the copper which is not introduced into the lattice has to be dispersed on the surface forming the BaOx-CuOx phase, which was detected by XRD and EDX, as copper content is higher for BFC4 (Table 1).

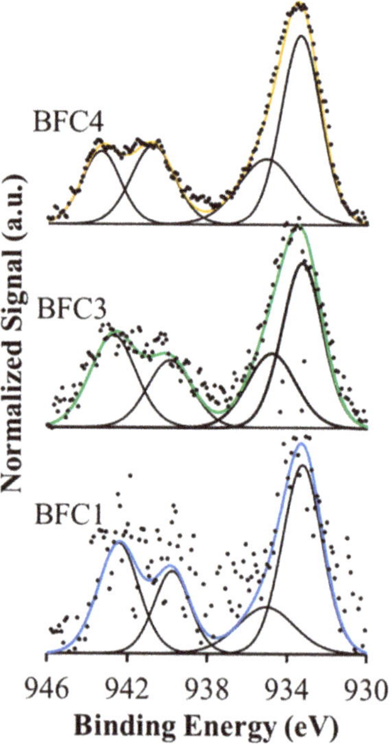

Figure 3. XPS spectra for Cu2p$^{3/2}$ transition.

Table 3. XPS characterization data of BaFe$_{1-x}$Cu$_x$O$_3$ catalysts.

Catalyst	Cu/ Ba+Fe+Cu (nominal)	Cu/ Ba+Fe+Cu (XPS)	O$_L$/ Ba+Fe+Cu (XPS)
FC0	-	-	1.30
BFC1	0.05	0.03	1.70
BFC3	0.15	0.09	1.10
BFC4	0.20	0.21	1.10

Figure 4 features the XPS spectra of the Fe 2p$_{3/2}$ for BaFe$_{1-x}$Cu$_x$O$_3$ catalysts and the corresponding to a Fe$_2$O$_3$ commercial sample use as reference. The maximum of the main XPS band for the four catalysts does not appear at exactly the same (BE) value than the corresponding to the reference suggesting the presence of Fe species with a different oxidation state or with the same oxidation state but in different proportion. The deconvolution of the main band shows two significant contributions at around 709 eV and 711 eV. Even though the identification of iron oxidation states by XPS is very difficult [46], according to literature [39,46–50], the first peak corresponds to Fe(III) species, and the second one could be assigned to Fe(IV) species [48–50]. It has been established that the position of the satellite peak is the key finger to detect the oxidation state of Fe [46,48]. Thus, the shake-up peak observed at 717 eV (which corresponds to the satellite peak of Fe(III)) supports the existence of this oxidation state [39,46,48,51]. However, the presence of Fe(IV) seems not to be supported by the XPS data, as the high BE peak at approximately 711 eV is not always unequivocally assigned to this oxidation state [46,48]. Thus, more evidence from other characterization techniques is needed to assume that Fe(IV) exits. The TPR-H$_2$ results (see below) indicate that Fe(IV) and Fe(III) oxidation states co-exist in the $BaFe_{1-x}Cu_xO_3$ catalysts, as it is observed that the experimental H_2 consumption is in between the nominal (calculated) H_2 consumption expected, considering that iron as Fe(III) or Fe(IV) is reduced to Fe(II). The presence of Fe(IV) in $BaFe_{1-x}Cu_xO_3$ catalysts is additionally supported by the well-known stabilization of high oxidation state for B cation, as Fe(IV), in perovskites [19,39,48–50]. On the basis of the $BaFeO_3$ stoichiometric formula, Fe(IV) must be the oxidation state for Fe in the perovskite, and, in the presence of copper, a rise in the Fe(IV) amount and /or the generation of additional oxygen vacancies into the perovskite structure would be expected to compensate the deficiency of positive charge due to the partial iron substitution [19]. In fact, the decrease in the lattice parameter observed by XRD for BFC1 with respect to BFC0 (Table 2) suggests an increase in the Fe(IV), which presents a lower ionic ratio that Fe(III). However, for BFC2 and BFC3, the lattice parameter (Table 2) increases revealing that the amount of Fe(IV) cannot be higher; thus, the generation of additional oxygen vacancies should be observed to balance the positive charge deficiency due to the increase of the copper content in the catalyst. This larger amount of oxygen vacancies has to cause the lattice expansion detected [41].

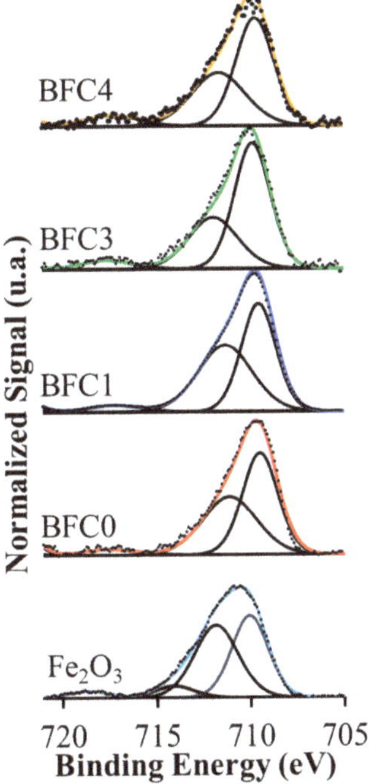

Figure 4. XPS spectra for $Fe2p^{3/2}$ transition.

Figure 5 presents the XPS spectra of the O1s transition for all catalysts, where three contributions are usually observed [36,42–44]: (i) at low BE (around 528 eV), corresponding to lattice oxygen (O_L) in metal oxides, (ii) at intermediate BE (between 529 and 531 eV), assigned to adsorbed oxygen species such as, O_2^{-2}, surface CO_3^{-2}, and/or OH^- groups, and (iii) at high BE (533 eV approximately) due to oxygen in adsorbed water [52–55]. The intensity of the bands is modified in the presence of copper revealing changes in the amount of oxygen species on the catalysts surface. The values of O_L/Cu+Ti+Ba ratio in Table 3 (determined from the peak area of O_L, $Fe2p^{3/2}$, $Ba3d^{3/2}$, and $Cu2p^{3/2}$ transitions) is higher for BFC1 than for the bare BFC0 perovskite, which means a lower amount of surface oxygen vacancies. This fact supports that the oxidation of Fe(III) to Fe(IV) occurs in the BFC1 perovskite to compensate the positive charge deficiency due to copper incorporation. However, for BFC3 and BFC4 catalysts, the lower O_L/Cu+Ti+Ba ratio with respect to the nominal value confirms the generation of additional oxygen vacancies to balance the positive charge deficiency due to partial iron substitution by copper. Additionally, these results justify the change in the values of lattice parameters observed, that is: (i) the lower lattice parameters values for BFC1 catalyst with respect to BFC0 (Table 2) are due to the decrease in the amount of oxygen vacancies (Table 3), as, for this catalysts, the oxidation of Fe(III) to Fe(IV) takes place and (ii) the larger values for BFC3 and BFC4 (Table 2) are due to the rise in the amount of oxygen vacancies with respect to BFC0 (Table 3).

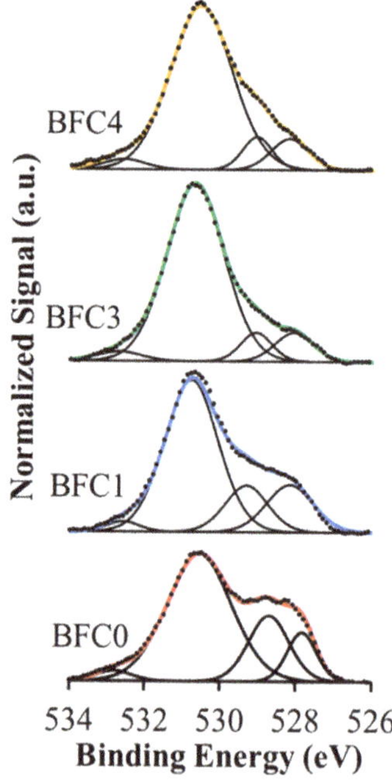

Figure 5. XPS spectra for O 1s transition.

3.1.3. Redox Properties

Reducibility and redox properties of the fresh $BaFe_{1-x}Cu_xO_3$ catalysts were analyzed by Temperature Programmed Reduction with H_2 (H_2-TPR), which are the H_2 consumption profiles shown in Figure 6. In Figure 7, the nominal (calculated) H_2 consumption (mL of H_2 per gram of catalysts) expected considering that iron, as Fe(III) or Fe(IV) in the perovskite, is reduced to Fe(II), is compared with the experimental H_2 consumption determined from the H_2-TPR profiles (Figure 6). It is observed that the experimental H_2 consumption is in between both nominal values revealing that Fe(IV) and Fe(III) oxidation states co-exist in the $BaFe_{1-x}Cu_xO_3$ catalysts. Furthermore, the experimental H_2 consumption data indicates that the amount of Fe(IV) increases in the presence of copper.

In the complex H_2 consumption profiles shown in Figure 6, three regions can be established [56–58]:

a) At low temperature, between approximately 200 °C and 550 °C, a broad H_2 consumption signal is observed for all the catalyst that, according to literature, can be ascribed to different reduction processes: (i) the Cu(II) [34,38] reduction, (ii) the Fe(IV) and Fe(III) reduction to Fe(III) and Fe(II), as was observed for Fe_3O_4, and iii) the reduction of weakly chemisorbed oxygen upon surface oxygen vacancies of perovskite (α-oxygen) [34].

b) From around 550 °C to 700 °C, the H_2 consumption peaks correspond to both the reduction of Fe(III) to Fe(II) as detected for the reduction of Fe_3O_4 to FeO and to the decomposition of surface oxygen species formed on oxygen vacancies (called α'-oxygen) [34], more strongly bonded to the perovskite than α-oxygen.

c) At high temperatures (T > 700 °C), broad TCD signals assigned to the reduction of Fe(II) to Fe(0) (causing the consequent destruction of the perovskite structure) could be found [56–58]. Nevertheless, the XRD data for catalysts after H_2-TPR (not shown) reveal that the perovskite structure is still present, thus, the reduction to Fe(0) is not taking place and, consequently, H_2 consumption is hardly observed at T > 700 °C. Therefore, the most relevant information related to the redox properties of the $BaFe_{1-x}Cu_xO_3$ catalysts is located at T < 700 °C.

BFC0 profile shows two peaks at temperature lower than 700 °C: a broad peak with maximum at ca 300 °C and a more defined peak with a maximum at around 670 °C. The H_2 consumption detected at temperature lower than 300 °C is usually related to the presence of Fe(IV) [56–58], supporting the existence of this oxidation state. The second H_2 consumption peak, with a maximum at 670 °C, corresponds to the reduction of Fe(III) to Fe(II) and to desorption/reduction of oxygen surface species formed on oxygen vacancies (α'-oxygen) [34].

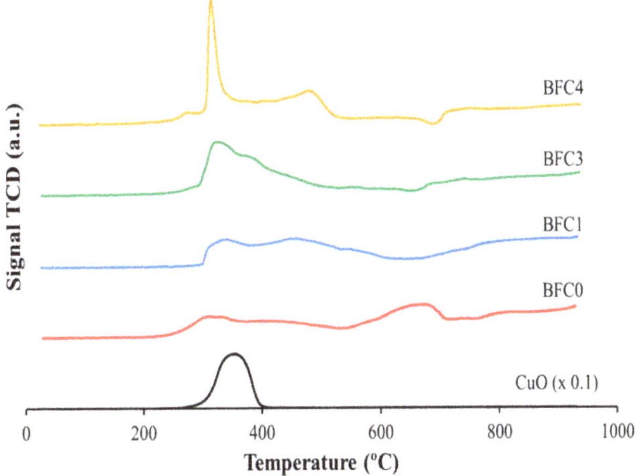

Figure 6. H_2-TPR profiles for $BaFe_{1-x}Cu_xO_3$ catalysts.

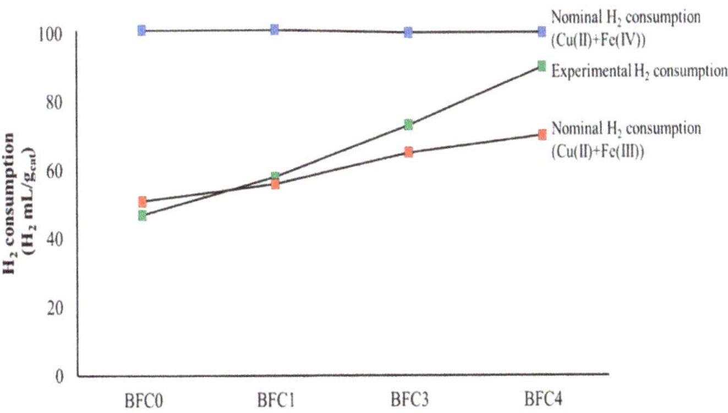

Figure 7. H_2 consumption (mL/g of catalyst).

In the H_2 consumption profile of BFC1 catalyst, a broad peak with two maxima, at approximately 350 °C and 450 °C, is identified. The first maximum is ascribed to the Cu(II) to Cu(0) reduction (appearing at lower temperature than the CuO used as a reference [38]) and also to the consumption due to the partial Fe(IV) and Fe(III) reduction to Fe(III) and Fe(II), respectively. The second maximum of this broad peak at 450 °C seems to correspond to: i) the reduction of Fe(III) to Fe(II), taking place at lower temperature than for the $BaFeO_3$ catalyst, due to the presence of reduced copper [26] and ii) the desorption/reduction of strongly bonded oxygen species ('-oxygen) [34].

In the H_2-TPR profile of the BFC3 catalyst, a broad peak between 300 and 500 °C is detected with a well-defined maximum at 320 °C, followed by a shoulder around 380 °C. The first maximum corresponds to the reduction of Cu(II) to Cu(0) and it is better defined than the corresponding to BFC1 due to the higher copper content. As for BFC1, the low H_2 consumption at T < 300 °C confirms the presence of Fe(IV). The shoulder at 380 °C has to be related with the reduction of Fe(III) to Fe(II) that seems to take place at lower temperature than for BFC1. This fact supports that the formation of metallic copper (which is more easily reduced than iron) promotes the reduction of Fe(III) to Fe(II), as was previously observed for the reduction of manganese species in $BaMn_{1-x}Cu_xO_3$ catalysts series [26].

Concerning BFC4, a sharp H_2 consumption peak with a maximum at 315 °C is found, followed by a low intensity peak with a maximum at around 475 °C. The presence of this well- defined peak, which is ascribed to Cu(II) to Cu(0) reduction, confirms the existence of copper oxide (II) species [34]. In fact, for this catalyst, BaO_x-CuO_x oxide has been detected by XRD and FESEM, thus, the sharp peak corresponds to the reduction of this copper oxide. The second peak has to be due to the Fe(III) reduction to Fe(II) that, for BFC1 and BFC3 catalysts, takes place at lower temperature that for BFC0.

After the analysis of the H_2-TPR profiles for three catalysts, it can be concluded that the Fe(III) reduction to Fe(II) takes place at similar temperature for BFC1 and BFC4 (450 °C and 475 °C, respectively); this happens at lower a temperature (380 °C) for the BFC3 catalyst, probably due to its higher content of lattice copper (see Table 3).

Concluding, H_2-TPR results indicate the co-existence of Fe(III) and Fe(IV) and suggest that copper incorporation promotes the reduction of Fe(III) to Fe(II).

3.1.4. O_2 Release During Heat-Treatment in He (O_2-TPD)

In the O_2 profiles evolved by perovskite mixed oxides during a heat treatment in He (O_2-TPD), three regions are usually observed [26,39,59–68]. The lower temperature region, at T < 400 °C corresponds to weakly chemisorbed oxygen upon surface-oxygen vacancies (denoted as oxygen). The intermediate region, between 400 °C and 700 °C, is ascribed to near-surface oxygen associated to lattice defects such as dislocations and grains frontiers (designed as α' oxygen). Therefore, the presence of α and α'-oxygen is directly linked with the presence of surface vacancies/defects of oxygen in the structure [64–66]. Finally, the oxygen evolved at temperature higher than 700 °C, named β oxygen, is generally related with the lattice oxygen (which comes from the reduction of B cation (Fe in this case) of the perovskite [66]) and it is related with the oxygen mobility and with the inner bulk oxygen vacancies.

Figure 8 shows the O_2 profiles for the $BaFe_{1-x}Cu_xO_3$ catalysts. The O_2–TPD profiles show that α and α'-oxygen are mainly evolved by most of the $BaFe_{1-x}Cu_xO_3$ catalysts [8,9]. BFC0, BFC3, and BFC4 exhibit a higher oxygen signals than BFC1 and, therefore, higher quantity of surface oxygen vacancies, agreeing with the XPS results (lower O_L/Cu+Ti+Ba ratio). Regarding α'-oxygen, BFC1 presents the highest signal, which evidences the great structure distortion (as exhibited by XRD) promoted by a small Cu incorporation. The total quantity of O_2, calculated from the area under the O_2 profiles, diminishes as copper content grows: 424 μmol/g cat (BFC0) > 333 μmol/g cat (BFC1) > 282 μmol/g cat (BFC3) > 275 μmol/g cat (BFC4). Thus, as it has been previously published [39], the addition of a dopant seems to stabilize the oxygen bonded to Fe and leads to a decrease in the desorbed O_2. For BFC4, a grown in the β oxygen has been detected, probably related to the structural modification and the presence of the BaOx-CuOx phase identified by XRD.

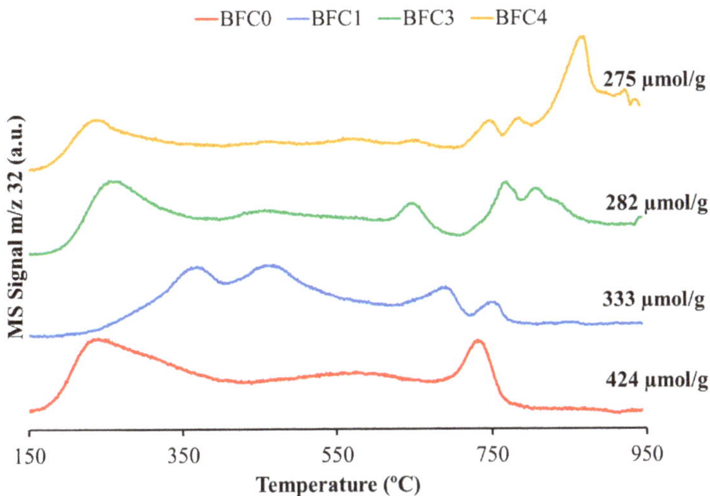

Figure 8. O_2-TPD profiles for $BaFe_{1-x}Cu_xO_3$ catalysts.

The phase composition of $BaFe_{1-x}Cu_xO_3$ catalysts after the O_2-TPD has been determined by XRD (Figure S3 in Supplementary Information). For BFC0 and BFC1 catalysts, the hexagonal perovskite structure is replaced by a monoclinic $BaFeO_{2.5}$ phase (with ordered oxygen vacancies) after losing a fraction of the lattice oxygen. On the contrary, BFC3 and BFC4 catalysts preserve the cubic perovskite structure after O_2 release. This founding agrees with the conclusions of Huang et al. [39], who pointed out the increase in the structure stability due to the presence of a dopant (copper in our case). Note that the most stable catalysts (BFC3 and BFC4) are those with ideal (cubic) perovskite structure. The higher structural stability in the presence of copper could be relevant for catalytic applications at high temperature.

3.2. Catalytic Activity

3.2.1. NO_2 Generation and Diesel Soot Oxidation

Figure 9 shows the NO_2 generation profiles obtained in TPR conditions for $BaFe_{1-x}Cu_xO_3$ catalysts including, as reference, the thermodynamic equilibrium profile. As observed for other perovskite-based catalysts [26,34–38,64,67], the thermodynamic equilibrium limits the NO_2 percentage at T > 500 °C. All catalysts accelerate the NO to NO_2 oxidation at temperature lower than 500 °C, being the copper-free catalyst (BFC0) the most active. In general terms, the NO_2 generation follows the same sequence than the amount of oxygen evolved during O_2-TPD, except for BFC4 catalyst. Note that the two catalysts evolving large amount of and '-oxygen, that is, BFC0 and BFC1, are also the catalysts generating more NO_2 at low temperature (T < 300 °C). This is in agreement with the relationship found by Onrubia et al. [64] between the amount of and ' oxygen evolved by the catalysts (Sr-doped $LaBO_3$ (B = Mn or Co perovskites) and the activity for the NO to NO_2 oxidation. Note that BFC4 shows a slightly higher NO_2 generation capacity than BFC3, which has to be related with the presence of copper species on the surface (BaOx-CuOx) that also catalyze the NO_2 production [34].

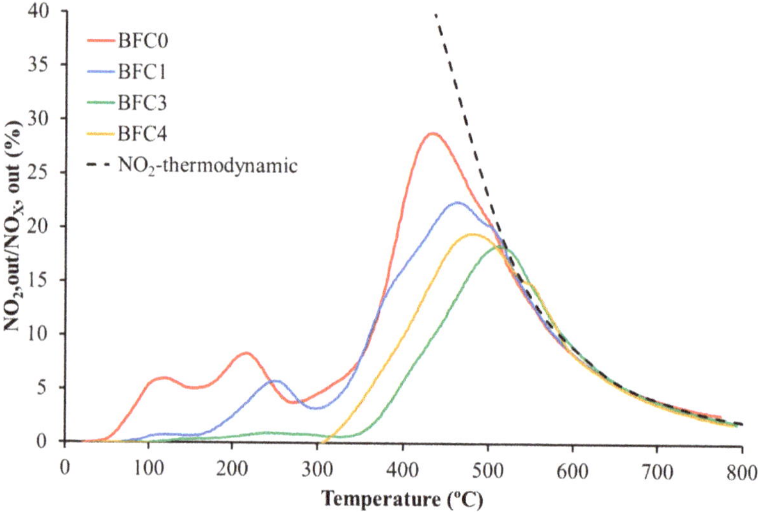

Figure 9. NO_2 generation profiles in TPR conditions s for $BaFe_{1-x}Cu_xO_3$ catalysts.

To evaluate the activity of the catalysts for NO_2-assisted diesel soot oxidation, Temperature Programmed Reactions in a NO/O_2 atmosphere (see Experimental Section for details) were carried out, and the TPR-NOx soot conversion profiles (calculated based on the amount of CO and CO_2 evolved) are featured in Figure 10. Relevant data, such as the ignition temperature ($T_{5\%}$), the temperature required to reach 50% of soot conversion ($T_{50\%}$), and the selectivity to CO_2, are included in Table 4. It can be concluded that all the catalysts shift the soot conversion profiles to lower temperatures compared to the uncatalyzed reaction (blank corresponding to bare soot) and, consequently, the $T_{5\%}$ and the $T_{50\%}$ are lower. In agreement with the NO_2 profiles (Figure 9), BFC0 is the most active catalyst for diesel soot oxidation as the addition of copper decreases the catalyst activity for soot conversion. Moreover, for BFC0 the $T_{50\%}$ value is close to 500 °C, thus, this perovskite could be used as potential catalyst for the soot removal from diesel engine exhaust [69]. The decrease in the activity for soot oxidation after the addition of a dopant (copper in our catalysts) was also observed by Huang et al. [39] for Ag-doped $LaFeO_3$ catalysts. These authors related the lower activity of Ag-perovskites for soot oxidation with the reduction in the amount of surface oxygen vacancies due to the anchorage of Ag nanoparticles. Furthermore, the reaction rate for methane combustion of a series of oxygen deficient $SrFeO_3$ perovskites was related with the quantity of oxygen vacancies in the structure [58]. In fact, a relationship between soot oxidation performance and oxygen vacancies has been published [70]. Thus, in $BaFe_{1-x}Cu_xO_3$ catalysts, the decrease in the total amount of O_2 evolved as copper content increases apparently leads to a decrease in the activity for both NO to NO_2 and soot oxidation. Note that BFC4 does not match this trend as it shows the lowest $T_{5\%}$ y $T_{50\%}$ values among the catalysts containing copper (BFC1, BFC3, and BFC4). This catalyst presents the highest fraction of surface copper species, which also catalyzes the NO_2/soot oxidation reaction [26,34,36], and hence, improves its catalytic performance. Therefore, the activity for NO_2 generation and the amount of surface copper species seem to determine the catalytic performance. Thus, the highest NO_2 generation capacity of BFC0 catalyst seems to justify its highest soot oxidation activity, while it is the largest fraction of surface copper species present in BFC4, which seems to justify its higher soot oxidation activity compared to BFC3.

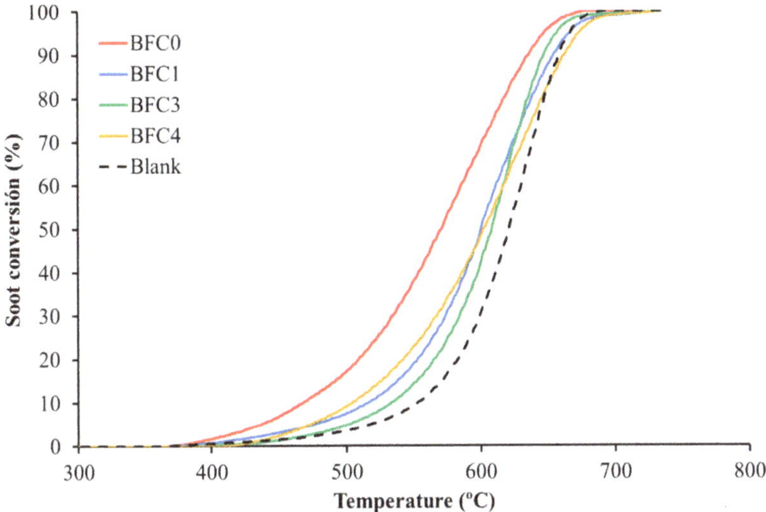

Figure 10. TPR soot conversion profiles in NOx for $BaFe_{1-x}Cu_xO_3$ catalysts.

Table 4. Data for NO_2-assisted diesel soot oxidation in TPR conditions $BaFe_{1-x}Cu_xO_3$ catalysts.

Catalysts	$T_{5\%}$ (°C)	$T_{50\%}$ (°C)	CO_2 Selectivity (%)
Bare soot	480	612	41
BFC0	430	543	51
BFC1	455	605	70
BFC3	480	605	66
BFC4	454	590	90

Additionally, the data in Table 4 reveal that, as could be expected [26,34,36,37], all the catalysts show a higher CO_2 selectivity than the uncatalyzed reaction (bare soot), with BFC4 being the most selective. Thus, CO_2 selectivity increases with the amount of surface copper species (Table 3) as this metal is a well-known catalyst for CO to CO_2 oxidation.

Due to its high activity, the performance of the BFC0 catalyst was deeply analyzed and five consecutives TPR-NOx soot oxidation cycles were carried out using the same portion of catalyst. As the $T_{50\%}$ values for the first (543 °C) and fifth cycle (561 °C) are still under those corresponding to the uncatalyzed reaction (612 °C), it can be concluded that the catalyst is not significantly deactivated. In fact, the XRD data (shown in Figure S4 in Supplementary Information) of this used catalyst (after five TPR-NOx cycles) reveal that the hexagonal perovskite structure is not significantly modified. This means that, in the presence of oxygen in the reaction atmosphere, the catalyst keeps its structure and, consequently, its activity for soot oxidation in TPR conditions.

Finally, to complete the BFC0 evaluation, its catalytic performance for NO_2-assisted diesel soot oxidation in isothermal conditions was determined by carrying out two consecutive soot oxidation experiments at 450 °C. The soot oxidation profiles at 450 °C (featured in Figure S5 in Supplementary Information) show that BFC0 catalyst is able to oxide soot without a significant deactivation and showing a high CO_2 selectivity (close to 80%). The soot oxidation rate was calculated at the beginning of the reaction, in order to avoid the effect of high soot consumption, as being 1.2 min^{-1}, which is not too far from a commercial model Pt/Al_2O_3 catalyst (1.8 min^{-1}) used in the same experimental conditions. Thus, it seems that $BaFeO_3$ perovskite could be a potential catalyst for diesel soot oxidation and, consequently, it could be used as an active phase for DPF.

3.2.2. Soot Oxidation in GDI Conditions

A preliminary study about the use of $BaFe_{1-x}Cu_xO_3$ perovskites to catalyze the oxidation reaction of soot under the highest demanding GDI exhaust requirements (regular stoichiometric GDI operation, i.e., 0% O_2) revealed that these oxides could be used as active phase for GPF [35]. It was concluded that the copper content has an essential role on the performance of the $BaFe_{1-x}Cu_xO_3$ catalysts for soot oxidation, agreeing with previous reports focused on diesel soot removal [10,26,34,36–38] and with the report for copper-supported ceria-zirconia catalysts for soot oxidation in GDI conditions [10]. These results confirm that the higher soot conversion presented by BFC4 with respect to the catalysts with lower copper content (BFC1 and BFC3) is linked both to the largest amount of β oxygen evolved by this catalyst, and with the presence of surface copper species (as BaO_x-CuO_x) [10,34,36–38].

To further analyze the performance of these $BaFe_{1-x}Cu_xO_3$ perovskites in GDI exhaust conditions, a study using 1% O_2 in He (which represents the named "fuel cuts" GDI exhaust conditions) has been developed [9]. Figure 12 shows the profiles of CO_2 and CO evolved during temperature programmed reaction experiments, including the profiles for the uncatalyzed reaction (blank corresponding to bare soot), as reference. Note that in the presence of $BaFe_{1-x}Cu_xO_3$ catalysts, the amount of CO decreases and the amount of CO_2 increases. This means that, as could be expected [26,34,36,37], and as was observed during soot conversion reaction in NOx atmosphere, the catalysts increase the selectivity to CO_2 from 56% for bare soot to 93% for BFC0, 70% for BFC1, 79% for BFC3, and 87% for BFC4. As for NO_2 assisted diesel soot oxidation, the BFC0 catalyst is the most active and the addition of copper seems to decrease the ability of the catalysts to improve the CO_2 selectivity. Additionally, as it has been deduced from soot conversion results in NOx atmosphere (see Table 4) that BFC4 presents the highest CO_2 selectivity among the copper containing catalysts due to the presence of the surface copper species.

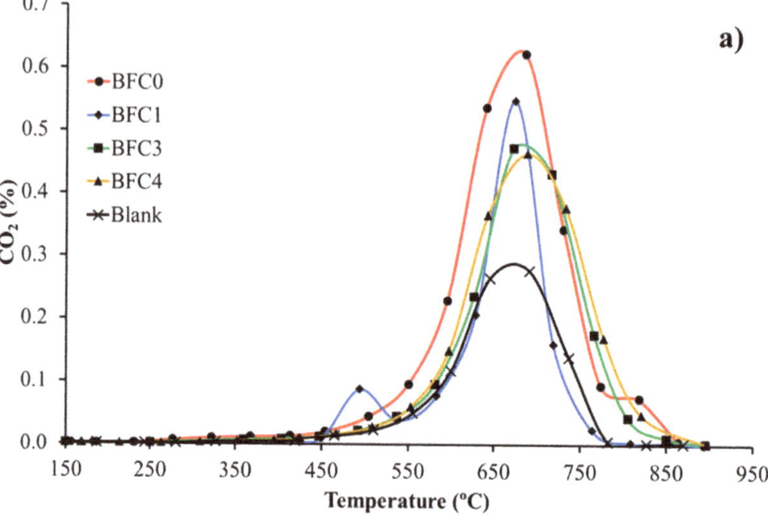

Figure 11. *Cont.*

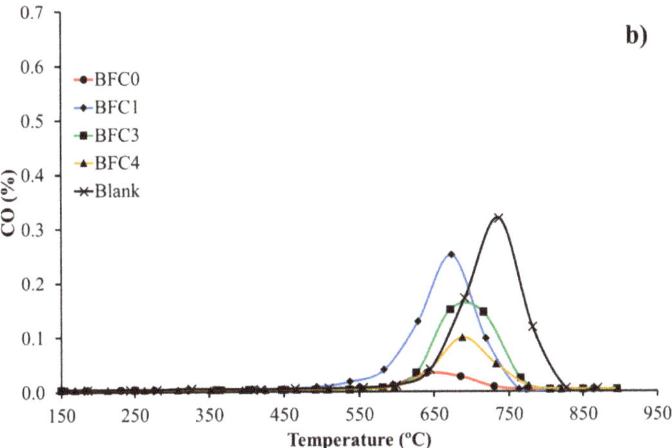

Figure 12. CO_2 (a) and CO (b) profiles during TPR soot oxidation in 1% O_2 for $BaFe_{1-x}Cu_xO_3$ catalysts.

Figure 13 shows the soot conversion profiles in 1% O_2, calculated based on the amount of evolved CO and CO_2 featured in Figure 12.

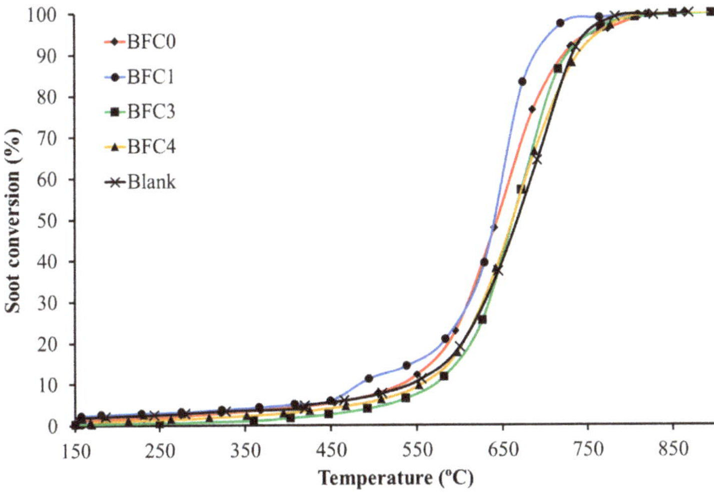

Figure 13. TPR soot conversion profiles in 1% O_2 for $BaFe_{1-x}Cu_xO_3$ catalysts.

In general terms, and in agreement with previous results for $BaMn_{1-x}Cu_xO_3$ catalysts [26], the catalytic effect is less relevant than the observed for NO_2-assisted diesel soot oxidation. Thus, only BFC0 and BFC1 catalyze the soot oxidation with oxygen as the conversion profiles are shifted to lower temperature with respect to bare soot for these two catalysts. As observed for NO_2 generation (see previous section), and in agreement with published conclusions [39,58,70], the soot conversion with oxygen follows the same trend than the amount of oxygen evolved during O_2-TPD (Figure 8) as the most active catalysts (BFC0 and BFC1) are those generating the largest amount of oxygen. In fact, the highest activity of BFC1 at low temperature, i.e., at T < 600 °C approximately, seems to be related with the largest amount of and ' evolved (Figure 8). Thus, it seems that a similar reaction pathway is followed by soot oxidation with oxygen and with NO_2-assisted diesel soot oxidation, even though the catalytic effect is more relevant for the latter reaction than for the former. Hence, it could be concluded that the $BaFe_{1-x}Cu_xO_3$ perovskites catalyze more effectively the NO_2-soot reaction than the O_2-soot reaction. Additionally, the comparison of these results with the obtained in the most demanding GDI conditions (0% O_2) [35] reveals that copper has an essential role on the performance of the $BaFe_{1-x}Cu_xO_3$ catalysts for soot oxidation only in the absence of oxygen in the reaction atmosphere.

Summarizing, the activity data above discussed reveals that the $BaFe_{1-x}Cu_xO_3$ perovskites catalyze both, the NO_2-assisted diesel soot oxidation (500 ppm NO, 5% O_2) and, to a lesser extent, the soot oxidation in the "fuel cut" GDI exhaust conditions (1% O_2). BFC0 is the most active catalyst as the activity seems to be mainly related with the amount of O_2 evolved during an O_2-TPD, which decreases with copper content.

4. Conclusions

The results obtained for the $BaFe_{1-x}Cu_xO_3$ ($x = 0$, 0.1, 0.3 and 0.4) catalyst series allows us to conclude that:

- Partial substitution of iron by copper in the lattice of a $BaFeO_3$ perovskite generates a distortion of the hexagonal perovskite structure for the lowest copper content catalyst (BFC1), and a change to cubic structure for the catalysts with higher copper content (BFC3 and BFC4).
- The amount of copper inserted into the perovskite framework achieve a maximum for the highest copper content catalyst (BFC4), which provokes the presence of BaOx-CuOx as a minority segregated phase.
- The positive charge deficiency due to the partial substitution of Fe by Cu seems to be balanced by the oxidation of Fe(III) to Fe(IV) in the BFC1 perovskite and by the generation of additional oxygen vacancies/defects, for BFC3 and BFC4 catalysts.
- $BaFe_1$-$xCuxO_3$ perovskite catalyze both the NO_2-assisted diesel soot oxidation (500 ppm NO, 5% O_2) and, to a lesser extent, the soot oxidation in the high demanding GDI exhaust conditions (1% O_2)
- BFC0 is the most active catalyst for both oxidation reactions. The activity seems to be mainly related with the amount of O_2 evolved during an O_2-TPD, which decreases with the copper content of the catalyst.

Supplementary Materials: The following are available online at http://www.mdpi.com/2079-4991/9/11/1551/s1, Figure S1: FESEM pictures for BaFe1-xCuxO3; Table S1: EDX data (atomic percentage) for BFC0 and BFC4 catalysts; Figure S2: Wagner (chemical state) plot for $BaFe_{1-x}Cu_xO_3$ catalysts; Figure S3: DRX patterns after O_2-TPD for $BaFe_{1-x}Cu_xO_3$ catalysts (dotted lines). As reference, XRD patterns of fresh catalysts (solid line) have been included; Figure S4: DRX patterns after TPR-NOx with soot for BFC0 catalyst (dotted line). As reference, XRD pattern of fresh catalyst (solid line) has been included; Figure S5: Soot conversion profiles at 450 °C for BFC0.

Author Contributions: Conceptualization, V.A.-F. and M.-J.I.-G.; methodology, V.A.-F. and M.-J.I.-G.; validation, V.A.-F., M.-S.S.-A. and M.-J.I.-G.; formal analysis, V.T.-R., C.M.-M., V.A.-F., M.-S.S.-A. and M.-J.I.-G.; investigation, V.T.-R. and C.M.-M. and V.A.-F.; resources, M.-S.S.-A. and M.-J.I.-G.; data curation, V.A.-F., M.-S.S.-A. and M.-J.I.-G.; writing—original draft preparation, V.T.-R., C.M.-M. and V.A.-F; writing—review and editing, M.-S.S.-A. and M.-J.I.-G.; visualization, C.M.-M., V.A.-F. and M.-S.S.-A.; supervision, V.A.-F. and M.-J.I.-G.; project administration, M.-J.I.-G.; funding acquisition, M.-S.S.-A. and M.-J.I.-G.

Funding: This research was funded by the Generalitat Valenciana (PROMETEO/2018/076 and Ph.D. Grant ACIF/2017/221 for V.Torregrosa-Rivero), Spanish Government (MINECO Project CTQ2015-64801-R), and EU (FEDER Founding).

Conflicts of Interest: The authors declare no conflict of interest.

References

1. Johnson, T.V. Vehicular Emissions in Review. *SAE Int. J. Engines* **2014**, *7*, 1207–1227. [CrossRef]
2. Mamakos, A.; Steininger, N.; Martini, G.; Dilara, P.; Drossinos, Y. Cost effectiveness of particulate filter installation on direct gasoline vehicles. *Atmos. Environ.* **2013**, *77*, 16–23. [CrossRef]

3. Guan, B.; Zhan, R.; Lin, H.; Huang, Z. Review of the state-of-the-art of exhaust particulate filter technology in internal combustion engines. *J. Environ. Manag.* **2015**, *154*, 225–258. [CrossRef] [PubMed]

4. Johnson, T.V. Vehicular Emissions in Review. *SAE Int. J. Engines* **2012**, *5*, 216–234. [CrossRef]

5. Kim, C.H.; Schmid, M.; Schmieg, S.J.; Tan, J.L.; Li, W. The Effect of Pt-Pd Ratio on Oxidation Catalysts Under Simulated Diesel Exhaust. *SAE Tech. Pap.* **2011**, *1*, 337–347.

6. Ad Hoc Working Group on Defining Critical Raw Materials. *Report on Critical Raw Materials for the EU*; Ad Hoc Working Group on Defining Critical Raw Materials: Brussels, Belgium, 2017.

7. Boger, T.; Rose, D.; Nicolin, P.; Gunasekaran, N.; Glasson, T. Oxidation of Soot (Printex®U) in Particulate Filters Operated on Gasoline Engines. *Emiss. Control Sci. Technol.* **2015**, *1*, 49–63. [CrossRef]

8. Hernández, W.Y.; Tsampas, M.N.; Zhao, C.; Bosselet, A.; Vernoux, P. La/Sr-Based Perovskites as Soot Oxidation Catalysts for Gasoline Particulate Filters. *Catal. Today* **2015**, *258*, 525–534. [CrossRef]

9. Hernández, W.Y.; López-González, D.; Ntais, S.; Zhao, C.; Boréave, A.; Vernoux, P. Ag-based catalysts for Diesel and gasoline soot oxidation. *Appl. Catal. B* **2018**, *226*, 202–212. [CrossRef]

10. Giménez-Mañogil, J.; Quiles-Díaz, S.; Guillén-Hurtado, N.; García-García, A. Catalyzed Particulate Filter Regeneration by Platinum Versus Noble Metal-Free Catalysts: From Principles to Real Application. *Top. Catal.* **2017**, *60*, 2–12. [CrossRef]

11. Aneggi, E.; De Leitenburg, C.; Trovarelli, A. Ceria-based formulations for catalysts for diesel soot combustion. In *Catalysis by Ceria and Related Materials*; Trovarelli, A., Fornaseiro, P., Eds.; Imperial College Press: London, UK, 2013; pp. 565–621.

12. Piumetti, M.; Bensai, S.; Russo, N.; Fino, D. Nanostructured ceria-based catalysts for soot combustion. Investigation on surface sensitivity. *Appl. Catal. B* **2015**, *154*, 742–751. [CrossRef]

13. Castoldi, L.; Matarrese, R.; Lietti, L.; Forzatti, P. Intrinsic reactivity of alkaline and alkaline-earth metal oxide catalysts for soot oxidation. *Appl. Catal. B* **2009**, *90*, 278–285. [CrossRef]

14. Obeid, E.; Tsampas, M.N.; Jonet, S.; Boréave, A.; Burel, L.; Steil, M.C.; Blanchard, G.; Pajot, K.; Vernoux, P. Isothermal catalytic oxidation of diesel soot on Ytria-stabilized zirconia. *Solid State Inonics* **2014**, *262*, 87–96.

15. Obeid, E.; Lizarraga, L.; Tsampas, M.N.; Cordier, A.; Boréave, A.; Steil, M.C.; Blanchard, G.; Pajot, K.; Vernoux, P. Continously regenerating diesel particulate filters based on ionically conducting ceramic. *J. Catal.* **2014**, *309*, 87–96. [CrossRef]

16. Ura, B.; Trawczynski, J.; Kotarba, A.; Bieniasz, W.; Illán-Gómez, M.J.; Bueno–López, A.; López-Suárez, F.E. Effect of potassium addition on catalytic activity of $SrTiO_3$ catalysts for diesel soot combustion. *Appl. Catal. B* **2011**, *101*, 169–175. [CrossRef]

17. Mgarajuan, S.K.; Rayalu, S.; Nishibori, M.; Teraoka, Y.; Labhsetwar, N. Effects of surface and bulk silver on $PrMnO_{3+\delta}$ perovskite for CO and soot oxidation: Experimental evidence for the chemical state of silver. *ACS Catal.* **2015**, *5*, 301–309. [CrossRef]

18. Wang, H.; Zhao, Z.; Xu, C.M.; Liu, J. Nanometric $La_{1-x}K_xMnO_3$ Perovskite-type oxides—Highly active catalysts for the combustion of diesel soot particle under loose contact conditions. *Catal. Lett.* **2005**, *102*, 251–256. [CrossRef]

19. Peña, M.A.; Fierro, J.L.G. Chemical Structures and performances of perovskite oxides. *Chem. Rev.* **2001**, *101*, 1981–2018. [CrossRef]

20. Shimokawa, H.; Kusaba, H.; Einaga, H.; Teraoka, Y. Effect of surface area of La–K–Mn–O perovskite catalysts on diesel particulate oxidation. *Catal. Today* **2008**, *139*, 8–14. [CrossRef]

21. Li, L.; Shen, X.; Wang, P.; Meng, X.; Song, F. Soot capture and combustion for perovskite La–Mn–O based catalysts coated on honeycomb ceramic in practical diesel exhaust. *Appl. Surf. Sci.* **2011**, *257*, 9519–9524. [CrossRef]

22. Wang, J.; Su, Y.; Wang, X.; Chen, J.; Zhao, Z.; Shen, M. The effect of partial substitution of Co in $LaMnO_3$ synthesized by sol–gel methods for NO oxidation. *Catal. Commun.* **2012**, *25*, 106–109. [CrossRef]

23. Peng, X.S.; Lin, H.; Shangguan, W.F.; Huang, Z. Surface Properties and Catalytic Performance of $La_{0.8}K_{0.2}Cu_xMn_{1-x}O_3$ for Simultaneous Removal of NO_x and Soot. *Chem. Eng. Technol.* **2007**, *30*, 99–104. [CrossRef]

24. Doggali, P.; Kusaba, S.; Teraoka, Y.; Chankapure, P.; Rayaluand, S.; Labhsetwar, N. $La_{0.9}Ba_{0.1}CoO_3$ perovskite type catalysts for the control of CO and PM emissions. *Catal. Commun.* **2010**, *11*, 665–669. [CrossRef]

25. Gao, Y.; Wu, X.; Liu, S.; Weng, D.; Zhang, H.; Ran, R. Formation of $BaMnO_3$ in Ba/MnO_x–CeO_2 catalyst upon the hydrothermal ageing and its effects on oxide sintering and soot oxidation activity. *Catal. Today* **2015**, *253*, 83–88. [CrossRef]

26. Torregrosa-Rivero, V.; Albaladejo-Fuentes, V.; Sánchez-Adsuar, M.S.; Illán-Gómez, M.J. Copper doped $BaMnO_3$ perovskite catalysts for NO oxidation and NO_2-assisted diesel soot removal. *RSC Adv.* **2017**, *7*, 35228–35238. [CrossRef]

27. Dai, H.X.; He, H.; Li, W.; Gao, Z.Z.; Au, C.T. Perovskite-type oxide $ACo_{0.8}Bi_{0.2}O_{2.87}$ (A = $La_{0.8}$ $Ba_{0.2}$): A catalyst for low-temperature CO oxidation. *Catal. Lett.* **2001**, *73*, 149–156. [CrossRef]

28. Borovskikh, L.; Mazo, G.; Kemnitz, E. Reactivity of oxygen of complex cobaltates $La_{1-x}Sr_xCoO_3$-δ and $LaSrCoO_4$. *Solid State Sci.* **2003**, *5*, 409–417. [CrossRef]

29. Isupova, L.A.; Rogov, V.A.; Yakovleva, I.S.; Alikina, G.M.; Sadykov, V.A. Oxygen States in Oxides with a Perovskite Structure and Their Catalytic Activity in Complete Oxidation Reactions: System La_{1-x} Ca_xFeO_{3-y} (x = 0–1). *Kinet. Catal.* **2004**, *45*, 446–453. [CrossRef]

30. Yakovleva, I.S.; Isupova, L.A.; Tsybulyaa, S.V.; Chernysh, N.N.; Boldyreva, G.; Alikina, G.M.; Sadykov, V.A. Mechanochemical synthesis and reactivity of $La_{1-x}Sr_xFeO_{3-y}$ perovskite (0 <x <1). *J. Mater. Sci.* **2004**, *39*, 5517–5521.

31. Yan, X.; Huang, Q.; Li, B.; Xu, X.; Chen, Y.; Zhu, S.; Shen, S. Catalytic performance of $LaCo_{0.5}M_{0.5}O_3$ (M = Mn, Cr, Fe, Ni, Cu) perovskite-type oxides and $LaCo_{0.5}Mn_{0.5}O_3$ supported on cordierite for CO oxidation. *J. Ind. Eng. Chem.* **2013**, *19*, 561–565. [CrossRef]

32. Yang, W.; Zhang, R.; Chen, B.; Bion, N.; Duprez, D.; Royer, S. Activity of perovskite-type mixed oxides for the low-temperature CO oxidation: Evidence of oxygen species participation from the solid. *J. Catal.* **2012**, *295*, 45–58. [CrossRef]

33. Seyfi, B.; Baghalha, M.; Kazemian, H. Modified $LaCoO_3$ nano-perovskite catalysts for the environmental application of automotive CO oxidation. *Chem. Eng. J.* **2009**, *148*, 306–311. [CrossRef]

34. Albaladejo-Fuentes, V.; López-Suárez, F.E.; Sánchez-Adsuar, M.S.; Illán-Gómez, M.J. Tailoring the properties of $BaTi_{0.8}Cu_{0.2}O_3$ catalyst selecting the synthesis method. *Appl. Catal. A* **2016**, *519*, 7–15. [CrossRef]

35. Moreno-Marcos, C.; Torregrosa-Rivero, V.; Albaladejo-Fuentes, V.; Sánchez-Adsuar, M.S.; Illán-Gómez, M.J. $BaFe_{1-x}Cu_xO_3$ Perovskites as Soot Oxidation Catalysts for Gasoline Particulate Filters (GPF): A Preliminary Study. *Top. Catal.* **2019**, *62*, 413–418. [CrossRef]

36. López-Suárez, F.E.; Parres-Esclapez, S.; Bueno-López, A.; Illán-Gómez, M.J.; Ura, B.; Trawczynski, J. Role of surface and lattice copper species in copper-containing (Mg/Sr)TiO_3 perovskite catalysts for soot combustion. *Appl. Catal. B* **2009**, *93*, 82–89. [CrossRef]

37. López-Suárez, F.E.; Bueno-López, A.; Illán-Gómez, M.J.; Trawczynski, J. Potassium-copper perovskite catalysts for mild temperature diesel soot combustión. *Appl. Catal. A* **2014**, *485*, 214–221. [CrossRef]

38. Albaladejo-Fuentes, V.; López-Suárez, F.E.; Sánchez-Adsuar, M.S.; Illán-Gómez, M.J. $BaTi_{1-x}Cu_xO_3$ perovskites: The effect of copper content in the properties and in the NOx storage capacity. *Appl. Catal. A Gen.* **2014**, *488*, 189–199. [CrossRef]

39. Huang, C.; Zhu, Y.; Wang, X.; Liu, X.; Wang, J.; Zhang, T. Sn promoted $BaFeO_{3-\delta}$ catalysts for N_2O decomposition: Optimization of Fe active centers. *J. Catal.* **2017**, *347*, 9–20. [CrossRef]

40. Langford, J.I.; Wilson, A.J.C. Scherrer after sixty years: A survey and some new results in the determination of crystallite size. *J. Appl. Cryst.* **1978**, *11*, 102–113. [CrossRef]

41. Chen, X.; Huang, L.; Wei, Y.; Wang, H. Tantalum stabilized $SrCoO_{3-\delta}$ perovskite membrane for oxygen separation. *J. Membr. Sci.* **2011**, *368*, 159–164. [CrossRef]

42. Ghijsen, J.; Tjeng, L.H.; Van Elp, J.; Eskes, H.; Westerink, J.; Sawatzky, G.A.; Czyzyk, M.T. Electronic structure of Cu_2O and CuO. *Phys. Rev. B* **1988**, *38*, 11322–11330. [CrossRef]

43. Espinós, J.P.; Morales, J.; Barranco, A.; Caballero, A.; Holgado, J.P.; González-Elipe, A.R. Interface Effects for Cu, CuO, and Cu_2O Deposited on SiO_2 and ZrO_2. XPS Determination of the Valence State of Copper in Cu/SiO_2 and Cu/ZrO_2 Catalysts. *J. Phys. Chem. B* **2002**, *106*, 6921–6929. [CrossRef]

44. Benoit, R. Centre de Recherche sur la Matière Divisée—CNRS. 2013. Available online: www.lasurface.com (accessed on 1 January 2019).

45. Biesinger, M.C. Advance analysis of copper X-ray photoelectron spectra. *Surf. Interface Anal.* **2017**, *49*, 1325–1334. [CrossRef]

46. Biesinger, M.C.; Payne, A.P.; Grosvenor, L.W.M.; Lau, F.; Wei, A.R.; Gerson, R.S.C. Smart resolving surface chemical states in XPS analysis of firts row transition metals oxides and hydroxides: Cr, Mn, Fe, Co and Ni. *Appl. Surf. Sci.* **2011**, *257*, 2717–2730. [CrossRef]

47. Wu, Y.; Cordier, C.; Berrier, E.; Nuns, N.; Dujardin, C.; Granger, P. Surface reconstructions of $LaCo_{1-x}Fe_xO_3$ at high temperature during N_2O decomposition in realistic exhaust gas composition: Impact on the catalytic properties. *Appl. Catal. B* **2013**, *140*, 151–163. [CrossRef]

48. Ghaffari, M.; Sahnnon, M.; Hui, H.; Tan, O.K.; Irannejad, A. Preparation, surface state and band structure studies of $SrTi_{(1-x)}Fe_{(x)}O_{(3-\delta)}$ (x = 0–1) perovskite-type nano structure by X-ray and ultraviolet photoelectron spectroscopy. *Surf. Sci.* **2012**, *606*, 670–677. [CrossRef]

49. Zhao, Z.; Dai, H.; Deng, J.; Du, Y.; Liu, Y.; Zhang, L. Preparation of three-dimensionally ordered macroporous $La_{0.6}Sr_{0.4}Fe_{0.8}Bi_{0.2}O_{3-\delta}$ and their excellent catalytic performance for the combustion of toluene. *J. Mol. Catal. A Chem.* **2013**, *366*, 116–125. [CrossRef]

50. Lee, E.-S. Characteristics of mixed conducting perovskites $(Ba_{1-x}Nd_x)Fe^{3+}_{1-t}Fe^{4+}_tO_{3-y}$. *J. Ind. Eng. Chem.* **2008**, *14*, 701–706. [CrossRef]

51. Tan, B.J.; Klabunde, K.J.; Sherwood, P.M.A. X-ray photoelectron spectroscopy studies of solvated metal atom dispersed catalysts. Monometallic iron and bimetallic iron-cobalt particles on alumina. *Chem. Mater.* **1990**, *2*, 186–191. [CrossRef]

52. He, H.; Dai, H.X.; Au, C.T. An investigation on the utilization of perovskite-type oxides $La_{1-x}Sr_xMO_3$ (M = $Co_{0.77}Bi_{0.20}Pd_{0.03}$) as three-way catalysts. *Appl. Catal. B* **2001**, *33*, 65–80. [CrossRef]

53. Nelson, A.E.; Schulz, K.H. Surface chemistry and microstructural analysis of $Ce_xZr_{1-x}O_{2-y}$ model catalyst surfaces. *Appl. Surf. Sci.* **2003**, *210*, 206–221. [CrossRef]

54. Merino, N.; Barbero, B.; Eloy, P.; Cadus, L. $La_{1-x}Ca_xCoO_3$ perovskite-type oxides: Identification of the surface oxygen species by XPS. *Appl. Surf. Sci.* **2006**, *253*, 1489–1493. [CrossRef]

55. Liao, J.X.; Yang, C.R.; Tian, Z.; Yang, H.G.; Jin, L. The influence of post-annealing on the chemical structures and dielectric properties of the surface layer of $Ba_{0.6}Sr_{0.4}TiO_3$ films. *J. Phys. D* **2006**, *39*, 2473–2479. [CrossRef]

56. Zhang, R.; Alamdari, H.; Kaliaguine, S. Fe-based perovskites substituted by copper and palladium for NO + CO reaction. *J. Catal.* **2006**, *242*, 241–253. [CrossRef]

57. Xian, H.; Zhang, X.; Li, X.; Li, L.; Zou, H.; Meng, M.; Li, Q.; Tan, Y.; Tsubaki, N. $BaFeO_{3-x}$ Perovskite: An Efficient NO_x Absorber with a High Sulfur Tolerance. *J. Phys. Chem. C* **2010**, *114*, 11844–11852. [CrossRef]

58. Falcón, H.; Barbero, J.A.; Alonso, J.A.; Martinez-López, M.J.; Fierro, J.L.G. $SrFeO_{3-\delta}$ Perovskite Oxides: Chemical Features and Performance for Methane Combustion. *Chem. Mater.* **2002**, *14*, 2325–2333. [CrossRef]

59. Zhang, J.Y.; Weng, X.I.; Wu, Z.B.; Liu, Y.; Wang, H.Q. Facile synthesis of highly active $LaCoO_3$/MgO composite perovskite via simultaneous co-precipitation in supercritical water. *Appl. Catal. B* **2012**, *126*, 231–238. [CrossRef]

60. Dhakad, M.; Rayalu, S.S.; Kumar, R.; Dpggali, P.; Bakardjieva, S.; Subst, J.; Mitsubashi, T.; Haneda, H.; Labhetwar, N. Low Cost Ceria Promoted Perovskite Type Catalysts for Diesel Soot Oxidation. *Catal. Lett.* **2008**, *121*, 137–143. [CrossRef]

61. Pereniguez, R.; Hueso, J.L.; Gaillard, F.; Holgado, J.P.; Caballero, A. Study of Oxygen Reactivity in $La_{1-x}Sr_xCoO_{3-\delta}$ Perovskites for Total Oxidation of Toluene. *Catal. Lett.* **2012**, *142*, 408–416.

62. Glisenti, A.M.; Pacella, M.; Guiotto, M.; Natile, M.M.; Canu, P. Largely Cu-doped $LaCo_{1-x}Cu_xO_3$ perovskites for TWC: Toward new PGM-free catalysts. *Appl. Catal. B.* **2016**, *180*, 94–105. [CrossRef]

63. Patcas, F.; Buciuman, F.C.; Zsako, J. Oxygen non-stoichiometry and reducibility of B-site substituted lanthanum manganites. *Thermochim. Acta* **2000**, *360*, 71–76. [CrossRef]

64. Onrubia, J.A.; Pereda-Ayo, B.; De-La-Torre, U.; González-Velasco, J.R. Key factors in Sr-doped $LaBO_3$ (B = Co or Mn) perovskites for NO oxidation in efficient diesel exhaust purification. *Appl. Catal. B* **2017**, *213*, 198–210. [CrossRef]

65. Najjar, H.; Lamonier, J.F.; Mentré, O.; Giraudon, J.M.; Batis, H. Optimization of thecombustion synthesis towards efficient $LaMnO_{3+y}$ catalysts inmethane oxidation. *Appl. Catal. B* **2011**, *106*, 149–159.

66. Merino, N.A.; Barbero, B.P.; Grange, P.; Cadús, L.E. $La_{1-x}Ca_xCoO_3$ perovskite-type oxides: Preparation, characterisation, stability, and catalytic potentiality for the total oxidation of propane. *J. Catal.* **2005**, *231*, 232–244. [CrossRef]

67. Torregrosa-Rivero, V.; Albaladejo-Fuentes, V.; Sánchez-Adsuar, M.S.; Illán-Gómez, M.J. Mecanismo de almacenamiento de NOx en el catalizador $BaTi_{0.8}Cu_{0.2}O_3$. In *Proceedings of the CICAT 2018 XXVI Congreso Ibero-Americano De Catálisis, Cohimbra, Portugal, 9–14 Septiembre 2018*; Gomes, H., Silva, A., Machado, B., Ribeiro, F., Fonseca, I., Faria, J., Pereira, M., Rocha, R., Eds.; Sociedade Portuguesa de Química: Coimbra, Portugal, 2018.

68. Ma, A.; Wang, S.; Liu, C.; Xian, H.; Ding, Q.; Guo, L.; Mung, M.; Tan, Y.; Tsubaki, N.; Zhaeng, J.; et al. Effects of Fe dopants and residual carbonates on the catalytic activities of the perovskite-type $La_{0.7}Sr_{0.3}Co_{1-x}Fe_xO_3$ NO_x storage catalyst. *Appl. Catal. B* **2014**, *146*, 24–34. [CrossRef]

69. Van Setten, B.; Makkee, M.; Moulijn, J. Science and technology of catalytic diesel particulate filters. *Catal. Rev.* **2001**, *43*, 489–564. [CrossRef]

70. Gao, Y.; Duan, A.; Liu, S.; Wu, X.; Liu, W.; Li, M.; Chen, S.; Wang, X.; Weng, D. Study of $Ag/Ce_xNd_{1-x}O_2$ nanocubes as soot oxidation catalysts for gasoline particulate filters: Balancing catalyst activity and stability by Nd doping. *Appl. Catal. B* **2017**, *203*, 116–126. [CrossRef]

Article

Catalytic Performance of Ni/CeO$_2$/X-ZrO$_2$ (X = Ca, Y) Catalysts in the Aqueous-Phase Reforming of Methanol

Daniel Goma [1,2], Juan José Delgado [1,2], Leon Lefferts [3], Jimmy Faria [3], José Juan Calvino [1,2] and Miguel Ángel Cauqui [1,2,*]

[1] Departamento de Ciencia de los Materiales e Ingeniería Metalúrgica y Química Inorgánica, Universidad de Cádiz, 11510 Puerto Real, Spain; dani.gomajimenez@gm.uca.es (D.G.); juanjose.delgado@uca.es (J.J.D.); jose.calvino@uca.es (J.J.C.)

[2] IMEYMAT, Instituto de Microscopía Electrónica y Materiales, 11510 Puerto Real, Spain

[3] Catalytic Processes and Materials group, University of Twente, P.O. Box 217, 7500 AE Enschede, The Netherlands; l.lefferts@utwente.nl (L.L.); j.a.fariaalbanese@utwente.nl (J.F.)

* Correspondence: miguelangel.cauqui@uca.es; Tel.: +34-956012747

Received: 3 October 2019; Accepted: 6 November 2019; Published: 8 November 2019

Abstract: In this study, we reported on the effect of promoting Ni/ZrO$_2$ catalysts with Ce, Ca (two different loadings), and Y for the aqueous-phase reforming (APR) of methanol. We mainly focused on the effect of the redox properties of ceria and the basicity provided by calcium or yttrium on the activity and selectivity of Ni in this reaction. A systematic characterization of the catalysts was performed using complementary methods such as XRD, XPS, TPR, CO$_2$-TPD, H$_2$ chemisorption, HAADF-STEM, and EDS-STEM. Our results reveal that the improvement in reducibility derived from the incorporation of Ce did not have a positive impact on catalytic behaviour thus contrasting with the results reported in the literature for other Ce-based catalytic compositions. On the contrary, the available Ni-metallic surface and the presence of weak basic sites derived from Ca incorporation seem to play a major role on the catalytic performance for APR of methanol. The best performance was found for a Ce-free catalyst with a molar Ca content of 4%.

Keywords: aqueous-phase reforming; nickel; ceria; zirconia; calcium; yttrium; methanol

---

1. Introduction

The future of the so-called hydrogen economy is linked to the possibility of developing economical clean and sustainable methods for both the production of H$_2$ and its subsequent conversion into energy [1]. Some of the applications adapted to this energy model would operate, for example, with electricity generated in a fuel cell powered by in situ produced hydrogen thus overcoming problems associated with storage, transport, and handling of hydrogen gas [2]. In this sense, methanol derived from biorefinery water fractions can be considered an interesting source of hydrogen, as it is an economic product with a considerable H$_2$ content (13%) and easily transportable (liquid at room temperature). The transformation of methanol into hydrogen can be achieved by reforming, with the aqueous-phase reforming (APR) proposed by Dumesic [3] being the most interesting alternative as long as it is carried out at low temperature and does not require prior vaporization of the reagents such as in the case of classic steam reforming processes [4].

The APR of methanol involves the decomposition of the organic compound to produce CO and H$_2$ (Equation (1)). In this case, C–C bond cleavage is not necessary, and CO and H$_2$ are produced through C–H and O–H bond cleavage. The CO is then converted into CO_2 and H_2 by the water–gas shift (WGS) reaction, for which H_2O dissociation is needed (Equation (2)).

$$CH_3OH \rightarrow CO + 2H_2 \tag{1}$$

$$CO + H_2O \rightarrow CO_2 + H_2 \tag{2}$$

However, under normal operating conditions (i.e., low temperature), the hydrogenation reactions of CO and CO_2 to produce methane and water (Equations (3) and (4)) are thermodynamically favoured, compromising the H_2 selectivity in APR.

$$CO + 3H_2 \rightarrow CH_4 + H_2O \tag{3}$$

$$CO_2 + 4H_2 \rightarrow CH_4 + 2H_2O \tag{4}$$

Thus, a suitable catalyst for APR should promote the reforming (Equation (1)) and WGS (Equation (2)) reactions and inhibit the methanation reactions (Equations (3) and (4)).

Davda et al. [5] reported a comparative study of the catalytic behaviour of a series of metals such as Pt, Ru, Rh, Pd, Ir, and Ni in APR reactions. The catalytic activities obtained for Pt and Ni were comparable and superior to those shown by the rest of the metals. Although nickel catalysts are economically preferable to Pt, they have two major drawbacks for APR: (1) they favour the methanation reactions therefore reducing the hydrogen selectivity [6,7] and (2) they have a significant tendency to deactivation by oxidation or sintering under hydrothermal conditions [8]. To become a real alternative to Pt in APR, these two limitations of Ni catalysts must be overcome. One of the most commonly used strategies consists of combining nickel with specific promoters or supports. Thus, for example, it has been reported that Ni/CeO_2 shows great potential to be used in the APR of glycerol, exhibiting higher H_2 selectivity than Ni/Al_2O_3 [9]. Improved catalytic performances were also obtained with Ni–Ce–O catalysts for aqueous-phase reforming of ethanol [10]. This is generally attributed to the singular redox properties of ceria (i.e., high oxygen storage capacity and oxygen mobility) which promote preferentially the WGS versus the methanation reaction and, thus, enhancing the hydrogen production [11,12]. The CeO_2–ZrO_2, CeO_2–La_2O_3 and CeO_2–TiO_2 mixed oxides have also been successfully used as support of Pt or Ni for APR, achieving a higher hydrothermal stability compared to pure CeO_2 [13–16].

The effect of the basicity of the support has also been investigated by several authors. Guo et al. [17] found that the catalytic performances of a series of Pt-supported catalysts followed the same trend than the basicity measured by CO_2-TPD. They concluded a correlation between the WGS ability, promoted in this case by basic supports, and the APR activity. Menezes et al. [18] reported that Pt/MgO performed better that Pt supported on alumina, zirconia or ceria, and attributed its success to the basicity of MgO.

In a previous work, we reported on the performance of a $Ni/CeO_2/YSZ$ (YSZ: yttrium-stabilized zirconia) catalyst for the CO_2 reforming of CH_4 [19]. This formulation exhibited not only high activity but also an outstanding stability under reaction conditions. The excellent redox properties provided by the support were found to be a key factor in preventing the accumulation of carbon deposits responsible for the deactivation of the catalyst. Based on that work, we here report on a series of $Ni/CeO_2/X$-ZrO_2 (X = Ca, Y) catalysts designed to simultaneously develop enhanced redox and basic properties, the latter resulting from the incorporation of calcium. These catalysts have been prepared and evaluated for hydrogen production by APR of methanol. The effect of these properties on the activity and selectivity of Ni has been specially addressed in order to determine their relative influence on the catalytic performance. This is a crucial point for tailoring more efficient catalysts for APR. A detailed characterization of the catalysts employing different techniques such as X-ray diffraction (XRD), X-ray photoelectron spectroscopy (XPS), temperature programmed reduction with H_2 (H_2-TPR), temperature programmed desorption of CO_2 (CO_2-TPD), high-angle annular dark-field scanning transmission electron microscopy (HAADF-STEM) and energy dispersive X-ray spectroscopy (EDS-STEM) was carried out to look for correlations between chemical properties and catalytic behaviour.

2. Materials and Methods

2.1. Catalysts Preparation

Calcia-stabilized zirconia (CSZ), yttria-stabilized zirconia (YSZ), and pure zirconia (Z) were used as supports for nickel and ceria-modified nickel catalysts. These oxides were prepared by a hydrothermal method from a 0.5 M zyrconil nitrate ($ZrO(NO_3)_2 \cdot 6H_2O$) (Sigma Aldrich, St. Louis, MO, USA) solution containing calculated amounts of calcium or yttrium nitrates (Sigma Aldrich, St. Louis, MO, USA) to obtain molar ratios of 4% and 14%, in the case of CSZ samples (4CSZ and 14CSZ), and 8% in the case of the YSZ (8YSZ). The pH was adjusted to 10 by adding a NaOH (Sharlau, Barcelona, Spain) 1 M solution under vigorous stirring. The hydrothermal synthesis was performed in PTFE vessels heated at 200 °C over 24 h. After cooling to room temperature, the powders were separated by centrifugation and washed with Milli-Q water until pH = 7. Subsequently, they were dried at 110 °C, grounded, sieved (300–600 µm fraction), and finally calcined at 500 °C for 1 h. Ceria (12% w/w) was incorporated by incipient wetness impregnation using a $Ce(NO_3)_3 \cdot 6H_2O$ (Alfa Aesar, Kandel, Germany) 2 M aqueous solution. After drying at 110 °C and calcination at 500 °C (1 h), the samples were further impregnated using a $Ni(NO_3)_2 \cdot 6H_2O$ (Sigma Aldrich, St. Louis, MO, USA) 2 M aqueous solution. After the second impregnation, the final powders were dried at 110 °C overnight and finally calcined at 500 °C (1 h). The nickel loading was always 6% w/w.

2.2. Catalysts Characterization

The metal contents of fresh and used catalysts were determined by inductively coupled plasma atomic emission spectroscopy (ICP-AES). The measurements were carried out using a Thermo Elemental plasma atomic emission spectrometer (model Intrepid, Thermo Scientific, Waltham, MA, USA).

The textural properties of catalysts were determined from nitrogen physisorption at −196 °C. The adsorption and desorption isotherms were obtained using a Quantachrome Autosorb IQ3 (Quantachrome Instruments, Boynton Beach, FL, USA). Prior to the physisorption experiments, the samples were degassed for 2 h at 200 °C under vacuum. Surface area was calculated with the BET method, and the pore volume and diameter were calculated by the BJH method using data from the desorption-isotherm.

The Ni-metallic surface area and average particle size were determined by hydrogen chemisorption. The experiments were performed at 35 °C in a ASAP 2020 equipment (Micromeritics Instrument Corp., Norcross, GA, USA). Prior to each adsorption, samples were reduced in 5% H_2/Ar flowing at 750 °C for 1 h and evacuated under vacuum at the same temperature. Estimated metal surface area and particle size values were based on spherical geometry and an H/Ni = 1 adsorption stoichiometry.

The structure of catalysts was determined by X-ray diffraction (XRD) in a diffractometer D8 Advance (Brucker, Billerica, MA, USA) using nickel-filtered CuKα (λ = 0.15418 nm) radiation with a step size of 0.02° and 0.2 s as the counting time. The crystallite diameter for NiO and Ni species in calcined and reduced samples, respectively, was estimated applying the Scherrer equation (k = 0.9), using the DIFFRAC.EVA software from Brucker. The reduction treatment consisted of heating in a flow of 5% H_2/Ar from room temperature to 750 °C (10 °C /min), followed by 1 h of isothermal treatment at this temperature. The gas was switched to He for 1 h, and then the samples were cooled down to −80 °C in He flow. Finally, they were heated from −80 °C up to room temperature under 5% O_2/He to avoid overheating of the reduced catalysts when exposed to air.

The reducibility of the catalysts was studied by H_2 temperature programmed reduction (H_2-TPR), using Autochem 2920 II equipment (Micromeritics Instrument Corp., Norcross, GA, USA). Before the TPR experiments, samples were treated in 5% O_2/He for 1 h at 500 °C, purged with He at the same temperature, and cooled down to room temperature. Then, they were heated (10 °C/min) from room temperature up to 950 °C in flowing 5% H_2/Ar. Hydrogen consumption throughout the reduction process was monitored by means of a thermal conductivity detector (TCD).

Temperature programmed desorption of CO_2 (CO_2-TPD) experiments were carried out to characterize the surface basic sites. Prior to the analyses, the catalysts were reduced in situ under a 5% H_2/Ar flow at 750 °C for 1 h, treated with He for an additional hour and cooled down to room temperature. Afterwards, they were contacted with a flow of pure CO_2 for one hour at 100 °C and finally cooled down again and purged with He at room temperature before starting the TPD experiment. CO_2 desorption was followed by monitoring the m/z = 44 signal using a quadrupole mass spectrometer model GSD301T1 (Pfeiffer Vacuum, Wetzlar, Germany). The calibration of this signal was carried out using standards of calcium oxalate diluted in alumina.

X-ray photoelectron spectroscopy (XPS) experiments were carried out in a Kratos Axis Ultra DLD (Kratos Analytical Ltd, Manchester, UK) equipped with a monochromatized Al-Kα X-ray source (1486.6 eV), operating with an accelerating voltage of 15 kV and 10 mA current. Spectra were acquired in a constant analyser energy mode, with a pass energy of 20 eV. Powder samples were analysed without any pre-treatment. Surface charging effects were corrected by adjusting the binding energy of the C(1s) peak at 248.8 eV. CasaXPS software (version 2.3.19) (Casa Software Ltd, Devon, UK) was used for the data analysis. Prior to the XPS experiments, the catalysts were reduced using the same protocol as that used for XRD measurements.

A JEOL-2010F microscope (Jeol Ltd., Tokyo, Japan) with a spatial resolution at Scherzer defocus conditions of 0.19 nm in TEM and operated at 200 kV was used in high-angle annular dark field scanning transmission electron microscopy (HAADF-STEM) mode. The microscope was equipped with an X-ray energy-dispersive spectrometer (X-EDS) model Xmax SSD (Oxford Instruments, Abingdon, UK), for composition analysis at the sub-nanometre scale. Samples for EM studies were prepared by depositing the powders onto holey carbon-coated Cu grids. Additionally, very high spatial resolution X-EDS maps were acquired using the ChemiSTEM capabilities of an FEI Titan Themis 60–300 microscope (FEI Company, Hillsboro, OR, USA).

Particle size distributions were obtained from the statistical analysis of at least 250 particles observed in the HAADF images. The values of mean particle size ($\bar{d}$) and surface area-weighted mean particle size ($\bar{d}_{sa}$) were calculated according to Equations (5) and (6), respectively. For a suitable consideration of the contribution of the larger particles, Ni dispersion (D%) was not estimated from the mean particle size, but as the ratio of total surface to bulk Ni atoms. These, in turn, were estimated by accumulating the surface or bulk Ni atoms corresponding to each of the particles considered in the analysis (assuming spherical morphology).

$$\bar{d} = \frac{\sum n_i\, d_i}{\sum n_i} \tag{5}$$

$$\bar{d}_{sa} = \frac{\sum n_i d_i^3}{\sum n_i d_i^2} \tag{6}$$

2.3. Catalytic Tests

Catalytic tests were carried out in a laboratory scale system as the one depicted in Figure S1 (Supporting Materials). The experiments were performed in a continuously operated stainless-steel tubular reactor with a cross-section of 6 mm using downward flow. The catalyst (without diluent) was loaded on the mid-section of the reactor on a fixed bed with a 5 μm pore mesh. Aqueous-phase reforming of methanol was carried out at 230 °C and 32 bar, using a flow rate of 0.338 mL/min of 5 wt. % aqueous solution of methanol as a feed (WSHV = 4 h^{-1}). The selected reaction time was 5 h (after 90 min to reach steady state).

The outlet stream from the reactor was cooled down, pressure-controlled by a back-pressure regulator, and separated (without dilution with inert gas) into the liquid and gas streams.

The composition of the gas products was analysed in situ by means of a micro gas chromatograph (GC, Varian CP4900, equipped with MS5 and PPQ columns) (Varian Inc., Palo Alto, CA, USA) and the liquid products were taken every 30 min and analysed offline by performing HPLC (RID-10A detector, Aminex HPX-87H column, 300 × 7.8 mm) (Shimadzu Corporation, Kyoto, Japan). Prior to the catalytic tests, samples were reduced at 750 °C for 1 h under pure hydrogen. After catalytic tests, the used samples were recovered and weighted to evaluate the catalyst loss during reaction. Methanol conversion, selectivity, and H_2 yield were obtained as follows:

$$\% \ CH_3OH \ conversion = \frac{[CH_3OH]_{In} - [CH_3OH]_{Out}}{[CH_3OH]_{In}} \times 100 \tag{7}$$

$$\% \ Selectivity \ to \ X = \frac{[X]_{Out}}{[H_2]_{Out} + [CO_2]_{Out} + [CO]_{Out} + [CH_4]_{Out}} \times 100 \tag{8}$$

$$H_2 \ yield \ (\%) = \frac{mol \ H_{2Out}}{mol \ CH_3OH_{In}} \times \frac{1}{3} \times 100 \tag{9}$$

3. Results and Discussion

3.1. Catalysts' Composition and Textural Characterization

The composition of the catalysts obtained from ICP analysis is shown in Table 1. As it can be seen, the elemental contents are in good agreement with the nominal values. Similar results were obtained in the analysis of the catalysts after reaction thus discarding the occurrence of lixiviation under reaction conditions (Table S1).

Table 1. Composition and textural properties.

Catalysts	Composition from ICP (% w/w)					S_{BET} $(m^2 \cdot g^{-1})$	Pore Volume $(cm^3 \cdot g^{-1})$	Average Pore Radius (nm)
	Ni	Ce	Ca	Y	Zr			
NiZr	7.0	-	-	-	60.0	30	0.178	9.4
NiCeZr	5.9	12.6	-	-	48.4	30	0.141	7.7
Ni4CSZ	5.0	-	1.20	-	60.0	48	0.214	5.7
NiCe4CSZ	5.9	13.0	1.80	-	47.6	47	0.123	3.9
Ni8YSZ	5.6	-	-	4.30	56.4	53	0.192	5.7
NiCe8YSZ	5.5	12.7	-	3.62	46.1	42	0.123	3.9
Ni14CSZ	6.9	-	4.90	-	53.0	54	0.220	4.8
NiCe14CSZ	5.5	13.5	4.00	-	43.3	67	0.133	2.8

The results obtained from the textural characterization of the catalysts using N_2 adsorption–desorption isotherms (Figure S2) are also gathered in Table 1. Samples exhibited mesoporosity, with surface areas values varying in the range 30–67 $m^2 \cdot g^{-1}$. The lower values (around 30 $m^2 \cdot g^{-1}$) were obtained for the catalysts using pure zirconia as support. Those systems based on structurally stabilised zirconia (CSZ or YSZ) exhibited both higher S_{BET} and pore volume values. This effect has previously been explained in the literature as being due to the existence of defects in the material which would impede the enlargement of grain boundaries and, hence, prevent growth of crystalline aggregates [20].

The incorporation of Ce did not have a clear influence on the surface area of the final catalyst; however, it provoked a decrease in both pore volumes and average pore sizes which, in principle, suggests that ceria was mainly blocking the largest pores of the support.

3.2. Structural Analysis by X-Ray Diffraction (XRD)

The XRD patterns of calcined and reduced catalysts are depicted in Figures 1 and 2. As expected, all the diffractograms were dominated by diffraction peaks characteristic of the corresponding support.

Thus, the catalyst supported on pure ZrO_2 presented peaks associated to the monoclinic structure of this oxide (PDF No. 37-1484). Doping with Ca^{2+} or Y^{3+} led to the appearance of additional peaks, indicating the partial stabilization of the cubic/tetragonal structure of zirconia (PDF No. 24-1074 and 83-0113, respectively). The degree of stabilization was higher in the case of the sample with 8% of Y^{3+} with respect to the sample doped with 4% of Ca^{2+}. We must recall that, according to the difference in oxidation states of Y and Ca, doping in these two samples introduced a similar concentration of oxygen vacancies in the ZrO_2 network. Increasing the doping amount up to 14% of Ca^{2+} led to an almost complete structural stabilization of the cubic/tetragonal phase, as deduced from the disappearance of the diffraction peaks associated to the monoclinic form. No peaks corresponding to segregated Ca-containing species were observed.

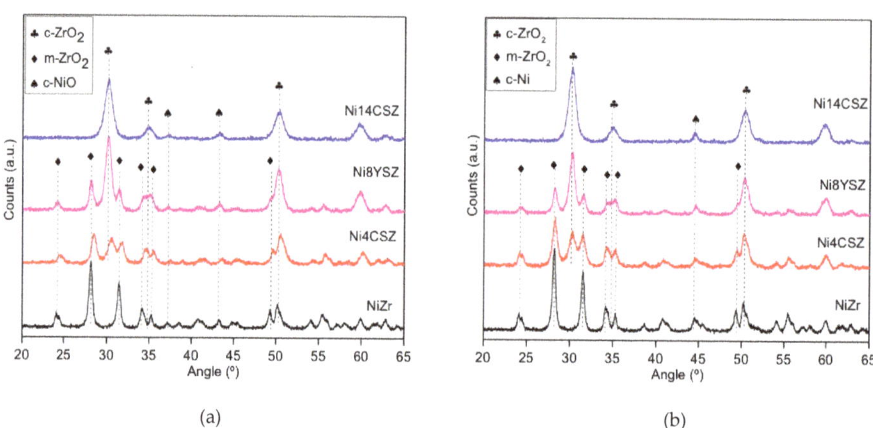

Figure 1. XRD patterns of the investigated catalysts (without Ce): (**a**) calcined; (**b**) reduced.

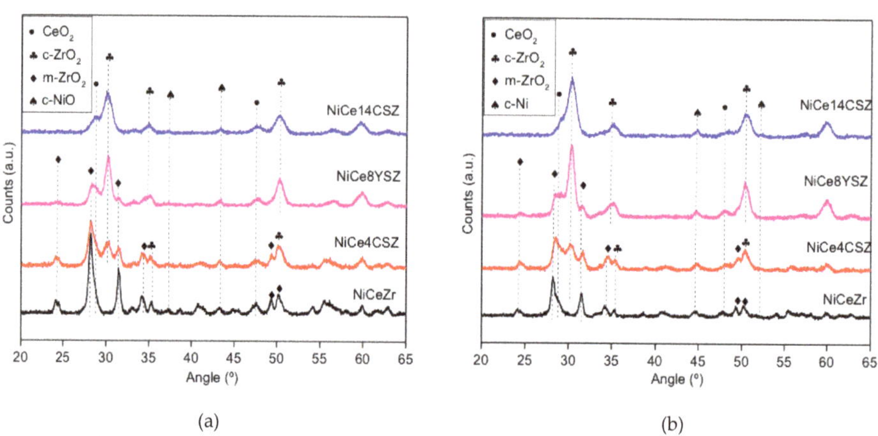

Figure 2. XRD patterns of the investigated catalysts (with Ce): (**a**) calcined; (**b**) reduced.

After the incorporation of Ce (Figure 2), some additional broad peaks at 2θ values of 28.5°, 33.1°, 47.5°, and 56.3° appeared, indicating the presence of small crystals of ceria with its typical fluorite-like structure (PDF No. 34-0394).

As for the Ni, all the calcined samples showed very small peaks at 37.2° and 43.2°, accounting for the presence of NiO (PDF No. 47-1049). After reduction at 750 °C, these peaks transformed into those corresponding to metallic Ni (Figures 2 and 3).

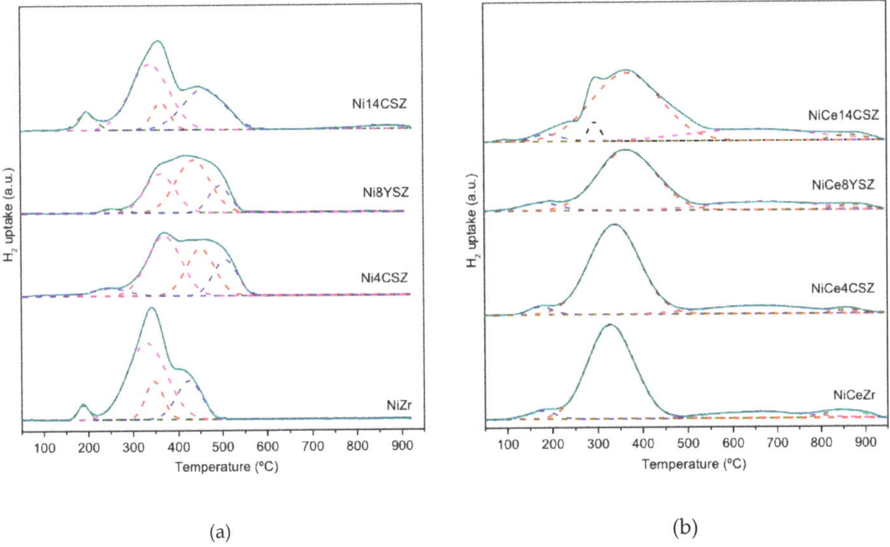

Figure 3. H_2-TPR profiles: (**a**) samples without Ce; (**b**) samples with Ce.

The average crystallite sizes of the NiO and Ni phases were estimated using XRD line-broadening and the Scherrer equation (Table 2). In particular, the peaks at 43.2° and 44.5° on the diffractograms corresponding to the calcined and reduced catalysts were used, respectively. Despite the low intensity of these peaks, the values obtained point to some influence of the support on the crystallite size of nickel species. In the case of the NiZr catalyst, values between 22 and 24 nm were obtained for NiO and Ni crystallites, respectively. The incorporation of either Ce or Ca to the catalysts slightly decreased the average crystallite size of nickel species, suggesting an improvement of the metal dispersion over the different Ce- or Ca-modified zirconia supports.

Table 2. Results from H_2 chemisorption, XRD line broadening, and H_2-TPR experiments.

Catalysts	Metal Surface [1] $(m^2 \cdot g^{-1})$	Ni Particle Size [1] (nm)	Crystallite Size (nm) [2]			H_2 Uptake [3] $(mmol \cdot g^{-1})$	Reduction Degree [3] (%)
			Ni	NiO	CeO$_2$		
NiZr	1.7	27.0 ± 1.0	24	22	-	1.16	97
NiCeZr	1.7	24.5 ± 2.0	10	17	22	1.53	105
Ni4CSZ	2.2	15.6 ± 0.5	17	15	-	0.94	110
NiCe4CSZ	2.1	19.3 ± 1.2	11	17	13	1.63	110
Ni8YSZ	1.2	32.5 ± 3.3	19	21	-	0.93	97
NiCe8YSZ	1.3	24.0 ± 1.6	16	18	15	1.28	92
Ni14CSZ	1.2	38.0 ± 1.4	21	18	-	1.24	105
NiCe14CSZ	1.4	25.7 ± 1.2	16	15	10	1.69	118

[1] From H_2 chemisorption; [2] from XRD line broadening. Crystallite sizes for NiO and CeO$_2$ were obtained from patterns corresponding to calcined catalysts, and, for Ni, they were obtained from patterns corresponding to the reduced catalysts. [3] From H_2-TPR experiments.

For samples containing ceria, the crystallite size for this phase was also estimated. As deduced from data in Table 2, the incorporation of Ca^{2+} or Y^{3+} into the zirconia support allows for the obtention of smaller (more dispersed) CeO_2 crystallites. This effect can be due to the higher surface area of the doped samples with respect to pure zirconia but also to a better structural coherence between the fluorite-type structure of ceria and the cubic/tetragonal structure stabilised in doped-zirconia supports.

3.3. Hydrogen Chemisorption and Temperature Programmed Reduction (TPR) Studies

To gain more insight into the Ni dispersion on the different supports, hydrogen chemisorption measurements were carried out. The results obtained are summarised in Table 2 in terms of Ni-metallic surfaces and Ni-particle sizes. As it can be seen, similar values were obtained for Ni-metallic surfaces, being only slightly higher in the case of the samples containing 4% of Ca. It should be noted that the values for particle size estimated from H_2 chemisorption were higher than those obtained by XRD (crystallite sizes). These fell around 20 nm, with a larger deviation in the case of the Ni14CSZ (38 nm) and Ni8YSZ (32 nm) catalysts. Results obtained from XRD line broadening were rather consistent with those obtained using electron microscopy, indicating that the estimations made from H_2 chemisorption measurements were likely affected by Ni–support interactions.

The TPR experiments were performed in order to obtain information about reducibility and also about Ni–support interactions in these catalysts. The reduction profiles are shown in Figure 3, and Table 2 includes the hydrogen consumption and estimated reduction degree values, assuming that, initially, Ce was completely Ce^{4+} and Ni as Ni^{2+}. Values close to 100% were obtained indicating that full reduction of these two species, from Ce^{4+} to Ce^{3+} and from Ni^{2+} to Ni^0, respectively, was achieved at the end of the TPR experiments. Values higher than 100% observed in some cases have also been reported by other authors for similar compositions, being generally attributed to (i) the existence of a small amount of Ni^{3+}-forming non-stoichiometric NiO_{1+x} species [21,22] and/or (ii) the reduction of labile oxygen adsorbed on vacancies in the ZrO_2 support generated as a consequence of doping with Ca^{2+} or Y^{3+} [23,24].

Concerning the structure of the TPR profiles, they were constituted by a combination of several peaks which account for the different reduction steps. For a better interpretation of the reduction sequence, the TPR profiles were deconvoluted into individual Gaussian functions as shown in Figure 2. These peaks are grouped into three general categories: low-temperature peaks (LT; 180–250 °C), intermediate-temperature peaks (IT; 300–500 °C), and high-temperature peaks (HT; >550 °C). The LT contributions were commonly low-intensity peaks assigned to the reduction of oxygen adsorbed on support vacancies and/or reduction of non-stoichiometric NiO_{1+x} species well dispersed on the catalyst surface [21,25]. This LT contribution appeared in the reduction scheme of all our catalysts, being slightly shifted to lower temperatures for catalysts containing Ce.

The IT part of the TPR profiles generally comprised several peaks which are commonly associated with the reduction of NiO species with different particle sizes and degrees of interaction with the support. Thus, it is widely assumed that NiO aggregates with little interaction with the substrate are reduced at temperatures around 400 °C while those with stronger interaction can be reduced at higher temperatures (approximately 550 °C) [26–29]. The reduction of large NiO aggregates in two consecutive steps ($Ni^{2+} \rightarrow Ni^{\delta+} \rightarrow Ni$) within this IT range has also been proposed by different authors [24].

In the case of catalysts without Ce, at least two contributions (centred at approximately 350 °C and 450 °C, respectively) are required to obtain a good fit of the curves in this IT range. Considering that no significant differences in NiO crystallite sizes were observed by XRD, we can assume that the intensity of the peak at higher temperature is related to the existence of NiO particles having a stronger interaction with the support [26]. Note that in the case of Ni8YSZ and Ni4CSZ catalysts, the first contribution was less intense, while the second was slightly shifted at higher temperature, suggesting the occurrence of a stronger NiO–support interactions in theses samples.

On the contrary, for the Ce-containing samples, a single peak was observed in the IT region of the TPR profiles, with the only exception being the NiCe14CSZ catalyst, which also showed a sharp shoulder at 300 °C. These simpler profiles indicate that CeO_2 improved the reducibility of the NiO phases in these catalysts. According to quantitative estimations, the amounts of H_2 involved in the IT part of the TPRs were higher than those required for the complete reduction of Ni, indicating that not only Ni^{2+} but also Ce^{4+} species were being reduced simultaneously at these temperatures.

High-temperatures peaks (>550 °C) also appearing in the TPR-profiles of Ce-containing samples would account for the reduction of bulk Ce species existing in larger (or not in contact with Ni) CeO_2 particles.

In summary, the TPR results indicate that NiO species in these catalysts were reduced mainly in a temperature range between 300 and 500 °C. For the catalysts supported on 4CSZ and 8YSZ, a shift to higher temperatures was observed presumably as a consequence of a stronger NiO-support interaction occurring in these cases. The incorporation of Ce improves the reducibility of NiO which seems to occur in a single step and simultaneously with the reduction of most of the CeO_2.

3.4. Thermal Programmed Desorption of CO_2 Basicity Studies

Thermal programmed desorption of CO_2 (CO_2-TPD) experiments were carried out to investigate the number and strength of surface basic sites. Desorption profiles are depicted in Figure 4. They were deconvoluted into three temperature ranges, as shown in Figure S3. The low temperature range (90–180 °C) contains information about weak basic sites; the intermediate temperature range (180–400 °C) gives information about moderate basic sites, and, finally, the high temperature range (>400 °C) accounts for the stronger basic sites. The amounts of CO_2 desorbed in each range as well as the total amount of CO_2 desorbed are gathered in Table 3.

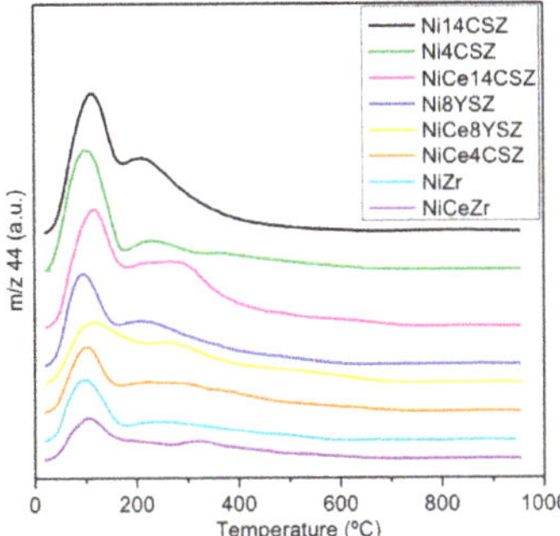

Figure 4. CO_2 desorption profiles.

Table 3. Amounts of CO_2 desorbed in the CO_2-TPD experiments. Values obtained by integration of the deconvoluted profiles in the low, medium and high temperature ranges.

Samples	Weak (90–180 °C) ($\mu mol \cdot g^{-1}$)	Intermediate (180–400 °C) ($\mu mol \cdot g^{-1}$)	Strong (>400 °C) ($\mu mol \cdot g^{-1}$)	Total ($\mu mol \cdot g^{-1}$)
NiZr	7.4	3.3	2.9	13.7
Ni4CSZ	21.1	1.5	3.1	25.6
Ni8YSZ	18.1	7.9	-	26.0
Ni14CSZ	29.0	7.4	-	36.5
NiCeZr	7.3	3.9	2.6	13.8
NiCe4CSZ	10.1	6.9	4.7	21.8
NiCe8YSZ	12.5	14.6	9.9	37.0
NiCe14CSZ	17.4	15.8	6.9	40.1

As can be seen in Figure 4, the CO_2-TPD profile of the NiZr sample showed a strong desorption peak at approximately 100 °C as well as a broad band extending up to 600 °C which can be deconvoluted into different contributions accounting for medium/strong basic sites. Taking this sample as a reference, we observed, as expected, that the incorporation of Ca results in a general increase in basicity [30], especially significant in the range of weak basic centres for the Ni4CSZ sample. When Y is used instead of Ca, the basicity with respect to the NiZr sample also increases, but more homogeneously for both weak- and intermediate-type centres. Thus, although the total amount of basic centres is almost the same for the Ni4CSZ and Ni8YSZ samples (25.6 and 26.0 μmol $CO_2 \cdot g^{-1}$, respectively), the latter exhibits fewer weak basic sites but more medium/strong basic centres. Increasing the calcium content by up to 14% results in an increase in the amounts of CO_2 desorbed over the whole temperature range. The incorporation of Ce has a minimal effect on the surface basicity of the NiZr sample, but, on the contrary, it greatly influences not only the quantity but mainly the nature of the basic centres of Ca(Y)-doped supports. In general, after the incorporation of Ce, the intensity of the low temperature peak (weak basicity) decreases, probably due to the partial covering of the support surface by CeO_2, while new contributions appeared in the intermediate/high temperature range, likely associated to the Ce-O sites [31].

3.5. Analysis of the Surface by XPS

X-Ray photoelectron spectroscopy was used to obtain information about the chemical state of elements on the surface of Ni catalysts after reduction treatment. Binding-energy values for Ni 2p, Ce 3d, and O 1s core levels are included in Table 4. Figure 5 gathers the core level Ni 2p XPS spectra corresponding to Ce-free and Ce-containing catalysts. The presence of Ni^0 was confirmed in all cases by the peak centred at approximately 852 eV. Moreover, two peaks at 855 eV and 860 eV were also observed in all the spectra. According to the literature, they can be assigned to Ni^{2+} (as $Ni(OH)_2$) and its satellite, respectively [32–34]. They have been labelled as Ni^{2+} (ii) in the spectra of Figure 5. The appearance of these peaks reveals a partial re-oxidation of the catalysts due to the exposure to atmospheric conditions during transfer from the preparation reactor to the XPS analysis chamber. It should be noted that, in the case of catalysts containing Ce, an additional component at around 853.5 eV, just in the middle of the positions corresponding to Ni^0 (852.6 eV) and Ni^{2+} (NiO, 854.6 eV), was observed. This peak, identified as Ni^{2+}(i) in the spectra, has been ascribed by some authors also to NiO [35]. However, its appearance exclusively in the case of Ce-containing samples suggests that it may be due to the cationic $Ni^{\delta+}$ species resulting of a Ni–Ce interaction [36] or simply Ni^{2+} incorporated into the CeO_2 lattice at the surface level [37,38].

Table 4. Binding-energy values of main peaks and the Ce^{3+} percentage from XPS.

Samples	Ni 2p	Ce 3d	O 1s	Ce^{3+} (%)
NiZr	852.7, -, 855.4	-	529.5, 531.2	-
Ni4CSZ	852.9, -, 855.3	-	529.7, 531.7	-
Ni8YSZ	852.8, -, 855.6	-	529.7, 531.7	-
Ni14CSZ	852.7, -, 855.3	-	529.8, 531.4	-
NiCeZr	852.2, 853.5, 855.4	885.0, 898.3	529.6, 531.3	9
NiCe4CSZ	851.9, 853.2, 855.2	884.7, 898.2	529.4, 531.0	19
NiCe8YSZ	851.9, 853.2, 855.1	885.2, 898.2	529.7, 531.7	34
NiCe14CSZ	852.1, 853.4, 855.5	884.7, 898.3	529.4, 531.1	18

The cerium-reduction degree was calculated fitting the experimental Ce 3d profiles (Figure S4) with two reference spectra corresponding to samples with either 100% Ce^{3+} or 100% Ce^{4+} according to the procedure described in Reference [39]. As observed in Table 4, the addition of Ca significantly increased the Ce^{3+} percentage with respect to the CeZr substrate, but a maximum of 18%–19% was obtained independently of the Ca loading (4% or 14%). The highest amount of reduced Ce (34%) was obtained for the NiCe8YSZ catalyst, suggesting that the surface concentration of oxygen vacancies over ceria particles was also higher in this catalyst compared to those using Ca-stabilized ZrO_2 supports.

From O 1s spectra (Figure S5), two types of oxygen species were found, with peaks at around 529 and 531 eV for all the catalysts. They can be ascribed to lattice oxygen and adsorbed OH^- groups as suggested in the literature [40–42].

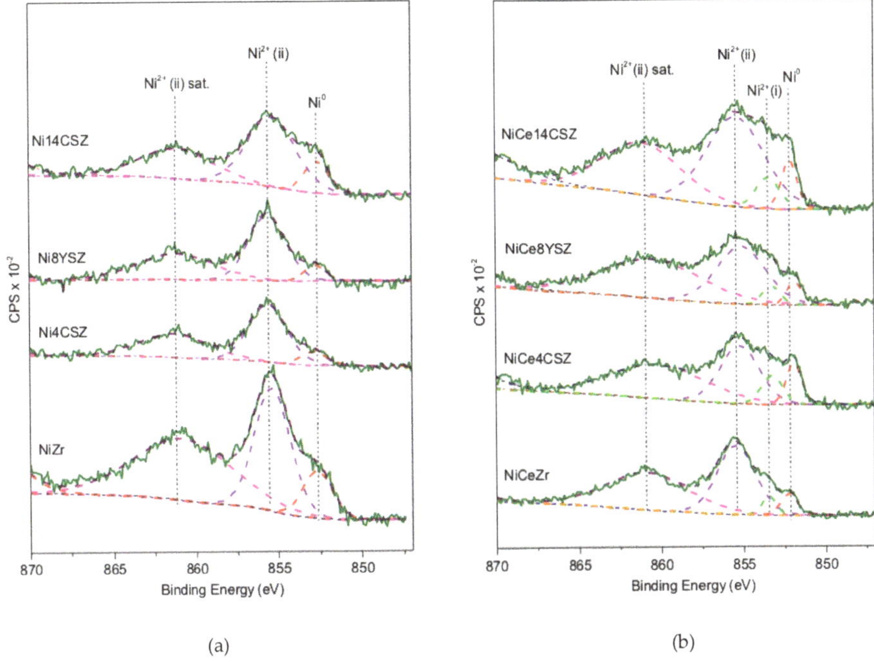

Figure 5. Nickel 2p XPS spectra: (**a**) samples without Ce; (**b**) samples with Ce.

3.6. Electron Microscopy Study of Ni Catalysts Using HAADF-STEM and EDS-STEM

Representative HAADF images recorded in STEM mode are shown in Figure 6. Moreover, and with the aim of exploring the spatial distribution of elements in the Ni catalysts, EDS analyses, also in STEM mode, were performed (for simplicity, only the images corresponding to NiZr, Ni4CSZ, Ni8YSZ, Ni14CSZ, and NiCe4CSZ are shown). As can be seen, the images and chemical maps indicate a rather homogeneous distribution of the different elements in all the investigated catalysts (maps for individual components are shown in Figure S6).

Particle size distributions obtained from the statistical analysis of the particles observed in the HAADF images are also included in Figure 6. Data corresponding to mean particle size ($\bar{d}$), surface area-weighted mean particle size ($\bar{d}_{sa}$), and Ni dispersion (D%) are gathered in Table 5. In the case of NiZr catalyst (Figure 6a), the EDS-STEM maps show that Ni particles were homogeneously distributed with a mean particle size of approximately 14 nm. A few particles with sizes in the range 40–84 nm were also observed. When the particle size distribution contains both very small and very large particles, the value of the surface area-weighted mean diameter is more representative in terms of metallic dispersion than that estimated from the mean particle size. As can be observed, the value of $\bar{d}_{sa}$ (27 nm) was similar to the averaged particle size estimated from H_2 chemisorption (27 nm) or XRD (24 nm).

Values between 20 and 23 nm were obtained for the Ni4CSZ, Ni8YSZ, and Ni14CSZ catalysts, which were also close to the crystallite sizes obtained by XRD. According to the values obtained by HAADF, the incorporation of Ca and Ce did not seem to have a significant influence on Ni particle size as suggested by the XRD results.

The accumulated dispersion values resulting from the particle size distribution curves were also very similar (5%–6%), being only slightly higher in the case of samples containing Ca.

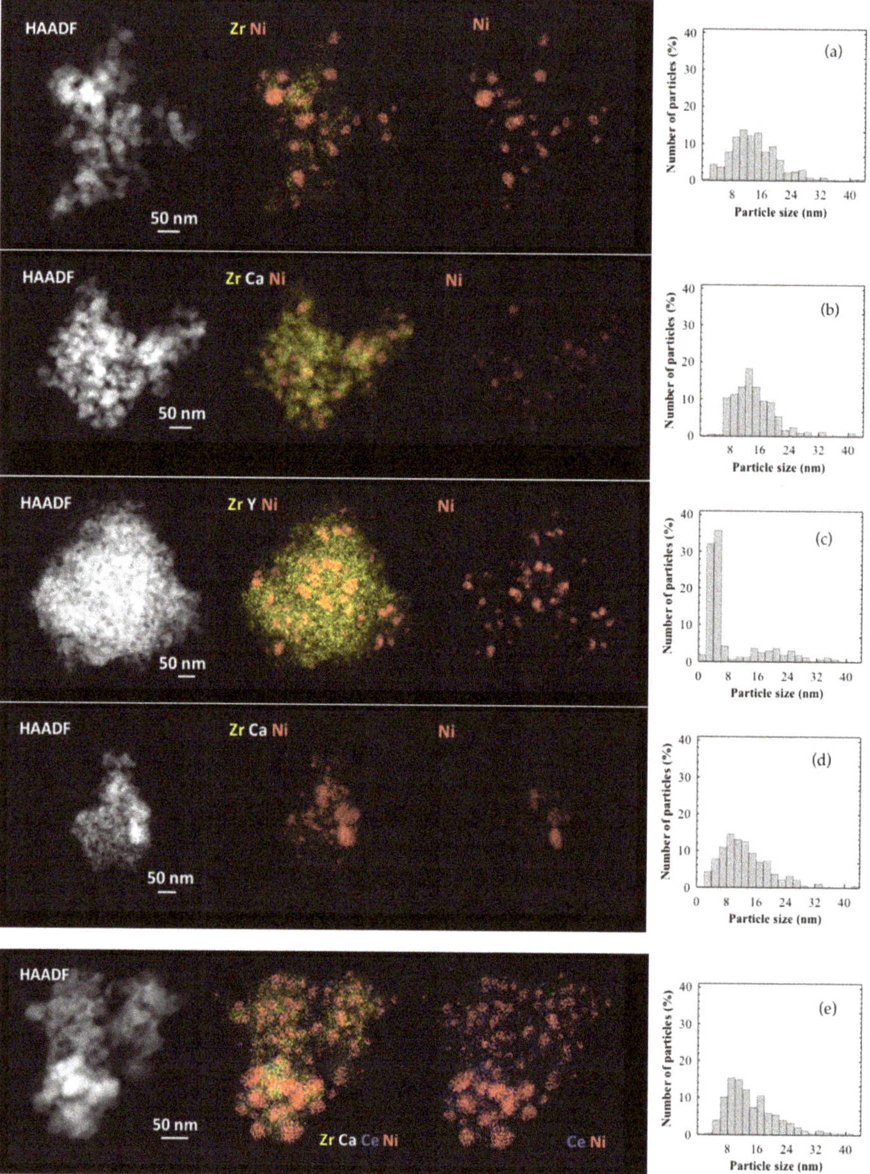

Figure 6. HAADF images, EDS mappings, and particle size distributions of: (**a**) NiZr; (**b**) Ni4CSZ; (**c**) Ni8YSZ; (**d**) Ni14CSZ; and (**e**) NiCe4CSZ.

It must be pointed out that, due to the atomic number of Ni, very small nickel nanoparticles cannot be easily identified in HAADF-STEM images. In order to evaluate this issue, EDS-STEM analyses were carried out at high magnifications, selecting areas of the material in which apparently no metallic particles but only a background signal was observed. Figure 7 illustrates this type of analysis. As can be seen in the EDS spectrum of the selected area, the presence of Ni was clearly detected which confirms the existence of highly dispersed forms of Ni, whose contribution in the particle size distribution analysis is not considered.

Table 5. Particle size results from electron microscopy analysis. For a better comparison, results from H_2 chemisorption and XRD line broadening are also included.

Sample	Particle Size from H_2 Chemisorption (nm)	Crystallite Size from XRD (nm)	$\bar{d}$(nm)	$\bar{d}_{sa}$(nm)	D (%) [1]
NiZr	27	24	14	27	5.2
Ni4CSZ	16	17	14	20	6.4
Ni8YSZ	32	19	9	23	5.4
Ni14CSZ	38	21	13	20	6.4
NiCe4CSZ	26	16	14	21	5.9

[1] From HAADF-STEM analysis.

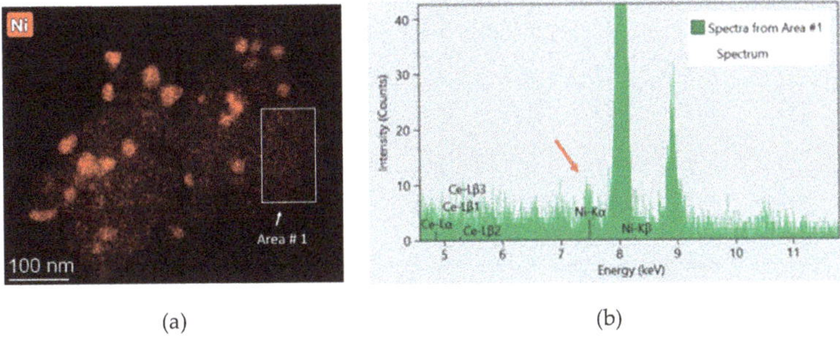

(a)   (b)

Figure 7. EDS analysis corresponding to Ni4CSZ: (**a**) Ni mapping; (**b**) spectrum from the area indicated in the image.

This technique also allowed us to obtain information about the relative spatial distribution of the different components of the catalyst at the nanoscopic level which is an important issue to understand the interactions among them, mainly in the case of the CeO_2-containing catalysts. As illustration, Figure 8 shows a detailed, high-magnification image of the NiCe4CSZ catalyst. As it can be clearly seen, CeO_2 is in direct contact with Ni, forming a layer at the interface between the metallic particle and the support. The interaction resulting from this direct contact may be responsible for the improvement in the reducibility of Ni particles observed in the TPR profiles corresponding to the catalysts containing CeO_2.

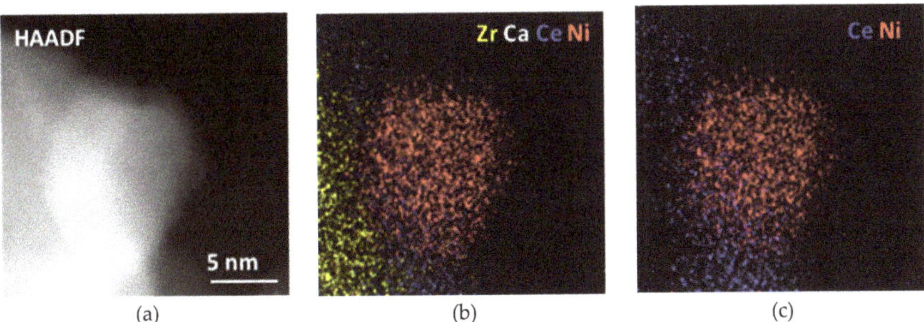

(a)   (b)   (c)

Figure 8. (**a**) HAADF image and (**b,c**) EDS mappings corresponding to the NiCe4CSZ catalyst.

3.7. Catalytic Activity Tests

Table 6 summarizes the performance of Ni catalysts in the APR of methanol at 230 °C and 32 bar after 5 h on stream. A blank experiment was carried out with the 4CSZ support, showing no activity under the same experimental conditions. Unconverted methanol in the liquid phase and CO_2, CO, H_2, and CH_4 in the gas phase were the only products found at the reactor outlet.

Table 6. Results from catalytic experiments.

Sample	Conversion (%)	Selectivity (%)				Products Ratio		H_2 Yield (%)
		H_2	CO_2	CO	CH_4	H_2/CO_2	CO_2/CO	
NiZr	48	72.9	20.2	4.6	2.2	3.6	4.4	40
Ni4CSZ	75	73.9	23.9	0.2	2.1	3.1	119.5	64
Ni8YSZ	46	75.5	21.5	1.5	1.5	3.5	14.3	36
Ni14CSZ	63	73.8	23.7	0.5	2.1	3.1	47.4	46
NiCeZr	40	76.8	20.5	1.8	0.9	3.7	11.4	34
NiCe4CSZ	68	73.8	23.4	0.6	2.2	3.2	39.0	57
NiCe8YSZ	54	74.1	21.3	2.2	2.5	3.5	9.7	40
NiCe14CSZ	44	75.7	21.6	1.9	0.9	3.5	11.4	33

Conversions higher than 40% were found with all the investigated Ni catalysts. Taking the NiZr sample as a reference, we observed that the incorporation of Ca had a positive effect on methanol conversion, which reached the highest value (75%) for the Ni4CSZ catalyst. In contrast, doping of ZrO_2 with Y had no significant impact on activity. As already mentioned, the doping of the zirconium oxide support with 4% Ca or 8% Y induced a stabilization of the tetragonal structure, generating in both cases a similar amount of oxygen vacancies. The different responses obtained with the Ni4CSZ and Ni8YSZ catalysts with respect to the NiZr sample suggests that there is not a straightforward relationship among these two parameters (oxygen vacancies concentration and APR activity). On the other hand, the conversion value obtained for the Ni14CSZ catalyst (63%) indicates that increasing the Ca content above 4% may have a negative effect on activity.

To explain the excellent behaviour exhibited by the Ni4CSZ sample, we should first consider its higher metallic surface area. For this purpose, we decided to compare the conversion and surface area of the different Ni catalysts. To facilitate the reading of the figure, we rationalized all the values with respect to the Ni4CSZ catalyst. As shown in Figure 9a, the higher conversions were obtained for Ni4CSZ and NiCe4CSZ which were also the catalysts with the larger Ni-surface areas. The metal surface is therefore a key factor determining the catalytic behaviour of these catalysts. However, the analysis of the whole set of values suggests that, in addition to the amount of Ni atoms exposed on the surface, some additional factor must be influencing the catalytic performance in this reaction.

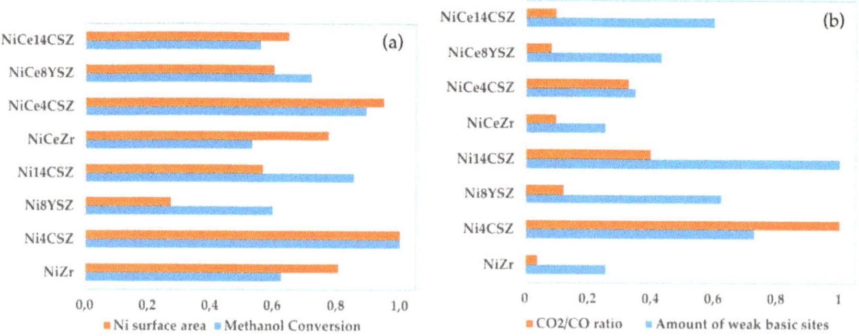

Figure 9. (**a**) Ni-surface area and methanol conversion (normalized values); (**b**): CO_2/CO ratio and amount of weak basic sites (normalized values).

One of the most significant differences resulting from the characterization of the catalysts was related to the surface basicity measured in the CO_2-TPD experiments. The influence of the basicity of the support on the APR reaction has been investigated in the literature and it has been proposed that basic sites promote a water–gas shift and further enhance the APR process [17]. As commented in a previous section, the incorporation of Ca or Y into the ZrO_2 lattice increased the total basicity of the samples (0.57 µmol $CO_2 \cdot g^{-1}$ for NiZr versus 1.07, 1.08, and 1.52 µmol$CO_2 \cdot g^{-1}$ for Ni4CSZ, Ni8YSZ, and Ni14CSZ, respectively). However, this increase in basicity did not have an identical incidence in all types of centres. Thus, in the case of the Ni4CSZ sample, the additional basicity was associated to the development of new weak basic sites, whereas, for Ni8YSZ, it resulted in an increase in the number of medium/strong centres. On the other hand, the situation for the Ni14CSZ sample can be described in terms of a simultaneous increase in both types of sites. The order of activity found for these samples suggests that weak centres have a positive influence on activity, while the influence of medium/strong sites seems to be negative for the APR reaction. The low metallic surface area of the Ni14CSZ may also be influencing its lower activity in comparison with Ni4CSZ.

Despite the differences found in methanol conversion, all catalysts showed very similar values of H_2 selectivity, ranging from 73% to 76%. The H_2/CO_2 ratio was close to the stoichiometric value ($H_2/CO_2 = 3$) in the case of Ca-containing catalysts and slightly higher in the case of NiZr (3.6) and Ni8YSZ (3.5), confirming that the WGS reaction was less favoured in the latter catalysts. Similar conclusions can be derived from data corresponding to CO selectivity (0.2 and 0.5 for Ni4CSZ and Ni14CSZ, respectively) or the CO_2/CO ratio.

To explain the influence of Ca in selectivity, we must recall that basic sites act promoting the transformation of CO to CO_2 in the WGS reaction [17]. However, sites with a high basic strength may have a negative influence avoiding the desorption of CO_2. The higher WGS activity showed by the Ni4CSZ catalyst could be related with the nature of the basicity exhibited by this sample, characterised by a high concentration of weak basic sites and a low concentration of medium/strong centres. The correlation between weak the basic sites and CO_2/CO ratio is illustrated in Figure 9b.

As for the effect of Ce on the APR performance of Ni, a decrease in conversion was generally observed in the catalysts containing Ce with respect to the homologous series without Ce, with the unique exception of Ni8YSZ, for which methanol conversion increased from 46% to 54% after the incorporation of Ce. Moreover, the activity order observed for this series was as follows: NiCe4CSZ (68) > NiCe8YSZ (54) > NiCe14CSZ (44) > NiCeZr (40). These results reveal that the improvement in reducibility derived from the incorporation of Ce (and evidenced in the TPR experiments) did not have a direct impact on catalytic behaviour thus contrasting with the results reported in the literature for other Ce-based catalytic compositions. For example, it has been reported that cerium oxide promotes catalyst activity in WGS reaction due to the fact of its well-known oxygen storage capacity and oxygen mobility throughout the lattice [11,12]. In order to explain this apparent disagreement, we must recall that Ce is not only changing the redox properties of the catalysts but also their surface basicity. According to CO_2 desorption measurements, new basic sites with intermediate strength appeared on catalyst surfaces after the addition of Ce. Simultaneously, the number of weaker basic sites decreased significantly, probably because these sites, initially located on the surface of the Ca(Y)–ZrO_2 supports, resulted in being partially covered by the CeO_2 layer. Changes in basic strength may be responsible for the decrease in methanol conversion after incorporation of Ce. These results also imply that the negative effects on APR of methanol resulting from changes in basicity prevail over the positive effects derived from changes in redox properties, both effects caused by the incorporation of Ce.

Another factor to be considered is the formation of cationic Ni species, presumably derived from the observed Ni–Ce interaction. These species, evidenced by XPS (peaks at around 853 eV) only in Ce-containing catalysts, would be inactive for the reaction. However, we must also bear in mind that reduction treatments prior to XPS measurements were not performed in situ and, therefore, this cationic species may result from air exposure of the pre-reduced catalysts.

As for selectivity, the most remarkable finding related with CeO_2 is a decrease in CO production for the NiCeZr sample compared to that of NiZr which can be explained considering that both catalysts showed a very similar basicity and, therefore, both conversion and selectivity values would depend mainly on other factors affected by Ce (e.g., reducibility).

4. Conclusions

In this work, a series of Ni/ZrO_2 catalysts were prepared, characterized, and tested in the aqueous-phase reforming of methanol reaction. The effects of doping zirconia with calcium or yttrium and the promotion with CeO_2 were investigated. The best catalytic performance in terms of activity and selectivity was found for the catalyst doped with 4% molar of Ca. This catalyst showed a higher metallic area and developed surface basicity characterized by the presence of a high concentration of weak basic centres and a low concentration of medium/strong basic sites. These two factors seem to be responsible for its excellent catalytic performance. The incorporation of ceria has two significant effects on the properties of Ni catalysts: (i) improves reducibility and (ii) increases the strength of the basic sites. The lower activity of catalysts promoted with CeO_2 suggests that the improvement in redox properties does not compensate for the negative effects on APR resulting from changes in surface basicity.

Supplementary Materials: The following are available online at http://www.mdpi.com/2079-4991/9/11/1582/s1, Figure S1: Experimental setup for APR of methanol, Figure S2: N_2 adsorption-desorption isotherms, Figure S3: CO_2-TPD profiles showing deconvoluted peaks, Figure S4: Ce 3d XPS spectra, Figure S5: O 1s XPS spectra, Figure S6: HAADF-STEM images and EDS-STEM maps for individual components.

Author Contributions: The manuscript was written through contributions by all authors. The individual contributions were as follows: conceptualization, D.G., M.Á.C., J.J.C. and L.L.; methodology, D.G., J.F. and M.Á.C.; investigation, D.G., J.F. and J.J.D.; writing—original draft preparation, D.G. and M.Á.C.; writing—review and editing, D.G., J.J.C., L.L., J.F., J.J.D. and M.Á.C.; supervision, M.Á.C., J.F., L.L. and J.J.C.; project administration, M.Á.C. and J.J.C.; funding acquisition, J.J.C, J.J.D. and M.Á.C.

Funding: This research was funded by MINECO/FEDER (Projects References: MAT2017-87579-R and ENE2017-82451-C3-2-R) and Junta de Andalucía (FQM334 and FQM110).

Acknowledgments: D.G. acknowledges the Spanish Government (MINECO) for his pre-doctoral grant. The authors are grateful for the help received from the "Plan Propio de I+D" of the University of Cadiz.

Conflicts of Interest: The authors declare no conflict of interest.

References

1. Muradov, N.Z.; Veziroglu, T.N. "Green" path from fossil-based to hydrogen economy: An overview of carbon-neutral technologies. *Int. J. Hydrogen Energy* **2008**, *33*, 6804–6839. [CrossRef]

2. Krummrich, S.; Llabres, J. Methanol reformer—The next milestone for fuel cell powered submarines. *Int. J. Hydrogen Energy* **2015**, *40*, 5482–5486. [CrossRef]

3. Cortright, R.D.; Davda, R.R.; Dumesic, J.A. Hydrogen from catalytic reforming of biomass-derived hydrocarbons in liquid water. *Nature* **2002**, *418*, 964–967. [CrossRef] [PubMed]

4. Coronado, I.; Stekrova, M.; Reinikainen, M.; Simell, P.; Lefferts, L.; Lehtonen, J. A review of catalytic aqueous-phase reforming of oxygenated hydrocarbons derived from biorefinery water fractions. *Int. J. Hydrogen Energy* **2016**, *41*, 11003–11032. [CrossRef]

5. Davda, R.R.; Shabaker, J.W.; Huber, G.W.; Cortright, R.D.; Dumesic, J.A. A review of catalytic issues and process conditions for renewable hydrogen and alkanes by aqueous-phase reforming of oxygenated hydrocarbons over supported metal catalysts. *Appl. Catal. B: Environ.* **2005**, *56*, 171–186. [CrossRef]

6. Shabaker, J.W.; Huber, G.W.; Dumesic, J.A. Aqueous-phase reforming of oxygenated hydrocarbons over Sn-modified Ni catalysts. *J. Catal.* **2004**, *222*, 180–191. [CrossRef]

7. Huber, G.W.; Shabaker, J.W.; Dumesic, J.A. Raney Ni-Sn catalyst for H-2 production from biomass-derived hydrocarbons. *Science* **2003**, *300*, 2075–2077. [CrossRef]

8. van Haasterecht, T.; Ludding, C.C.I.; de Jong, K.P.; Bitter, J.H. Stability and activity of carbon nanofiber-supported catalysts in the aqueous phase reforming of ethylene glycol. *J. Energy Chem.* **2013**, *22*, 257–269. [CrossRef]

9. Manfro, R.L.; da Costa, A.F.; Ribeiro, N.F.P.; Souza, M.M.V.M. Hydrogen production by aqueous-phase reforming of glycerol over nickel catalysts supported on CeO_2. *Fuel Process. Technol.* **2011**, *92*, 330–335. [CrossRef]

10. Roy, B.; Leclerc, C.A. Study of preparation method and oxidization/reduction effect on the performance of nickel-cerium oxide catalysts for aqueous-phase reforming of ethanol. *J. Power Sources* **2015**, *299*, 114–124. [CrossRef]

11. Rahman, M.M.; Church, T.L.; Minett, A.I.; Harris, A.T. Effect of CeO_2 addition to Al_2O_3 supports for Pt catalysts on the aqueous-phase reforming of glycerol. *ChemSusChem* **2013**, *6*, 1006–1013. [CrossRef] [PubMed]

12. Hilaire, S.; Wang, X.; Luo, T.; Gorte, R.J.; Wagner, J. A comparative study of water-gas-shift reaction over ceria-supported metallic catalysts. *Appl. Catal. A: Gen.* **2004**, *258*, 271–276. [CrossRef]

13. Chen, A.P.; Guo, H.J.; Song, Y.M.; Chen, P.; Lou, H. Recyclable CeO_2-ZrO_2 and CeO_2-TiO_2 mixed oxides based Pt catalyst for aqueous-phase reforming of the low-boiling fraction of bio-oil. *Int. J. Hydrogen Energy* **2017**, *42*, 9577–9588. [CrossRef]

14. Jeon, S.; Park, Y.M.; Saravanan, K.; Han, G.Y.; Kim, B.W.; Lee, J.B.; Bae, J.W. Aqueous phase reforming of ethylene glycol over bimetallic platinum-cobalt on ceria-zirconia mixed oxide. *Int. J. Hydrogen Energy* **2017**, *42*, 9892–9902. [CrossRef]

15. Larimi, A.S.; Kazemeini, M.; Khorasheh, F. Aqueous phase reforming of glycerol using highly active and stable $Pt_{0.05}CeXZr_{0.95}$-XO_2 ternary solid solution catalysts. *Appl. Catal. A: Gen.* **2016**, *523*, 230–240. [CrossRef]

16. Stekrova, M.; Rinta-Paavola, A.; Karinen, R. Hydrogen production via aqueous-phase reforming of methanol over nickel modified Ce, Zr and La oxide supports. *Catal. Today* **2018**, *304*, 143–152. [CrossRef]

17. Guo, Y.; Azmat, M.U.; Liu, X.H.; Wang, Y.Q.; Lu, G.Z. Effect of support's basic properties on hydrogen production in aqueous-phase reforming of glycerol and correlation between WGS and APR. *Appl. Energy* **2012**, *92*, 218–223. [CrossRef]

18. Menezes, A.O.; Rodrigues, M.T.; Zimmaro, A.; Borges, L.E.P.; Fraga, M.A. Production of renewable hydrogen from aqueous-phase reforming of glycerol over Pt catalysts supported on different oxides. *Renew. Energy* **2011**, *36*, 595–599. [CrossRef]

19. Muñoz, M.A.; Calvino, J.J.; Rodríguez-Izquierdo, J.M.; Blanco, G.; Arias, D.C.; Pérez-Omil, J.A.; Hernández-Garrido, J.C.; González-Leal, J.M.; Cauqui, M.A.; Yeste, M.P. Highly stable ceria-zirconia-yttria supported Ni catalysts for syngas production by CO_2 reforming of methane. *Appl. Surf. Sci.* **2017**, *426*, 864–873. [CrossRef]

20. Mercera, P.D.L.; van Ommen, J.G.; Doesburg, E.B.M.; Burggraaf, A.J.; Roes, J.R.H. Stabilized tetragonal zirconium oxide as a support for catalysts Evolution of the texture and structure on calcination in static air. *Appl. Catal.* **1991**, *78*, 79–96. [CrossRef]

21. Jiménez-González, C.; Boukha, Z.; de Rivas, B.; Delgado, J.J.; Cauqui, M.Á.; González-Velasco, J.R.; Gutiérrez-Ortiz, J.I.; López-Fonseca, R. Structural characterisation of Ni/alumina reforming catalysts activated at high temperatures. *Appl. Catal. A: Gen.* **2013**, *466*, 9–20. [CrossRef]

22. Drozdz, E.; Lacz, A.; Spalek, Z. Deposition of NiO on 3 mol% yttria-stabilized zirconia and $Sr_{0.96}Y_{0.04}TiO_3$ materials by impregnation method. *J. Therm. Anal. Calorim.* **2017**, *130*, 291–299. [CrossRef]

23. Wei, Y.G.; Wang, H.; Li, K.Z.; Zhu, X.; Du, Y.P. Preparation and characterization of $Ce1-xNixO_2$ as oxygen carrier for selective oxidation methane to syngas in absence of gaseous oxygen. *J. Rare Earths* **2010**, *28*, 357–361. [CrossRef]

24. Shan, W.J.; Luo, M.F.; Ying, P.L.; Shen, W.J.; Li, C. Reduction property and catalytic activity of $Ce_{1-x}NiXO_2$ mixed oxide catalysts for CH_4 oxidation. *Appl. Catal. A: Gen.* **2003**, *246*, 1–9. [CrossRef]

25. Tang, C.J.; Li, J.C.; Yao, X.J.; Sun, J.F.; Cao, Y.; Zhang, L.; Gao, F.; Deng, Y.; Dong, L. Mesoporous NiO-CeO_2 catalysts for CO oxidation: Nickel content effect and mechanism aspect. *Appl. Catal. A: Gen.* **2015**, *494*, 77–86. [CrossRef]

26. Mori, H.; Wen, C.J.; Otomo, J.; Eguchi, K.; Takahashi, H. Investigation of the interaction between NiO and yttria-stabilized zirconia (YSZ) in the NiO/YSZ composite by temperature-programmed reduction technique. *Appl. Catal. A: Gen.* **2003**, *245*, 79–85. [CrossRef]

27. Wang, Y.; Zhu, A.M.; Zhang, Y.Z.; Au, C.T.; Yang, X.F.; Shi, C. Catalytic reduction of NO by CO over NiO/CeO$_2$ catalyst in stoichiometric NO/CO and NO/CO/O-2 reaction. *Appl. Catal. B: Environ.* **2008**, *81*, 141–149. [CrossRef]

28. Luisetto, I.; Tuti, S.; Di Bartolomeo, E. Co and Ni supported on CeO$_2$ as selective bimetallic catalyst for dry reforming of methane. *Int. J. Hydrogen Energy* **2012**, *37*, 15992–15999. [CrossRef]

29. Chagas, C.A.; de Souza, E.F.; Manfro, R.L.; Landi, S.M.; Souza, M.M.V.M.; Schmal, M. Copper as promoter of the NiO-CeO$_2$ catalyst in the preferential CO oxidation. *Appl. Catal. B: Environ.* **2016**, *182*, 257–265. [CrossRef]

30. Liu, S.; Ma, J.; Guan, L.; Li, J.; Wei, W.; Sun, Y. Mesoporous CaO–ZrO$_2$ nano-oxides: A novel solid base with high activity and stability. *Microporous Mesoporous Mater.* **2009**, *117*, 466–471. [CrossRef]

31. Radlik, M.; Adamowska-Teyssier, M.; Krztoń, A.; Kozieł, K.; Krajewski, W.; Turek, W.; Da Costa, P. Dry reforming of methane over Ni/Ce$_{0.62}$Zr$_{0.38}$O$_2$ catalysts: Effect of Ni loading on the catalytic activity and on H2/CO production. *Comptes Rendus Chim.* **2015**, *18*, 1242–1249. [CrossRef]

32. Grosvenor, A.P.; Biesinger, M.C.; Smart, R.S.; McIntyre, N.S. New interpretations of XPS spectra of nickel metal and oxides. *Surf. Sci.* **2006**, *600*, 1771–1779. [CrossRef]

33. Biesinger, M.C.; Payne, B.P.; Lau, L.W.M.; Gerson, A.; Smart, R.S.C. X-ray photoelectron spectroscopic chemical state Quantification of mixed nickel metal, oxide and hydroxide systems. *Surf. Interface Anal.* **2009**, *41*, 324–332. [CrossRef]

34. Carley, A.F.; Jackson, S.D.; O'Shea, J.N.; Roberts, M.W. The formation and characterisation of Ni^{3+}—An X-ray photoelectron spectroscopic investigation of potassium-doped Ni(110)–O. *Surf. Sci.* **1999**, *440*, L868–L874. [CrossRef]

35. Löfberg, A.; Guerrero-Caballero, J.; Kane, T.; Rubbens, A.; Jalowiecki-Duhamel, L. Ni/CeO$_2$ based catalysts as oxygen vectors for the chemical looping dry reforming of methane for syngas production. *Appl. Catal. B: Environ.* **2017**, *212*, 159–174. [CrossRef]

36. Kugai, J.; Subramani, V.; Song, C.; Engelhard, M.H.; Chin, Y.-H. Effects of nanocrystalline CeO$_2$ supports on the properties and performance of Ni–Rh bimetallic catalyst for oxidative steam reforming of ethanol. *J. Catal.* **2006**, *238*, 430–440. [CrossRef]

37. Zhao, J.; Xu, X.Y.; Li, M.R.; Zhou, W.; Liu, S.M.; Zhu, Z.H. Coking-resistant Ce$_{0.8}$Ni$_{0.2}$O$_2$-delta internal reforming layer for direct methane solid oxide fuel cells. *Electrochim. Acta* **2018**, *282*, 402–408. [CrossRef]

38. Sun, K.; Lu, W.; Wang, M.; Xu, X. Characterization and catalytic performances of La doped Pd/CeO$_2$ catalysts for methanol decomposition. *Appl. Catal. A: Gen.* **2004**, *268*, 107–113. [CrossRef]

39. Fernandez-Garcia, S.; Jiang, L.; Tinoco, M.; Hungria, A.B.; Han, J.; Blanco, G.; Calvino, J.J.; Chen, X. Enhanced Hydroxyl Radical Scavenging Activity by Doping Lanthanum in Ceria Nanocubes. *J. Phys. Chem. C* **2016**, *120*, 1891–1901. [CrossRef]

40. Skårman, B.; Grandjean, D.; Benfield, R.E.; Hinz, A.; Andersson, A.; Reine Wallenberg, L. Carbon monoxide oxidation on nanostructured CuO$_x$/CeO$_2$ composite particles characterized by HREM, XPS, XAS, and high-energy diffraction. *J. Catal.* **2002**, *211*, 119–133. [CrossRef]

41. Requies, J.; Cabrero, M.A.; Barrio, V.L.; Güemez, M.B.; Cambra, J.F.; Arias, P.L.; Pérez-Alonso, F.J.; Ojeda, M.; Peña, M.A.; Fierro, J.L.G. Partial oxidation of methane to syngas over Ni/MgO and Ni/La$_2$O$_3$ catalysts. *Appl. Catal. A: Gen.* **2005**, *289*, 214–223. [CrossRef]

42. Francisco, M.S.P.; Mastelaro, V.R.; Nascente, P.A.P.; Florentino, A.O. Activity and characterization by XPS, HR-TEM, Raman spectroscopy, and bet surface area of CuO/CeO$_2$-TiO$_2$ catalysts. *J. Phys. Chem. B* **2001**, *105*, 10515–10522. [CrossRef]

Article

Coupling Plasmonic and Cocatalyst Nanoparticles on N–TiO$_2$ for Visible-Light-Driven Catalytic Organic Synthesis

Yannan Wang, Yu Chen, Qidong Hou, Meiting Ju and Weizun Li *

College of Environmental Science and Engineering, Nankai University, Tianjin 300350, China; wangyannan@nankai.edu.cn (Y.W.); chenyu0870@gmail.com (Y.C.); houqidong@nankai.edu.cn (Q.H.); jumeit@nankai.edu.cn (M.J.)
* Correspondence: liweizun@nankai.edu.cn; Tel.: +86-135-1221-2566

Received: 29 January 2019; Accepted: 1 March 2019; Published: 7 March 2019

Abstract: The use of the surface plasmon resonance (SPR) effect of plasmonic metal nanocomposites to promote photocarrier generation is a strongly emerging field for improving the catalytic performance under visible-light irradiation. In this study, a novel plasmonic photocatalyst, AuPt/N–TiO$_2$, was prepared via a photo-deposition–calcination technique. The Au nanoparticles (NPs) were used herein to harvest visible-light energy via the SPR effect, and Pt NPs were employed as a cocatalyst for trapping the energetic electrons from the semiconductor, leading to a high solar-energy conversion efficiency. The Au$_2$Pt$_2$/N–TiO$_2$ catalyst, herein with the irradiation wavelength in the range 460–800 nm, exhibited a reaction rate ~24 times greater than that of TiO$_2$, and the apparent quantum yield at 500 nm reached 5.86%, indicative of the successful functionalization of N–TiO$_2$ by the integration of Au plasmonic NPs and the Pt cocatalyst. Also, we investigated the effects of two parameters, light source intensity and wavelength, in photocatalytic reactions. It is indicated that the as-prepared AuPt/N–TiO$_2$ photocatalyst can cause selective oxidation of benzyl alcohol under visible-light irradiation with a markedly enhanced selectivity and yield.

Keywords: plasmonic photocatalyst; metal nanoparticle; N–TiO$_2$; nanocomposites; photocatalytic selective oxidation

---

1. Introduction

Titanium dioxide (TiO$_2$) was extensively studied in the past two decades as a photocatalyst because it can eliminate environmental pollutants, purify air, and produce clean hydrogen energy through the efficient utilization of solar energy [1]. Nevertheless, owing to the rapid recombination rate of the photogenerated electron–hole pairs and limited visible-light response, the application of pure TiO$_2$ is restricted. Appropriate modification, such as doping non-metals, is essential for TiO$_2$ to allow the further utilization of solar energy [2]. Nonetheless, the reported reactivity and quantum efficiency of TiO$_2$ derivative materials remains extremely low to meet the requirements of practical applications.

In order to sufficiently improve photocatalytic efficiency, both the visible-light absorption region and electron–hole separation of the photocatalyst should be optimized. Recently, semiconductor nanomaterials decorated with noble-metal nanoparticles (NPs) were recognized as a promising method for boosting the performance of photocatalysts [3–8]. Coupling semiconductors with noble metals (such as platinum and palladium) as the cocatalyst can form a Schottky barrier, serving as the "electron trapper" to improve charge migration and separation [9,10]. Plasmonic metals (gold and silver) nanoparticles with attractive SPR properties under visible-light excitation can be used as antennas for converting light energy into a local electric field [7,11,12], and improve the photocarrier generation/separation via plasmon-induced resonance energy transfer (PIRET) and the hot-electron injection mechanism [13,14]. Consequently, the combination of multi-functional metal NPs in a noble metal/semiconductor nanostructure might effectively enhance the generation of photo-carriers and strengthen charge migration and separation.

For the hot-electron injection effect, the so-called SPR-sensitization effect, the plasmonic metal nanoparticles act as a dye molecule in dye-sensitized solar cells; as excited by the incident high-energy photon, the SPR effect of the plasmonic metal causes confined free electrons oscillating with incident light to generate excitation of hot electrons via non-radiative decay, so-called "plasmonic hot-electron generation". Furthermore, these hot electrons with energy high enough to overcome the Schottky barrier can inject into the adjacent semiconductor's conduction band [15–17]. To facilitate the PIRET process, the existence of intra-bandgap level-related defects can play a crucial role in the promotion of PIRET [18,19]. For instance, a TiO_2 photoanode based on N-doped exhibits enhanced photocurrent behavior induced by a PIRET water-splitting reaction [20]. N-doped TiO_2 introduces a new intra-bandgap level, which causes the absorption range of the semiconductor photocatalyst to overlap with the extinction wavelength of the plasmonic material, thereby obtaining a sufficient resonance interaction. In this case, the energy of the plasmonic oscillation is transferred from the plasmonic material to the semiconductor photocatalyst by an electromagnetic field or a dipole–dipole interaction.

In order to fully understand the excellent photocatalytic activity of the bifunctional noble-metal-modified N-doped TiO_2 under visible-light excitation, a detailed comparative study of light intensity and light wavelength is required. Herein, we integrated the plasmonic effect and a Schottky junction into one nanostructure by forming bifunctional plasmonic photocatalyst co-decorated with Au and Pt NPs and N–TiO_2. The activities of the AuPt/N–TiO_2 samples were evaluated by photocatalytic oxidation of benzyl alcohol. By strictly limiting the effects of other factors, we observed a direct correlation between photocatalytic activity and the irradiation parameter, which is essential for the design to improve the efficiency of the photocatalytic reaction. The results obtained in this paper are expected to contribute to the rational design and development of multifunctional metal nanoparticles for applications targeting solar energy conversion.

2. Materials and Methods

2.1. Synthesis

In a typical procedure, TiO_2-supported nanocrystals were prepared according to our previous paper [21]. The prepared TiO_2 product was mixed and ground with urea (1:4), and then the mixture was heated in air at 400 °C for two hours to obtain N–TiO_2 [22]. Finally, the noble metals were deposited on the N–TiO_2 via a typical photo-deposition calcination method, and the as-prepared catalysts were annealed in air for further use. All experimental methods are fully reported in the Supplementary Materials.

2.2. Sample Characterization

X-ray diffraction (XRD) patterns of the samples were recorded using a PANalytical X'pert MRD system (Almelo, Netherlands). Diffuse-reflectance ultraviolet–visible light (UV–Vis) spectra (DRS) of the samples were recorded on a Shimadzu 3600 UV–Vis spectrophotometer (Kyoto, Japan) in the air against $BaSO_4$. Transmission electron microscopy (TEM) images and scanning electron microscopy (SEM) images were taken by a FEI Tecnai F20 microscope (Hillsboro, OR, USA) and Hitachi S4800 microscope (Tokyo, Japan). X-ray photoelectron spectra (XPS) of the samples were recorded on a Thermo-Fisher Scientific ESCALAB 250XI system (Waltham, OR, USA). The steady-state photoluminescence (PL) spectrum was recorded by a Hitachi F-7000 fluorescence spectrophotometer (Tokyo, Japan). Photocurrent and electrochemical impedance spectroscopy measurements of the photocatalyst were performed on a CHI 760D workstation (Shanghai, China). All electrochemical measurements were made at room temperature.

3. Results and Discussion

SEM and TEM images (Figure 1 and Figure S1) were recorded to observe the morphology of as-prepared Au_2Pt_2/N–TiO_2. As shown in the TEM images, gold nanoparticles with an average size of ~25 nm can be observed instead of the presence of bimetallic Au–Pt alloy. It is suggested that the size of the metal nanoparticles is critical for modifying the chemical composition of the resulting nanomaterials, and the bimetallic alloy usually can be observed in the case of small gold (~7 nm) and platinum (~2 nm) nanoparticles [23]. Here, in this case, the gold nanoparticle size was larger than 20 nm, the formation of the bimetallic alloy NPs was not thermodynamically favored, and the segregation of gold and platinum nanoparticles was maintained. On the other hand, the platinum metal NPs was observed with a mean size of 2 nm (Figure 1h), which was uniformly decorated on the N–TiO_2 support. Furthermore, interplanar distances of 0.235 and 0.225 nm for gold and platinum NPs were observed in the high-resolution TEM images (Figure 1f,g), which were indexed to the lattice spacings of Au(111) and Pt(111) planes of face-centered cubic (fcc) structures, respectively [24]. Another type of lattice fringe (~0.352 nm) can be indexed to the (101) plane of anatase TiO_2. The results obtained from the energy-dispersive X-ray (EDX) spectrum showed that the noble metal's composition in Au_2Pt_2/N–TiO_2 (list in Table S1, Supplementary Materials) was consistent with the nominal load. Furthermore, the Brunauer–Emmett–Teller (BET) characterization results (Figure S2, Supplementary Materials) indicated that all catalysts exhibited similar surface areas, while pore volume and pore size decreased after the loading of noble-metal nanoparticles.

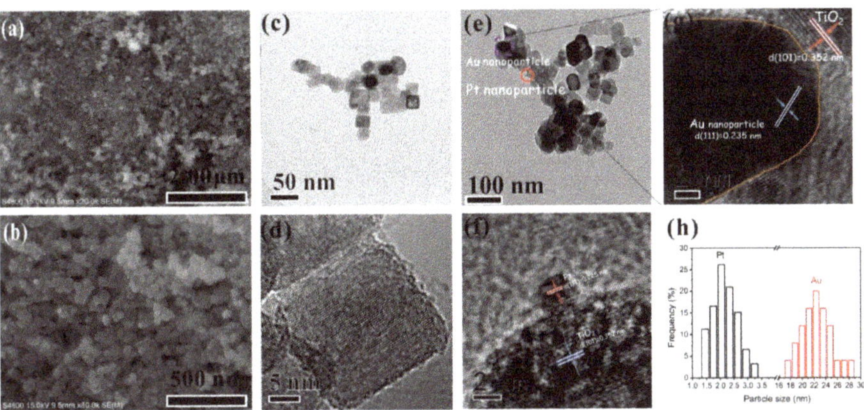

Figure 1. SEM image of Au_2Pt_2/N–TiO_2 (**a**,**b**) and TEM and high-resolution TEM (HRTEM) images of N–TiO_2 (**c**,**d**) and Au_2Pt_2/N–TiO_2 (**e**–**g**). Nanoparticle size distributions of Au and Pt in Au_2Pt_2/N–TiO_2 (**h**). Scale bar (**e**,**f**): 2 nm.

In the XRD characterization, all catalysts exhibited diffraction peaks dominated by TiO_2 (Figure 2a). Concerning the Au NPs, some additional weak peaks observed at 38° corresponded to Au; however, diffraction peaks for Pt were not found in the XRD patterns, possibly related to the line broadening caused by the quantum-size effects of small-sized Pt NPs [25]. Figure 2b shows the UV–Vis absorption spectra (DRS) of the as-prepared catalyst. The absorption band of the N–TiO_2 sample in the visible region of 400–500 nm corresponds to the presence of nitrogen. This effect is related to nitrogen doping, which possibly leads to the formation of hybridized states at the top of the valence band of the nitrogen $2p$ states and oxygen $2p$ states or an N-induced intermediate gap level [26]. XPS profiles were recorded to investigate the localization of nitrogen. N was mainly located at the interstitial atom on the Ti–O–N bonds (Figure S3, Supplementary Materials). By introducing impurity levels into the TiO_2 lattice, more overlapping portions of the absorption spectrum can be obtained between the TiO_2 and Au nanoparticles. In this way, the near-field electromagnetic resonance of the SPR effect can collect enough energy to stimulate the generation of electron–hole pairs through the PIRET process. Compared to N–TiO$_2$, Au$_2$Pt$_2$/N–TiO$_2$ exhibited a stronger absorption feature around 550 nm in Figure 2b, corresponding to the SPR peak of Au NPs [27]. As reported previously [6,28–31], plasmonic nanoparticles (gold, silver) loaded on a semiconductor, with a broad absorption cross-section, are capable of absorbing visible light and generating hot electrons through intraband transitions. These high-energy hot electrons then overcome the Schottky barrier and inject into the conduction band of the semiconductor. In this way, the SPR effect of the metal nanoparticles leads the photon energy transfer to the adjacent semiconductor or molecular complex, which in turn drives the chemical reaction.

Also, the photoluminescence emission (PL) of the samples was recorded to understand the behavior of the electrons and holes generated by light in catalysts. Here, the steady-state fluorescence emission spectrum (Figure 2c) showed a substantial attenuation of the PL signal owing to the deposition of noble metal (Pt), indicating that Pt NPs effectively form the Schottky barrier at the metal/N–TiO$_2$ heterojunction. This Schottky barrier, in turn, reduces electron–hole (e$^-$–h$^+$) pair recombination and increases the number of photoreactive photo-carriers available for photoreaction [32]. To further determine the role of the noble-metal nanoparticles in illumination, the photoelectrochemical properties of catalyst were characterized (Figure 2d), and it is demonstrated that the photocurrent intensity of Au$_2$Pt$_2$/N–TiO$_2$ is considerably higher than that of N–TiO$_2$. Such an apparent transient photocurrent enhancement is primarily associated with the available gold NPs, which absorb visible light and promote photocarrier generation through the SPR effect. Subsequently, we used the electrochemical impedance spectra (EIS) experiments to investigate the generation of the electron. The results of charge transport characteristics (Figure 2d inset) revealed that the radius of Au$_2$Pt$_2$/N–TiO$_2$ in the middle-frequency region is smaller than the radius of N–TiO$_2$, which demonstrates that the photoinduced electron–hole separation efficiency is higher, and the interface charge can be transferred to the electron donor more quickly.

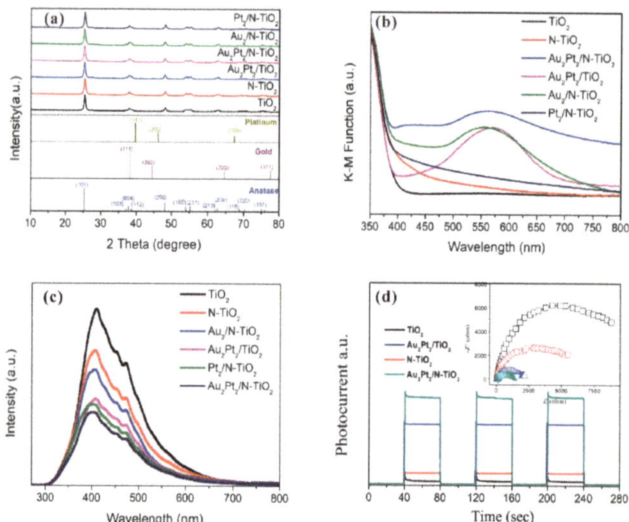

Figure 2. X-ray diffraction (XRD) patterns (**a**), ultraviolet–visible light (UV–Vis) diffuse reflectance spectra (DRS) (**b**), and photoluminescence spectra (**c**) of prepared photocatalysts. Photocurrent transient response (**d**) and electrochemical impedance spectroscopy (EIS) Nyquist plots (inset) of the sample electrodes of TiO$_2$, N–TiO$_2$, Au$_2$Pt$_2$/TiO$_2$, Au$_2$Pt$_2$/N–TiO$_2$ under visible-light irradiation.

For investigating the photocatalytic performance, we used the selective oxidation of benzyl alcohol as a probe reaction to study the photocatalytic activity of $Au_2Pt_2/N–TiO_2$ catalyst for visible-light-driven organic catalytic synthesis [33–35]. Figure 3a summarizes the reaction parameters such as conversion, yield, and selectivity data. After 2.5 h, benzaldehyde formed over bare TiO_2 (yield: 3.37%) under visible-light irradiation. Considering that bare TiO_2 does not absorb visible light, this visible-light catalytic reactivity can be ascribed to the ligand-to-metal charge transfer resulting from the surface complex formed by the adsorption of benzyl alcohol on the surface of TiO_2 [36–38]. Moreover, $N–TiO_2$ does not significantly increase the activity, and the selectivity in TiO_2 and $N–TiO_2$ cases was low (~70%). Hence, the loading of metal NPs can significantly improve reaction efficiency compared with TiO_2. Among all samples, the $Au_2Pt_2/N–TiO_2$ composite presented the highest photocatalytic performance, and its yield was 24 times that of TiO_2. On the contrary, the yield of the Au_2Pt_2/TiO_2 photocatalyst was only 70% of the $Au_2Pt_2/N–TiO_2$, which suggests that the overlapped intrinsic absorption of $N–TiO_2$ with plasmonic material may boost the PIRET process. Moreover, the yields over $Au_2/N–TiO_2$ and $Pt_2/N–TiO_2$ increased by 5.5 and 19 times, respectively, indicating that the Schottky barrier formed between Pt nanoparticles and TiO_2 is crucial for the improvement in the catalyst efficiency. It is interesting to note that, after the loading of noble-metal NPs, the selectivity increased from 73.8% to greater than 95%, which means that, when noble-metal nanoparticles are used as photocatalysts for selective oxidation of benzyl alcohol, the photolysis of the reaction is negligible.

To better understand the factors affecting the performance of $Au_2Pt_2/N–TiO_2$ photocatalyst, we tuned and investigated the light-source wavelength and intensity in the photocatalytic reaction. The most significant enhancement in the yield of the reaction was observed by irradiation of 460–560 nm over $Au_2Pt_2/N–TiO_2$ photocatalyst (Figure 3b), accounting for 81.26% of the strengthening of the total light irradiation. Also, we used a multiple-wavelength laser light source to confirm the effect of illumination wavelength (Figure 3d and Figure S4, Supplementary Materials). $Au_2Pt_2/N–TiO_2$ exhibited an exceptionally high apparent quantum yield at two wavelengths (500 nm and 532 nm), with 5.86% at 500 nm and 4.57% at 532 nm. Moreover, it is believed that there are two possible mechanisms that may affect the performance of the photocatalytic activity, namely hot-electron injection and PIRET. Upon irradiation of visible light, following light absorption and SPR excitation in these nanostructures, electromagnetic decay takes place on a femtosecond timescale non-radiatively by transferring the energy to hot electrons; then, these "hot enough" electrons with high energy would inject into the $N–TiO_2$ conduction band. In this manner, the apparent quantum yield will fit well with the pattern of the plasmonic metal absorption spectrum, which is consistent with an observation reported in previous literature [39]. On the other hand, in this case, nitrogen doping introduces a new intra-bandgap level above the TiO_2 valence band, which can resonate with the electromagnetic field generated by the gold SPR effect, and the electromagnetic field is then able to improve the generation of photocarriers from intra-bandgap levels to the TiO_2 conduction band through the PIRET process. As a result, it will further increase the photocatalytic efficiency. Therefore, a high apparent quantum yield (AQY) was observed in the band that was contributed by the hot-electron injection mechanism caused by plasmonic absorption and the PIRET mechanism. Subsequently, further analysis (Figure 3c) showed that the correlation between the dependence of light enhancement on optical irradiance of all photocatalysts indicates that photoexcitation intensity is a crucial factor for the photo-enhancing activity, which is consistent with previously published literature [40,41]. The above analysis suggested that the enhanced activity on the $Au_2Pt_2/N–TiO_2$ catalyst is dominated by the specific illumination wavelength and irradiation intensity. Through the analysis of PL and photoelectrochemical tests, we believe the following two factors can describe this: (1) through the loading with Au NPs (plasmonic nanoparticles), on the one hand, plasmonic photocatalysts can utilize a specific wavelength of photons (especially in the visible-light range) to extract hot electrons from the plasmonic metals and more efficiently generate electron–hole pairs. On the other hand, due to the strong near-field electromagnetic resonance caused by the surface plasmons, the rate of generation of photocarriers in TiO_2 is enhanced by the PIRET between the electromagnetic field and the resonance electronic energy levels of TiO_2; (2) the integration of a cocatalyst such as Pt NPs can result in the formation of a Schottky barrier that acts as an "electron trapper" for improving photoinduced charge transport and separation.

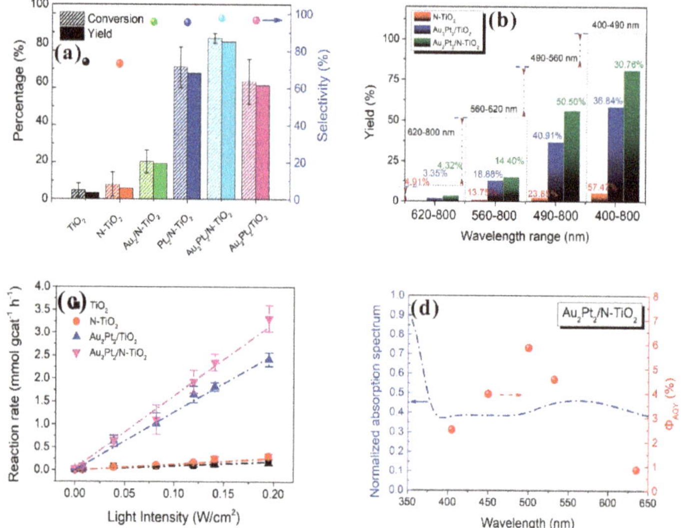

Figure 3. (**a**) Conversion, yield, and selectivity for the photo-oxidation of benzyl alcohol to benzaldehyde over prepared photocatalysts. (**b**) The dependence of yield and irradiation wavelength over photocatalysts for the selective oxidation of benzyl alcohol. (**c**) The rate of photocatalytic reaction over TiO_2, N–TiO_2, Au_2Pt_2/TiO_2, and Au_2Pt_2/N–TiO_2 as a function of irradiance intensity. (**d**) Diffuse-reflectance UV–Vis spectra of Au_2Pt_2/N–TiO_2 photocatalyst and the quantum yield for the formation of benzaldehyde under a multiple-wavelength laser light source. The apparent quantum yield was calculated using the equation $\Phi AQY = (Y_{vis} - Y_{dark})/N \times 100\%$, where Y_{vis} and Y_{dark} are the yields of photocatalytic reaction under irradiation or dark conditions, and N is the number of incident photons in the reaction vessel.

The reaction mechanism involved in the photocatalytic oxidation of benzyl alcohol on Au_2Pt_2/N–TiO_2 was inspected by a control experiment using different radical scavengers and by electron spin resonance (ESR) spectroscopy measurement using spin trapping and labeling [42]. As shown in Figure 4a, there was no significant change in the reaction process for the hydroxyl (·OH) radicals scavenged by TBA. However, when ammonium oxalate, silver nitrate, and benzoquinone were separately added to capture photogenerated holes, electrons, and superoxide (·O^{2-}) radicals, the yield of the reaction was significantly reduced. This observation indicated that, in addition to the hydroxyl (·OH) radicals, radicals such as photogenerated holes, electrons, and superoxide radicals are involved in the process of visible-light photooxidation of benzyl alcohol. Furthermore, ESR measurement using spin trapping and labeling (Figure 4b) indicated that oxygen can be used to capture photogenerated electrons, providing superoxide ·O^{2-}) radicals, which played a vital role in the photocatalytic process. It is a known fact that the ·OH radical is a highly reactive intermediate which can oxidize substrate molecules indiscriminately without selectivity [43]; however, superoxide (·O^{2-}) radicals are well-known oxidants for selective oxidation reactions [44,45]. Thus, the specific oxidation behavior of the superoxide species in the system and the absence of hydroxyl (·OH) radicals can advantageously favor the selective oxidation of benzyl alcohol and can be a significant cause of high reaction selectivity.

The possible reaction mechanism is illustrated in Figure 4c. Under specific wavelength and intensive visible-light irradiation, the incident photons excite the SPR of the gold nanoparticles. The electrons collectively oscillating by localized surface plasmons decay non-radiatively through intraband or interband excitations on Au NPs, generating hot electrons with high enough energy, then finally transfer into the N–TiO$_2$ conduction band. Meanwhile, in this case, N-doping introduces a new intra-bandgap level above the TiO$_2$ valence band, which can resonate with the electromagnetic field generated by the gold SPR effect, and the electromagnetic field is then able to improve the generation of photocarriers from intra-bandgap levels to the TiO$_2$ conduction band through the PIRET process. After the contact of Pt and N–TiO$_2$, a Schottky junction can be established at the interface, wherein the conduction and valence band are bent upward to the N–TiO$_2$ interface. The electrons on the N–TiO$_2$ conduction band are enriched by the Pt NPs via the Schottky barrier between cocatalyst and N–TiO$_2$ and then captured by oxygen molecules, affording superoxide ($\cdot O^{2-}$) species [11,46]. The superoxide ($\cdot O^{2-}$) species may attract the hydrogen atom of the substrate (benzyl alcohol) to form an alkoxide intermediate; after that, the transient alkoxide intermediate undergoes rapid hydride transfer, resulting in the elimination of proton hydrogen, and ultimately resulting in benzaldehyde. The local electromagnetic field generated by the gold nanoparticles under visible light enhances the excitation probability of the photogenerated electron and hole pairs of the N–TiO$_2$ support material, resulting in more photogenerated carriers, and then these photogenerated electrons migrate to the cocatalyst nanoparticles across the Schottky barrier between Pt–TiO$_2$ and react with oxygen to form superoxide radicals.

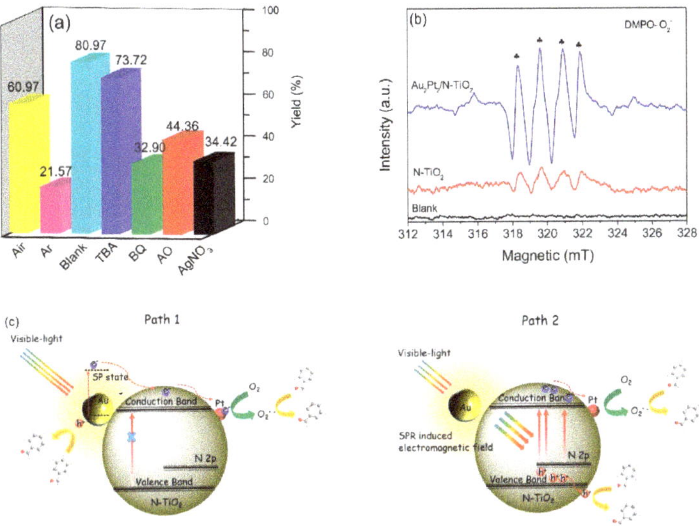

Figure 4. (**a**) Controlled experiment using different radical scavengers; (**b**) ESR spectra collected using 5,5-dimethyl-1-pyrroline-N-oxide ((DMPO-$\cdot O_2^-$)) as the spin trap. (**c**) A plausible mechanism for the photo-oxidation of benzyl alcohol over Au$_2$Pt$_2$/N–TiO$_2$ under visible-light irradiation.

As listed in Table 1, we further investigated the photocatalytic oxidation of various aromatic alcohols over Au$_2$Pt$_2$/N–TiO$_2$. As expected, Au$_2$Pt$_2$/N–TiO$_2$ has not only high activity for oxidation of aromatic alcohols, but also has excellent selectivity for carbonyl compounds. Furthermore, the conversions of different aromatic alcohol substrates were significantly different; for example, the substitution of *para*-substituted benzyl alcohol with an electron-donating group (–OCH$_3$ and –CH$_3$) can increase the efficiency of the reaction, while substitution with an electron-withdrawing group (–Cl) lowers the activity. Furthermore, the durability of the photocatalyst is also a crucial parameter for its further application. As shown in Figures S5 and S6 (Supplementary Materials), no significant decrease in the photocatalytic activity was observed after five cycles, indicating that $Au_2Pt_2/N–TiO_2$ maintained highly durability in the photocatalytic reaction.

Table 1. Photocatalytic selective oxidation of various aromatic alcohols on $Au_2Pt_2/N–TiO_2$ [a].

Entry	Substrate	Product	Yield [b] (%)	Selectivity [c] (%)
1	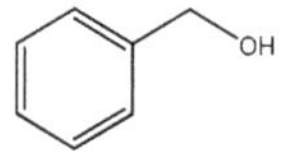	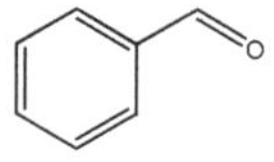	80.9	97.7
2	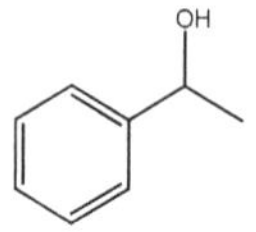	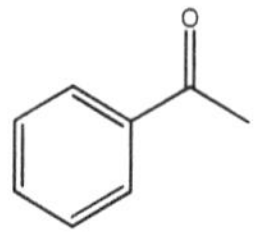	78.1	79.6
3 [d]	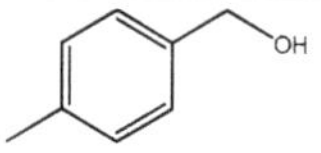	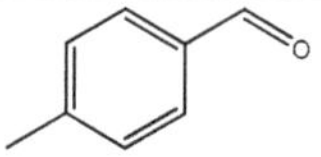	87.9	97.9
4	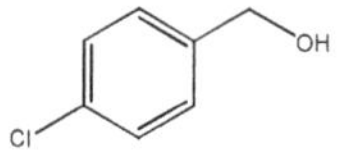	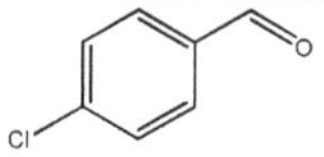	63.8	76.9
5 [e]	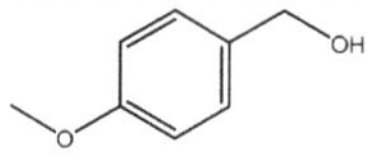	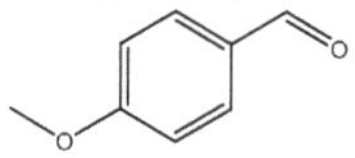	91.2	100
6	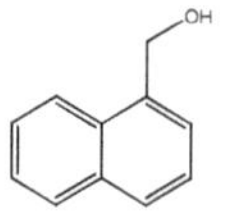	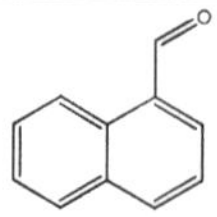	50.1	54.2
7	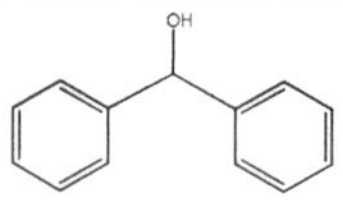	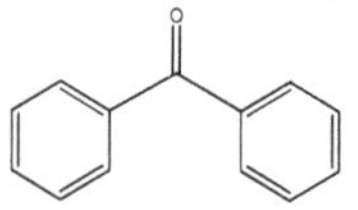	52.2	75.1

[a] Solvent: 1.5 mL of trifluorotoluene; substrate: 0.1 mmol aromatic alcohol; photocatalyst: 10 mg of $Au_2Pt_2/N–TiO_2$; 1 atm of O_2, reaction temperature was maintained at 30 °C, for 2.5 h of reaction time. [b] Yield of aromatic aldehyde; [c] selectivity of aldehyde or ketone; [d] 1 h of reaction time; [e] 1 h of reaction time.

4. Conclusions

In this study, we successfully synthesized the plasmonic photocatalyst $Au_2Pt_2/N–TiO_2$ and investigated its catalytic performance for photo-oxidation of aromatic alcohol. A combination of bifunctional metal NPs was demonstrated for their dual properties related to plasmonic absorption, as well as efficient electron trapping. Coupling a semiconductor with Pt NPs as the cocatalyst can form a Schottky barrier interface, serving as the "electron trapper" to improve charge migration and separation. The Au nanoparticles can be used as an antenna for converting light energy into a local electric field, and hot-electron injection and PIRET mechanisms improve photocarrier generation. The intra-bandgap states of N-doped TiO_2 take a crucial part in improving both hot-electron injection and PIRET from plasmonic metal nanoparticles to the semiconductor. As a result, the reaction rate of the $Au_2Pt_2/N–TiO_2$ catalyst, herein, is ~24 times than that of TiO_2, and the AQY at 500 nm reaches 5.86%, indicative of the successful functionalization of $N–TiO_2$ via the integration of Au plasmonic NPs and the Pt cocatalyst. Furthermore, it is indicated that the intensity and wavelength of the illumination source and the choice of the light source have a significant impact on the activity of photocatalytic reaction. This modification of multifunctional metal NPs demonstrates promise for visible-light-driven catalytic applications.

Supplementary Materials: The following are available online at http://www.mdpi.com/2079-4991/9/3/391/s1: Figure S1: TEM images of Au_2Pt_2/N–TiO_2 photocatalyst; Figure S2: (a) N_2 adsorption/desorption isotherm of as-prepared catalysts; (b) the corresponding Barrett–Joyner–Halenda (BJH) desorption pore size distribution; Figure S3: Fine XP spectra of (a) Au *4f*, (b) Pt *4f*, (c) N *1s*, and (d) O *1s* obtained from Au_2Pt_2/N–TiO_2; Figure S4: (a–d) Diffuse-reflectance UV–Vis spectra of TiO_2, N-TiO_2, Au_2Pt_2/TiO_2, and Au_2Pt_2/N-TiO_2 catalysts and the quantum yield for the formation of benzaldehyde under laser irradiation of different wavelengths, such as 405 nm, 450 nm, 500 nm, 532 nm, and 635 nm. The apparent quantum yield was calculated using the equation $\Phi_{AQY} = (Y_{vis} - Y_{dark})/N \times 100\%$, where Y_{vis} and Y_{dark} denote the yield of benzaldehyde under light and dark conditions, respectively. N denotes the number of incident photons in the reaction vessel; Figure S5: Recyclability tests of Au_2Pt_2/N–TiO_2 for the selective oxidation of benzyl alcohol; Figure S6: (a,b) TEM images of Au_2Pt_2/N–TiO_2 catalyst after photocatalysis reaction; (c,d) the EDX analysis of photocatalysis before and after photocatalysis reaction; Table S1: Summarized physical and chemical data for TiO_2, N–TiO_2, Au_2Pt_2/TiO_2, and Au_2Pt_2/N–TiO_2 photocatalysts.

Author Contributions: W.L. and M.J. guided the project and designed the experiments; Y.W. and Y.C. performed the nanomaterial preparation and characterization; Y.C. contributed to the analysis; Y.W. wrote the manuscript; W.L. and Q.H. revised the manuscript. All authors read and approved the manuscript.

Funding: This research was supported by the National Natural Science Foundation of China (21878163), the National key Research Project (2018YFD0800103), the Fundamental Research Funds for the Central Universities, and the Natural Science Foundation of Tianjin, China (17JCZDJC39500).

Conflicts of Interest: The authors declare no conflicts of interest.

References

1. Fujishima, A.; Honda, K. Electrochemical photolysis of water at a semiconductor electrode. *Nature* **1972**, *238*, 37–38. [CrossRef] [PubMed]

2. Asahi, R.; Morikawa, T.; Ohwaki, T.; Aoki, K.; Taga, Y. Visible-light photocatalysis in nitrogen-doped titanium oxides. *Science* **2001**, *293*, 269–271. [CrossRef] [PubMed]

3. Linic, S.; Aslam, U.; Boerigter, C.; Morabito, M. Photochemical transformations on plasmonic metal nanoparticles. *Nat. Mater.* **2015**, *14*, 567–576. [CrossRef] [PubMed]

4. Chen, Y.; Wang, Y.; Li, W.; Yang, Q.; Hou, Q.; Wei, L.; Liu, L.; Huang, F.; Ju, M. Enhancement of photocatalytic performance with the use of noble-metal-decorated TiO_2 nanocrystals as highly active catalysts for aerobic oxidation under visible-light irradiation. *Appl. Catal. B Environ.* **2017**, *210*, 352–367. [CrossRef]

5. Zhang, N.; Han, C.; Fu, X.; Xu, Y.-J. Function-Oriented Engineering of Metal-Based Nanohybrids for Photoredox Catalysis: Exerting Plasmonic Effect and Beyond. *Chem* **2018**, *4*, 1832–1861. [CrossRef]

6. Sarina, S.; Zhu, H.; Jaatinen, E.; Xiao, Q.; Liu, H.; Jia, J.; Chen, C.; Zhao, J. Enhancing catalytic performance of palladium in gold and palladium alloy nanoparticles for organic synthesis reactions through visible light irradiation at ambient temperatures. *J. Am. Chem. Soc.* **2013**, *135*, 5793–5801. [CrossRef] [PubMed]

7. Fang, C.; Jia, H.; Chang, S.; Ruan, Q.; Wang, P.; Chen, T.; Wang, J. (Gold core)/(titania shell) nanostructures for plasmon-enhanced photon harvesting and generation of reactive oxygen species. *Energy Environ. Sci.* **2014**, *7*, 3431–3438. [CrossRef]

8. Zhang, N.; Han, C.; Xu, Y.-J.; Foley Iv, J.J.; Zhang, D.; Codrington, J.; Gray, S.K.; Sun, Y. Near-field dielectric scattering promotes optical absorption by platinum nanoparticles. *Nat. Photonics* **2016**, *10*, 473–482. [CrossRef]

9. Tanaka, A.; Nakanishi, K.; Hamada, R.; Hashimoto, K.; Kominami, H. Simultaneous and stoichiometric water oxidation and Cr(VI) reduction in aqueous suspensions of functionalized plasmonic photocatalyst Au/TiO_2-Pt under irradiation of green light. *ACS Catal.* **2013**, *3*, 1886–1891. [CrossRef]

10. Zhang, P.; Wang, T.; Chang, X.; Gong, J. Effective Charge Carrier Utilization in Photocatalytic Conversions. *Acc. Chem. Res.* **2016**, *49*, 911–921. [CrossRef] [PubMed]

11. Christopher, P.; Xin, H.; Linic, S. Visible-light-enhanced catalytic oxidation reactions on plasmonic silver nanostructures. *Nat. Chem.* **2011**, *3*, 467–472. [CrossRef] [PubMed]

12. Li, B.; Gu, T.; Ming, T.; Wang, J.; Wang, P.; Wang, J.; Yu, J.C. (Gold Core)@(Ceria Shell) Nanostructures for Plasmon-Enhanced Catalytic Reactions under Visible Light. *ACS Nano* **2014**, *8*, 8152–8162. [CrossRef] [PubMed]

13. Cushing, S.K.; Li, J.; Meng, F.; Senty, T.R.; Suri, S.; Zhi, M.; Li, M.; Bristow, A.D.; Wu, N. Photocatalytic activity enhanced by plasmonic resonant energy transfer from metal to semiconductor. *J. Am. Chem. Soc.* **2012**, *134*, 15033–15041. [CrossRef] [PubMed]

14. Li, J.; Cushing, S.K.; Meng, F.; Senty, T.R.; Bristow, A.D.; Wu, N. Plasmon-induced resonance energy transfer for solar energy conversion. *Nat. Photonics* **2015**, *9*, 601–607. [CrossRef]

15. Mubeen, S.; Lee, J.; Singh, N.; Krämer, S.; Stucky, G.D.; Moskovits, M. An autonomous photosynthetic device in which all charge carriers derive from surface plasmons. *Nat. Nanotechnol.* **2013**, *8*, 247–251. [CrossRef] [PubMed]

16. Zhang, Y.; He, S.; Guo, W.; Hu, Y.; Huang, J.; Mulcahy, J.R.; Wei, W.D. Surface-Plasmon-Driven Hot Electron Photochemistry. *Chem. Rev.* **2018**, *118*, 2927–2954. [CrossRef] [PubMed]

17. Zhang, P.; Wang, T.D.; Gong, J. Mechanistic Understanding of the Plasmonic Enhancement for Solar Water Splitting. *Adv. Mater.* **2015**, *27*, 5328–5342. [CrossRef] [PubMed]

18. Naldoni, A.; Fabbri, F.; Altomare, M.; Marelli, M.; Psaro, R.; Selli, E.; Salviati, G.; Dal Santo, V. The critical role of intragap states in the energy transfer from gold nanoparticles to TiO_2. *Phys. Chem. Chem. Phys.* **2015**, *17*, 4864–4869. [PubMed]

19. Wang, X.; Long, R.; Liu, D.; Yang, D.; Wang, C.; Xiong, Y. Enhanced full-spectrum water splitting by confining plasmonic Au nanoparticles in N-doped TiO_2 bowl nanoarrays. *Nano Energy* **2016**, *24*, 87–93. [CrossRef]

20. Liu, Z.; Hou, W.; Pavaskar, P.; Aykol, M.; Cronin, S.B. Plasmon resonant enhancement of photocatalytic water splitting under visible illumination. *Nano Lett.* **2011**, *11*, 1111–1116. [CrossRef] [PubMed]

21. Chen, Y.; Li, W.; Wang, J.; Gan, Y.; Liu, L.; Ju, M. Microwave-assisted ionic liquid synthesis of Ti3+ self-doped TiO_2 hollow nanocrystals with enhanced visible-light photoactivity. *Appl. Catal. B Environ.* **2016**, *191*, 94–105. [CrossRef]

22. Wang, W.; Lu, C.; Ni, Y.; Su, M.; Xu, Z. A new sight on hydrogenation of F and N-F doped {001} facets dominated anatase TiO_2 for efficient visible light photocatalyst. *Appl. Catal. B Environ.* **2012**, *127*, 28–35. [CrossRef]

23. Usón, L.; Sebastian, V.; Mayoral, A.; Hueso, J.L.; Eguizabal, A.; Arruebo, M.; Santamaria, J. Spontaneous formation of Au–Pt alloyed nanoparticles using pure nano-counterparts as starters: A ligand and size dependent process. *Nanoscale* **2015**, *7*, 10152–10161. [CrossRef] [PubMed]

24. Liu, L.; Dao, T.D.; Kodiyath, R.; Kang, Q.; Abe, H.; Nagao, T.; Ye, J. Plasmonic janus-composite photocatalyst comprising Au and C-TiO_2 for enhanced aerobic oxidation over a broad visible-light range. *Adv. Funct. Mater.* **2014**, *24*, 7754–7762. [CrossRef]

25. Al-Azri, Z.H.N.; Chen, W.T.; Chan, A.; Jovic, V.; Ina, T.; Idriss, H.; Waterhouse, G.I.N. The roles of metal co-catalysts and reaction media in photocatalytic hydrogen production: Performance evaluation of M/TiO_2 photocatalysts (M = Pd, Pt, Au) in different alcohol-water mixtures. *J. Catal.* **2015**, *329*, 355–367. [CrossRef]

26. Li, X.; Liu, P.; Mao, Y.; Xing, M.; Zhang, J. Preparation of homogeneous nitrogen-doped mesoporous TiO_2 spheres with enhanced visible-light photocatalysis. *Appl. Catal. B Environ.* **2015**, *164*, 352–359. [CrossRef]

27. Murdoch, M.; Waterhouse, G.I.N.; Nadeem, M.A.; Metson, J.B.; Keane, M.A.; Howe, R.F.; Llorca, J.; Idriss, H. The effect of gold loading and particle size on photocatalytic hydrogen production from ethanol over Au/TiO_2 nanoparticles. *Nat. Chem.* **2011**, *3*, 489–492. [CrossRef] [PubMed]

28. Pan, X.; Xu, Y.J. Defect-mediated growth of noble-metal (Ag, Pt, and Pd) nanoparticles on TiO_2 with oxygen vacancies for photocatalytic redox reactions under visible light. *J. Phys. Chem. C* **2013**, *117*, 17996–18005. [CrossRef]

29. Sakamoto, H.; Ohara, T.; Yasumoto, N.; Shiraishi, Y.; Ichikawa, S.; Tanaka, S.; Hirai, T. Hot-Electron-Induced Highly Efficient O_2 Activation by Pt Nanoparticles Supported on Ta_2O_5 Driven by Visible Light. *J. Am. Chem. Soc.* **2015**, *137*, 9324–9332. [CrossRef] [PubMed]

30. Jiang, D.; Wang, W.; Sun, S.; Zhang, L.; Zheng, Y. Equilibrating the plasmonic and catalytic roles of metallic nanostructures in photocatalytic oxidation over Au-modified CeO_2. *ACS Catal.* **2015**, *5*, 613–621. [CrossRef]

31. Lu, D.; Ouyang, S.; Xu, H.; Li, D.; Zhang, X.; Li, Y.; Ye, J. Designing Au Surface-Modified Nanoporous-Single-Crystalline $SrTiO_3$ to Optimize Diffusion of Surface Plasmon Resonance-Induce Photoelectron toward Enhanced Visible-Light Photoactivity. *ACS Appl. Mater. Interfaces* **2016**, *8*, 9506–9513. [CrossRef] [PubMed]

32. Shiraishi, Y.; Tsukamoto, D.; Sugano, Y.; Shiro, A.; Ichikawa, S.; Tanaka, S.; Hirai, T. Platinum nanoparticles supported on anatase titanium dioxide as highly active catalysts for aerobic oxidation under visible light irradiation. *ACS Catal.* **2012**, *2*, 1984–1992. [CrossRef]

33. Verma, S.; Baig, R.B.N.; Nadagouda, M.N.; Varma, R.S. Selective Oxidation of Alcohols Using Photoactive VO@g-C3N4. *ACS Sustain. Chem. Eng.* **2016**, *4*, 1094–1098. [CrossRef]

34. Sugano, Y.; Shiraishi, Y.; Tsukamoto, D.; Ichikawa, S.; Tanaka, S.; Hirai, T. Supported Au-Cu bimetallic alloy nanoparticles: An aerobic oxidation catalyst with regenerable activity by visible-light irradiation. *Angew. Chem. Int. Ed.* **2013**, *52*, 5295–5299. [CrossRef] [PubMed]

35. Jiang, T.; Jia, C.; Zhang, L.; He, S.; Sang, Y.; Li, H.; Li, Y.; Xu, X.; Liu, H. Gold and gold-palladium alloy nanoparticles on heterostructured TiO$_2$ nanobelts as plasmonic photocatalysts for benzyl alcohol oxidation. *Nanoscale* **2015**, *7*, 209–217. [CrossRef] [PubMed]

36. Higashimoto, S.; Kitao, N.; Yoshida, N.; Sakura, T.; Azuma, M.; Ohue, H.; Sakata, Y. Selective photocatalytic oxidation of benzyl alcohol and its derivatives into corresponding aldehydes by molecular oxygen on titanium dioxide under visible light irradiation. *J. Catal.* **2009**, *266*, 279–285. [CrossRef]

37. Higashimoto, S.; Suetsugu, N.; Azuma, M.; Ohue, H.; Sakata, Y. Efficient and selective oxidation of benzylic alcohol by O$_2$ into corresponding aldehydes on a TiO$_2$ photocatalyst under visible light irradiation: Effect of phenyl-ring substitution on the photocatalytic activity. *J. Catal.* **2010**, *274*, 76–83. [CrossRef]

38. Kobayashi, H.; Higashimoto, S. DFT study on the reaction mechanisms behind the catalytic oxidation of benzyl alcohol into benzaldehyde by O$_2$ over anatase TiO$_2$ surfaces with hydroxyl groups: Role of visible-light irradiation. *Appl. Catal. B Environ.* **2015**, *170–171*, 135–143. [CrossRef]

39. Tanaka, A.; Sakaguchi, S.; Hashimoto, K.; Kominami, H. Preparation of Au/TiO$_2$ with metal cocatalysts exhibiting strong surface plasmon resonance effective for photoinduced hydrogen formation under irradiation of visible light. *ACS Catal.* **2013**, *3*, 79–85. [CrossRef]

40. Tanaka, A.; Hashimoto, K.; Kominami, H. Preparation of Au/CeO$_2$ exhibiting strong surface plasmon resonance effective for selective or chemoselective oxidation of alcohols to aldehydes or ketones in aqueous suspensions under irradiation by green light. *J. Am. Chem. Soc.* **2012**, *134*, 14526–14533. [CrossRef] [PubMed]

41. Xiao, Q.; Liu, Z.; Bo, A.; Zavahir, S.; Sarina, S.; Bottle, S.; Riches, J.D.; Zhu, H. Catalytic transformation of aliphatic alcohols to corresponding esters in O$_2$ under neutral conditions using visible-light irradiation. *J. Am. Chem. Soc.* **2015**, *137*, 1956–1966. [CrossRef] [PubMed]

42. Fu, H.; Zhang, L.; Zhang, S.; Zhu, Y.; Zhao, J. Electron spin resonance spin-trapping detection of radical intermediates in N-doped TiO$_2$-assisted photodegradation of 4-chlorophenol. *J. Phys. Chem. B* **2006**, *110*, 3061–3065. [CrossRef] [PubMed]

43. Li, A.; Wang, T.; Chang, X.; Cai, W.; Zhang, P.; Zhang, J.; Gong, J. Spatial separation of oxidation and reduction co-catalysts for efficient charge separation: Pt@TiO$_2$@MnOx hollow spheres for photocatalytic reactions. *Chem. Sci.* **2016**, *7*, 890–895. [CrossRef] [PubMed]

44. Salmistraro, M.; Schwartzberg, A.; Bao, W.; Depero, L.E.; Weber-Bargioni, A.; Cabrini, S.; Alessandri, I. Triggering and monitoring plasmon-enhanced reactions by optical nanoantennas coupled to photocatalytic beads. *Small* **2013**, *9*, 3301–3307. [CrossRef] [PubMed]

45. Primo, A.; Corma, A.; García, H. Titania supported gold nanoparticles as photocatalyst. *Phys. Chem. Chem. Phys.* **2011**, *13*, 886–910. [CrossRef] [PubMed]

46. Christopher, P.; Xin, H.; Marimuthu, A.; Linic, S. Singular characteristics and unique chemical bond activation mechanisms of photocatalytic reactions on plasmonic nanostructures. *Nat. Mater.* **2012**, *11*, 1044–1050. [CrossRef] [PubMed]

Article

Comparative Study of Different Acidic Surface Structures in Solid Catalysts Applied for the Isobutene Dimerization Reaction

José M. Fernández-Morales [1], Eva Castillejos [2,*], Esther Asedegbega-Nieto [1,*], Ana Belén Dongil [3], Inmaculada Rodríguez-Ramos [3] and Antonio Guerrero-Ruiz [1]

1   Dpto. Química Inorgánica y Técnica, Facultad de Ciencias, UNED, c/Senda del Rey No. 9, 28040 Madrid, Spain; jmfernandez@ccia.uned.es (J.M.F.-M.); aguerrero@ccia.uned.es (A.G.-R.)
2   Dpto. Ingeniería Química, Facultad de Ciencias, UCM, Avda. Complutense s/n, 28040 Madrid, Spain
3   Instituto de Catálisis y Petroleoquímica, CSIC, c/Marie Curie No. 2, Cantoblanco, 28049 Madrid, Spain; a.dongil@csic.es (A.B.D.); irodriguez@icp.csic.es (I.R.-R.)
*   Correspondence: castillejoseva@ccia.uned.es (E.C.); easedegbega@ccia.uned.es (E.A.-N.)

Received: 1 June 2020; Accepted: 22 June 2020; Published: 25 June 2020

Abstract: Dimerization of isobutene (IBE) to C_{8s} olefins was evaluated over a range of solid acid catalysts of diverse nature, in a fixed bed reactor working in a continuous mode. All catalytic materials were studied in the title reaction performed between 50–250 °C, being the reaction feed a mixture of IBE/helium (4:1 molar ratio). In all materials, both conversion and selectivity increased with increasing reaction temperature and at 180 °C the best performance was recorded. Herein, we used thermogravimetry analysis (TGA) and temperature programmed desorption of adsorbed ammonia (NH_3-TPD) for catalysts characterization. We place emphasis on the nature of acid sites that affect the catalytic performance. High selectivity to C_{8s} was achieved with all catalysts. Nicely, the catalyst with higher loading of Brønsted sites displayed brilliant catalytic performance in the course of the reaction (high IBE conversion). However, optimum selectivity towards C_8 compounds led to low catalyst stability, this being attributed to the combined effect between the nature of acidic sites and structural characteristics of the catalytic materials used. Therefore, this study would foment more research in the optimization of the activity and the selectivity for IBE dimerization reactions.

Keywords: catalysts; dimerization; isobutene; olefins

1. Introduction

The energy dependence of fuels obtained from fossil sources continues to be a serious problem in many countries that do not count with such natural reserves. Without going any further, in 2017, gross imports of massive energy into the European Union (EU) stood at 87% [1], highlighting transport as the sector that consumes the most energy (33%) [2]. In addition, this sector is deeply dependent on fossil fuels, since 95% of the energy it uses is derived from these sources [3]. This energy dependency is further troubling when taking into consideration that we are currently running out of these sources and, in addition, the fuels that are extracted are the main contributors to climate change. Therefore, there is a need to progressively replace the non-renewable energy sources by inexhaustible ones. One of these renewable sources for fuel production is biomass [4], which is a sustainable carbon resource with neutral CO_2 emissions.

In the last decade, important studies have been conducted with the aim of producing fuels, using butanol as an intermediate, which in turn is a byproduct of biomass. This compound can be dehydrated to butene and subsequently undergo oligomerization reactions to produce from C_8 to C_{16}, and after subsequent hydrogenation yield the desired fuels [5,6]. One of the most used hydrocarbons in the fuel market is isooctane, because it is a non-aromatic compound, has a high-octane rating, has a low sulfur content and low volatility [7]. This compound can be obtained from isobutene (IBE), which is a relevant molecule employed in the synthesis of polymers, as well as gasoline additives such as methyltertbutyl ether (MTBE), by reaction with an alcohol [8], or isooctane itself. Industrially, isooctane can be obtained from the alkylation reaction of IBE in liquid phase with strong acids such as HF or H_2SO_4. However, the use of these acids as homogeneous catalysts make their regeneration very expensive, apart from the environmental problems that they can provoke due to their corrosive capacity and water solubility [9,10]. Therefore, for years, an alternative route has been studied with the use of solid acid catalysts, which make safer handling possible and provide a lower environmental impact. In the literature among the solid acid catalysts used for this reaction we find ion exchange resins [6,11–13], sulfated metal oxides [14–18], heteropolyacids [19–21], zeolites [22–26] or Metal-Organic Frameworks (MOFs) [27,28].

The studied reaction in this work consists on IBE to produce isooctane C8 such as 2,4,4-trimethylpen-1-ene (TP1) and 2,4,4-trimethylpen-2-ene (TP2). This alkylation reaction of IBE needs an acidic medium or a metal capable of breaking the double bond, which generates a tertiary carbocation that enables interaction with another IBE molecule present in the medium. It is an acknowledged fact that the surface acidity of the catalyst plays a relevant role on the IBE dimerization reaction. However, the acidity or the presence of metal can also promote chain polymerization reactions, which could be minimized working at low conversions, between 20% and 60%, or decreasing the contact time with the catalytic bed. Thus, catalysis systems achieving high catalytic activities and optimum selectivity towards dimers in the oligomerization of IBE still remains a challenge.

In this reaction we have used the following catalysts: an ion exchange resin (Amberlyst 15), a carbon supported heteropolyacid, a sulfated zirconium oxide and a NiO/Al_2O_3 catalyst. The common denominator of these catalysts is their surface acidity, since in the structure of the Amberlyst we can find sulfonic groups ($-SO_3H$) [6], in that of the heteropolyacids there is the presence of protons attached to the metal cation that makes them considered superacids [21] and in the sulfated zirconium oxide the presence of sulfate groups bound to the oxide structure [15–18]. On the other hand, the incorporation of NiO into the alumina structure was consistent with the fact that the metal atom played the role of reaction intermediate, being able to interact with the electrons of the double bond in the IBE molecule; thereby allowing the formation of the carbocation and the attack of a new IBE molecule for alkylation to occur [29–31].

A competent method is pursued in order to orientate the catalytic behavior in reactions based on solid acid-catalysts. From another perspective, this research would decipher the relevant role of adequate textual properties of the materials required in these particular reactions. A great deal of effort is undertaken in order to reveal the nature of active site which is responsible for IBE dimerization. All these tests employing different catalysts have the objective of accomplishing the IBE dimerization reaction selectively and, therefore, avoiding polymerization reactions that generate carbonaceous species which would lead to the deactivation of the catalyst.

On the other hand, another motivation for this study with solid acid catalysts is that most of the research on oligomerization of butenes is performed in liquid phase at moderately low temperatures, though, high pressure is required. Therefore, additional specialized components are needed to cautiously carry out the reaction. Contrary to this, in our work, these requirements are ruled out as IBE dimerization reaction is carried out under atmospheric pressure conditions in a gas-phase continuous flow reactor.

2. Materials and Methods

2.1. Catalysts Preparation

The Amberlyst 15 catalyst (denoted A15) was supplied by Alfa-Aesar, Kandel, Germany (S_{BET} = 15 m^2/g, CAS: 39389-20-3). Amberlyst 15 presents high acidity due to its content of sulfonic acid groups, equal to 4.7 mmol/g (https://www.sigmaaldrich.com/catalog/product/sial/06423?lang=es®ion=ES). Before its use, the catalyst A15 was purified. The commercial resin beads (0.5 mm) were incessantly washed with acetone, dried at 110 °C during 24 h and kept in a desiccator. Prior to all catalytic tests, the grains were loaded into the reactor, which was thereafter heated at 180 °C for 2 h under helium flow.

The second catalyst, denoted STA/HSAG$_{100}$, was a heteropolyacid or polyoxometalates supported on graphite HSAG$_{100}$. STA ($H_4SiW_{12}O_{40} \cdot nH_2O$, CAS: 12027-43-9) was supplied by Sigma-Aldrich (Darmstadt, Germany). Commercial high surface area graphite HSAG$_{100}$ supplied by TIMCAL (Bodio, Switzerland, CAS: 7782-42-5) with a specific surface area of 100 m^2/g (particle size <125 µm) was chosen as the support for STA. This catalyst containing an STA loading of 15 wt.%, was synthesized by the incipient impregnation of HSAG$_{100}$, employing as solvent for the STA solid a mixture of ethanol/water (1:1 ratio) as reported previously [32]. Before each experiment, STA/HSAG$_{100}$ was inserted into the reactor and heated prior to the reaction at 180 °C for 2 h under a helium flow of 20 mL/min.

Commercial ZrO$_2$/S Zirconium Hydroxide (Sulfate doped, CAS: 34806-73-0) Grade XZ0682/01 was purchased from Mel (Darmstadt, Germany). ZrO$_2$/S Zirconium Hydroxide precursor was calcined in flowing air at 600 °C for 5 h to obtain active sulfated zirconia (ZS) with a specific surface area of 116 m^2/g [33]. This catalyst was also treated prior to its use in a helium flow of 20 mL/min at 180 °C.

NiO/Al$_2$O$_3$ catalyst was prepared by incipient impregnation of γ-Al$_2$O$_3$ (CK-300 from Cyanamid Ketjen, S$_{BET}$ = 180 m^2/g, CAS: 1344-28-1) with an aqueous solution of NiSO$_4 \cdot 6H_2O$ (also provided by Sigma-Aldrich, CAS: 10101-97-0). The theoretical metal (Ni) amount incorporated was 5% W/W. Subsequently, the catalyst was dried at 100 °C overnight and calcined at 500 °C during 5 h. The final sample, named NiO/Al$_2$O$_3$, was also treated prior to its use under a 20 mL/min helium flow at 180 °C. This catalyst was also treated in situ with pure hydrogen flow in order to reduce the Ni ions to their metallic state.

2.2. Supports and Catalysts Characterization

All the samples were characterized by thermogravimetric analyses (TGA). TGA were conducted under helium using TA Instruments (New Castle, DE, USA), model SDT Q600 TA System. About 5 mg of the sample was deposited in an alumina crucible and the temperature was raised until to 800 °C (starting from room temperature) at a rate of 10 °C/min. In order to remove H$_2$O, on reaching 100 °C, the temperature was kept constant for 30 min. The measured weight loss gives information on the changes in the solid sample composition and thermal stability features. A derivative weight loss curve (DTG) can be used to determine the temperature at which the rates of weight changes are maximum. Gases issued at the outlet of the TGA equipment are CO$_2$, CO, H$_2$, O$_2$ and SO$_2$. This was evidenced by a Mass Quadrupole Spectrometer (Pfeiffer Vacuum Omnistar™ GSD 301, Asslar, Germany) coupled to the TGA, this is the technique abbreviated as temperature programmed desorption (TPD)-MS. On the other hand, total acidity of the samples was estimated with the aid of static ammonia adsorption using an ASAP 2010 Micromeritics unit (Norcross, GA, USA). The adsorption was carried out at 50 °C and 50 mg of material were analyzed. The samples were first pretreated in helium at 230 °C (5 °C min^{-1}) for 60 min after which it was cooled to 50 °C and then NH$_3$ was adsorbed during 30 min. After flushing with He the amount of desorbed NH$_3$ was determined ramping (10 °C/min) from 50 °C until 230 °C (NH$_3$-TPD). This final temperature was maintained for 30 min and from the analysis of the gas evolved by a thermal conductivity detector (TA Instruments, New Castle, DE, USA) the amount of chemisorbed ammonia can be detected.

2.3. Catalytic Evaluation in Gas Phase Isobutene Dimerization

The IBE dimerization reaction was carried out in a continuous flow fixed-bed tubular reactor contained within a PID Eng&Tech system (Alcobendas, Spain), equipped with a furnace using PID control and counting on mass flow controllers mod. EL-Flow Select (Bronkhorst High-Tech B.V., Ruurlo, The Netherlands). Before reaction, the catalyst (50 mg) was pre-treated under He atmosphere

(20 mL/min) at 180 °C for 2 h. Any substantial mass-transfer limitation was hindered applying high linear velocity and the adequate catalysts grain size (sieve fraction 0.25–0.5 mm). Thereafter, the system was cooled and the catalytic tests were carried out feeding the reactor with IBE:He in a 1:4 molar ratio counting with a total gas flow of 8 cm^3/min^{-1}. The catalytic measurements were done once the catalyst was brought to the required reaction temperature (180 °C) under He. Thereafter, the gas stream was switched to the reaction mixture. Conversion measurements were conducted every 10 min after stabilizing the samples. The IBE conversions were derived upon the on-line analysis of the exit gas with the aid of a Varian CP-3800 chromatograph (Agilent, Santa Clara, CA, USA) equipped with a FID detector (200 °C, detection range 10) and a Supelco alumina sulfate PLOT capillary column (30 m × 0′53 mm/10 μm). The programmed temperature of the column is as follows: it was maintained at 120 °C during 5 min; then a raised until 180 °C at a rate of 15 °C min^{-1} and kept constant at 180 °C for 10 min. IBE conversions, activity and selectivity were estimated as:

$$\text{Conv IBE (\%)} = (([\text{IBE}]^\circ - [\text{IBE}])/[\text{IBE}]^\circ) \times 100 \tag{1}$$

where Conv IBE (%) (1) is IBE conversion percentage, $[\text{IBE}]^\circ$ and $[\text{IBE}]$ are the inlet and outlet IBE concentrations, respectively.

$$\text{Selec TP1 (\%)} = (2 \times [\text{TP1}])/(2 \times \Sigma_i\,[\text{products}]_i) \times 100 \tag{2}$$

where Selec TP1 (%) (2) is percentage of 2,4,4-trimethylpent-1-ene produced, $[\text{TP1}]$ represents the concentration of TP1 produced and $\Sigma_i\,[\text{products}]_i$ is total concentration of products.

$$\text{Selec TP2 (\%)} = (2 \times [\text{TP2}]_i)/(2 \times \Sigma_i\,[\text{products}]_i) \times 100 \tag{3}$$

where Selec TP2 (%) (3) is percentage of 2,4,4-trimethylpent-2-ene produced, $[\text{TP2}]$ represent the concentration of TP2 produced and $\Sigma_i\,[\text{products}]_i$ is total concentration of products.

$$\text{Selec C8s (\%)} = (2 \times [\text{C}_8\text{s}]_i)/(2 \times \Sigma_i\,[\text{products}]_i) \times 100 \tag{4}$$

where Selec C8s (%) (4) is percentage of another C_{8s} produced, $[C_{8s}]$ represent the concentration of C_{8s} produced and $\Sigma_i\,[\text{products}]_i$ is total concentration of products.

$$\text{Activity (μmol IBE converted/g}_{\text{cat}}) = (\text{Conv IBE (\%)} \times [\text{IBE}]^\circ/\text{g}_{\text{cat}}) \tag{5}$$

where Activity (5) is catalytic activity, Conv IBE is IBE conversion percentage, $[\text{IBE}]^\circ$ are inlet IBE concentration and g_{cat} is mass of catalysts.

In order to verify that the solids exhibited a reproducible behavior the catalytic tests were carried out in duplicate.

3. Results

3.1. Catalysts Characterization

The acidity of catalysts was studied by NH$_3$-TPD and their profiles are shown in Figure 1. As it is well-known, the temperature at which NH$_3$ desorbs can be related to the strength of acid sites. However, NH$_3$ is a strong non-site-specific base that can be adsorbed on both Brønsted and Lewis acid sites, and simultaneous desorption from several sites can occur [34]. The experimental conditions employed in the present work allow us to disregard the contribution of gaseous or adsorbed NH$_3$ as well as those from species evolved from the material itself and to consider, exclusively, contributions ascribed to acid sites. In addition, under these low temperatures, it is unlikely that surface reduction by ammonia and, hence, change in the acidic properties might take place. Moreover, none of the catalysts hold micropores, so ammonia desorption temperatures should not be influenced by diffusion constrains.

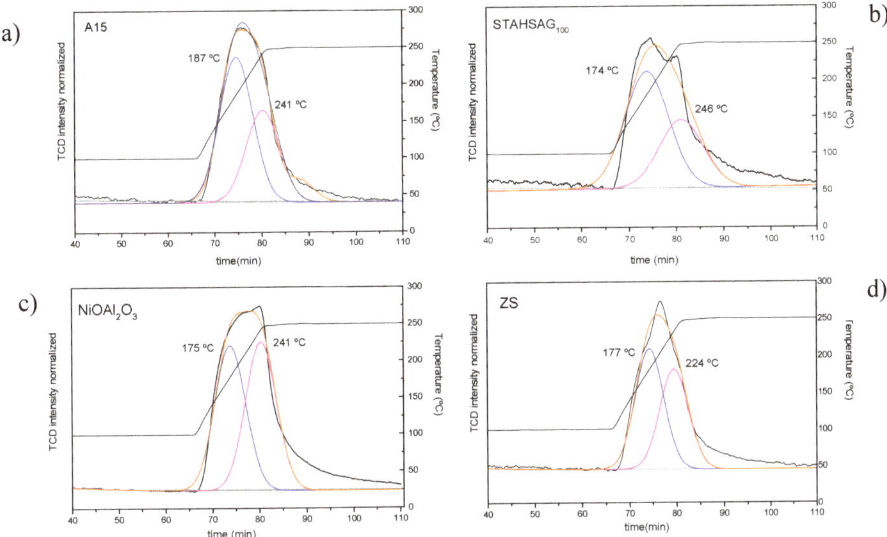

Figure 1. Temperature programmed desorption of adsorbed ammonia (NH_3-TPD) profiles and deconvolution after adsorption of NH_3 at 250 °C on the catalysts: (**a**) A15; (**b**) STA/HSAG$_{100}$; (**c**) NiO/Al$_2$O$_3$; (**d**) sulfated zirconia (ZS).

Overall, it can be observed (Figure 1) that the temperature range in which NH_3 desorption occurs is similar for all the samples, and that the maximum is shifted to higher temperatures following the order: NiO/Al$_2$O$_3$ > ZS > A15 > STA/HSAG$_{100}$. Moreover, the intensity of the STA/HSAG$_{100}$ TPD-NH_3 profile is significantly lower compared with that of the other samples. This was expected since HSAG$_{100}$ is a graphitic support that barely holds any acid site; hence the acidity of this catalyst is given mostly by the STA which is in a 15 wt.%. However, it is worth highlighting that the heteropolyacid also retains its acidity despite being supported, and this suggests a weak interaction between STA and the graphite as observed previously in silica-supported systems [35]. In contrast, the other tested catalysts are either bulk catalysts, A15 and ZS, or NiO 5 wt.% supported on alumina, so the total acidity is obviously higher as it emerges from the structure and the surface groups themselves.

In order to analyze the results, the NH_3-TPD profiles were fitted by two symmetrical peaks. The resulting peaks, with maxima in the range of 174–187 °C and 224–246 °C, represent sites of increasing acidity strength. The first contribution is observed at quite similar maximum temperature for all the samples except for A15 that displays the maximum at higher temperature, ca. 187 °C. The second contribution is observed at the same temperature for NiO/Al$_2$O$_3$ and A15, ca. 241 °C, while it is slightly shifted to higher temperature for STA/HSAG$_{100}$, ca. 246 °C and to significantly lower temperature for the sample ZS, ca. 224 °C.

The ratio of these sites LT/HT (Low T: LT, and High T: HT) was also estimated (in Figure 2), and results indicate that it follows the trend ZS > NiO/Al$_2$O$_3$ > A15 > STA/HSAG$_{100}$. These values indicate that, even though ZS presented less acid sites in the HT range, as refers to the desorption temperature, its LT/HT ratio is the highest among the samples, i.e., 1.21. In contrast, both A15 and STA/HASG$_{100}$, despite having stronger acid sites than ZS as evidenced in this higher temperature of desorption, hold significantly lower LT/HT ratios, i.e., 0.64 and 0.57, respectively.

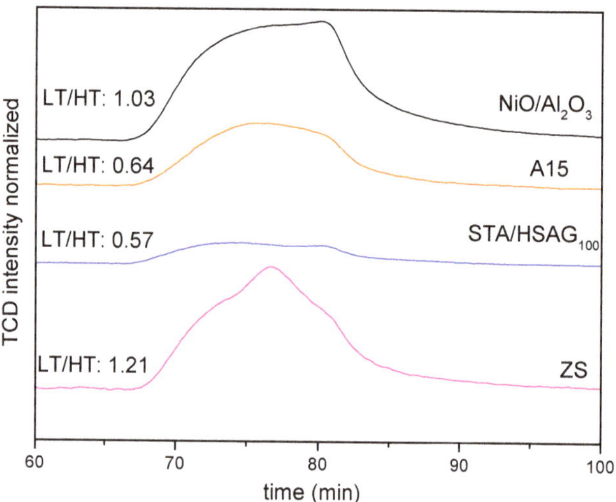

Figure 2. TPD profiles after adsorption of NH$_3$ at 250 °C on the catalysts.

We have evaluated four samples with acid sites of different nature, both Brønsted (mainly SO$_3$H, H$^+$ and -OH from the support surface) and Lewis (Ni, Zr, Al), all of them contributing to weak and strong acid sites [36]. The evaluated samples have surface -OH groups that could be responsible for the acidity given by the first desorption peak. The shift to higher temperatures of the LT peak observed in A15 could be tentatively ascribed to the contribution of the H$^+$ of this cation exchange resin, which would be more acidic than the -OH groups. The second contribution could also include the desorption of NH$_3$ from -OH groups along with desorption from more acidic sites like Lewis, e.g., Al and Ni$^{+\delta}$, as observed previously [37,38]. The contribution derived from –SO$_3$H groups of the ZS and A15, which correspond to Brønsted acid sites, have been reported to occur at higher temperatures [39].

Table 1 and Figure 3 summarizes the total weight loss at 600 °C for the studied catalysts. The TGA technique offers information on the stability of the catalysts as refers to the loss of water and their subsequent thermal decomposition. Generally, the catalysts present a low weight loss being stable at the reaction temperature range. It should be noted that the catalysts had been pretreated prior to the reaction at 180 °C for several hours, assuring the removal of water species in all cases. The analysis of TPD-MS showed the evolution of m/z 64 ascribed to desorption of SO$_2$ on ZS, between 350–390 °C associated with sulfonic acids groups. While STA/HSAG$_{100}$ and NiO/Al$_2$O$_3$ reveal a weight loss due to decomposition at higher temperatures (>350 °C). Finally, the TGA of A15 has been extensively reported [40,41] revealing an intensive loss of water at temperatures near 200 °C. The corresponding TGA and DTG curves are presented in Figure 3. The mass loss observed within 100 °C is associated with the removal of physically adsorbed molecular water. However, the significant loss of water at temperatures near 200 °C of A15, leads to an unclear comparison among samples. Therefore, the exact values are presented in the Table 1 that can be taken into account for comparison purposes.

Table 1. Total weight loss determined gravimetrically at 600 °C.

Catalysts	Weight Loss
A15	41.3%
STA/HSAG$_{100}$	2.4%
ZS	1.7%
NiO/Al$_2$O$_3$	3.2%

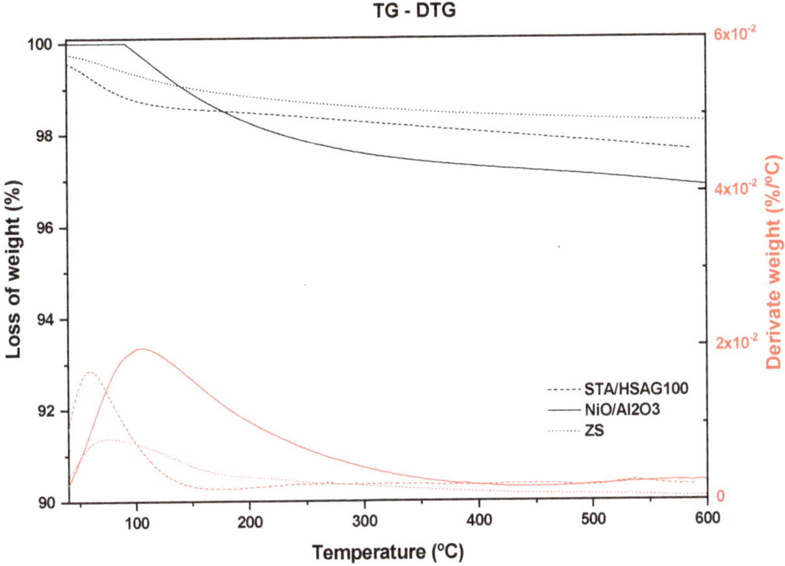

Figure 3. Weight loss % of STA/HSAG$_{100}$, NiO/Al$_2$O$_3$ and ZS samples.

3.2. Catalytic Results

After the analysis of the precedent information, it seemed appropriate to investigate the dimerization of IBE over catalytic materials that possess these properties. IBE oligomerization is representative of acid-catalyzed reactions where the main defiance involves procuring high selectivity to C$_8$ (Scheme 1), inhibiting the formation of heavier olefinic products (C$_{12}$ and C$_{16+}$) and improving the catalyst lifespan at the same time.

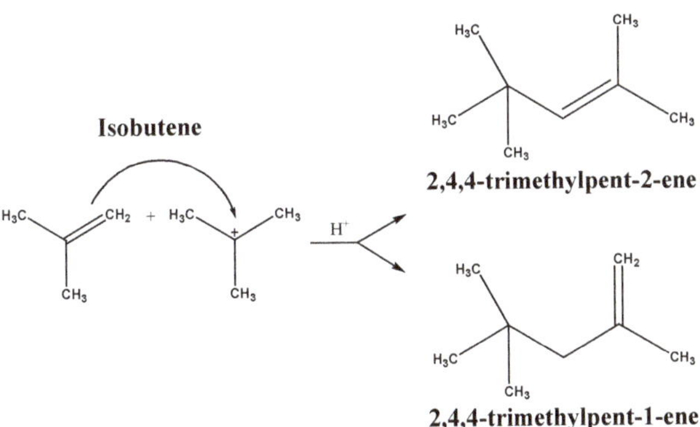

Scheme 1. Dimerization of the isobutene.

Figure 4 depicts the catalyst activity (5) results, over the different samples at 180 °C with respect to time on stream. As portrayed, the highest initial activity for IBE transformation was found with A15 and STA/HSAG$_{100}$ catalysts. However, a noteworthy diminution of this parameter within the first four hours was observed (about 60–75%). It is interesting to notice the catalyst deactivation with time on stream, especially at the early stages of the reaction (rapid initial deactivation). Thereafter, the catalysts' activities were maintained constant for another 24 h of the process (not shown). The IBE activity was lower for ZS and NiO/Al$_2$O$_3$ than that obtained with A15 and STA/HSAG$_{100}$.

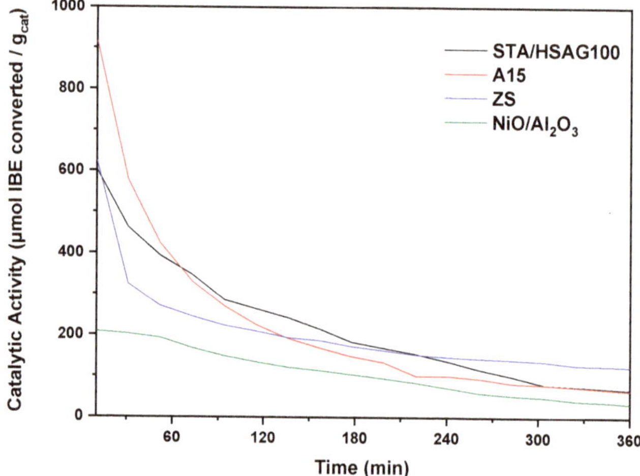

Figure 4. Catalytic activity at 180 °C for all samples.

Figure 5 describes the results which emphasize the influence of the reaction time on the selectivity towards C_8 olefins, produced with the aid of the different catalysts. The on-stream analysis of reaction products evidenced that C8s, mainly TP2 (**3**) and TP1 (**2**), were the only products obtained. High molecular weight oligomers were not identified at any of the studied temperatures. Nonetheless, compounds of the sort are surely formed during the process, though may not be accounted for as they remain confined at the catalysts' surface (see the discussion on the catalyst deactivation). STA/HSAG$_{100}$, ZS and NiO/Al$_2$O$_3$ presented high selectivity to TP2 (**3**) (around 70%) being the remaining 25% to TP1 (**2**). It should be stressed that the IBE conversion to TP2 offered the lowest values in the case of A15, see Figure 5a, and the formation of other C8s olefins (**4**) (not reported in Figure 5) was significant higher (20% at 150 min) compared with STA/HSAG$_{100}$, ZS and NiAl$_2$O$_3$ (less 10%). Overall, all catalysts exhibited high activity toward the selective production of TP2 (**3**) olefin. The high selectivity efficiency of the catalysts in the IBE dimerization disclosed in this work is of immense relevance because, as far as we can tell, such high selectivities to C_8 products have not been divulged, so far, by other researchers under continuous flow reaction conditions.

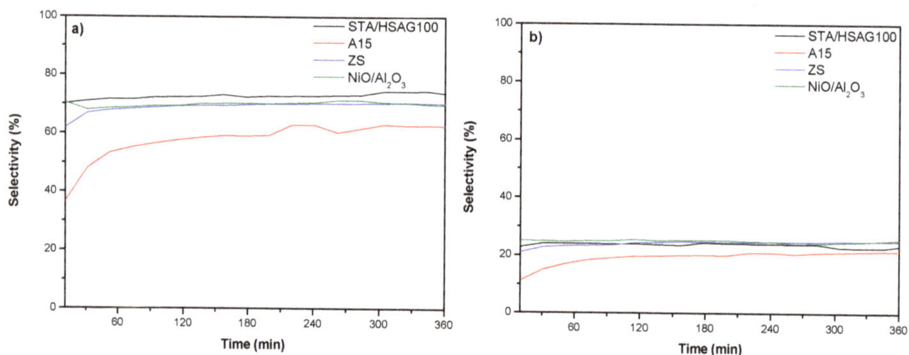

Figure 5. Selectivity to C_8 olefins formed in the isobutene (IBE) dimerization reaction: (**a**) 2,4,4-trimethylpen -2-ene (TP2) and (**b**) 2,4,4-trimethylpen-1-ene (TP1).

The effect of reaction temperature on IBE conversion (**1**) has been previously studied within the 50–180 °C temperature interval (see Figure S1, Supplementary Materials). The reason behind this is, in the first place, concerning the fact that this oligomerization reaction is highly exothermic [42,43]. The temperature effect on IBE dimerization conversion as well as in subsequent oligomerizations, was considered in order to maximize the amount of C8s obtained in the reaction. The chosen studied temperatures were selected after numerous preliminary experiments. In addition, some tests were conducted in the absence of catalyst, seeking the reactant's activity at high reaction temperatures. In addition, employing the same reaction conditions, blank experiments of IBE reaction with just the bare supports ($HSAG_{100}$ and Al_2O_3) and pure STA were carried out in order to check their inherent catalytic activity. Furthermore, external mass transfer resistance and pore diffusion are found to be non-limiting factors in this catalytic application of these materials.

4. Discussion

The IBE dimerization is a complex, extremely exothermic process whereupon a succession of consecutive and parallel reactions unfolds. Due to this, without neglecting the formation of high molecular weight derivatives, such as C_{12} and C_{16} olefins or even higher (C_{20+}), the selectivity to a specific C_8 olefins is the key parameter.

Overall, the literature described reaction processes of IBE dimerizations via carbenium ion mechanism. As an example, Schmidl [44] explains how this reaction advances through this mentioned path (carbenium ion way). The process conditions and the relative stability of the intermediate carbenium ions will determine the reaction equilibria and hence the products obtained. The catalytic cycle of the solid acid isomerization includes three steps: 1—chain initiation to form the first active carbenium ions, 2—rearrangement of the carbenium ions and 3—chain propagation. Solid acid catalysts release a proton, which favors alkene protonation to form an active carbenium ion. The dimerization of light olefins, such as IBE, in the solid acid catalyst presence, is based on the consecutive reaction sequence going on through carbocation intermediates. The addition of a proton (from the solid acid catalyst) to an olefin leads to a t-butyl cation formation, which then combines with another C_4 to give the corresponding carbocation with eight atoms of carbon. This C_8 may isomerize via hydride transfer and methyl shifts to form more stable cations. The presence of suitable amounts of Brønsted and Lewis acid sites is ascribed to favorable activities and selectivities. However, it is a current challenge to relate acid (Lewis and Brønsted) sites with activity or selectivity.

The resin material Amberlyst "A15" used as catalyst in this study has been selected due to its structural and chemical complexity. It consists of a styrene-divinylbenzene-based support where the active sites are sulfonic groups. Hence, and as it was detected earlier on (see Section 3.1) by ammonia adsorption, the surface acid capacity is provided by sulfonic acid groups $-SO_3H$ and the distance between these acid sites. SO_3 with an adjacent Brønsted acid site (H^+) is regarded as the active site. In fact, Xiaolong Zhou et al. [12] reported that the surface acidity capacity of A15 is about 4.60 mmol H^+/g. These acid sites of A15 determine the IBE conversion (1) and dimerization selectivity, where the density of sites or distance between them plays a decisive role.

It can be deduced on viewing Table 2 that the initial conversion of IBE (1) (for A15) was near 50% and that the principal dimer (close to 60%, w/w, Figure 5a) is TP2. It can be derived from the data given in Table 2 that IBE conversion (1) is almost quantitative for weight–time values reaching 300 min.

Table 2. Isobutene conversion (1) values at initial, 100 and 300 min for all the samples.

Catalysts	$Conv_0$	$Conv_{100}$	$Conv_{300}$
A15	49	17	5
$STA/HSAG_{100}$	36	14	5
ZS	17	6	4
NiO/Al_2O_3	7	5	2

As was already mentioned, initial high conversion was obtained over commercial Amberlyst A15. It is evident that IBE conversion (1) rises with increasing of active sites and with the optimal distance between acid sites. Since A15 presents mostly sulfonic acid groups -SO$_3$H, it would seem logical that the catalytic results are mainly due to these Brønsted acidic sites. However, especially at the beginning of the process, the catalyst activity rapidly decreased. The conversion of IBE over A15 gradually diminished achieving a ~90% of conversion decrease, making obvious the deactivation behavior. Apparently, a higher amount of Brønsted sites or a not suitable enough distance led to the fast-initial deactivation of the catalyst. When the number of the Brønsted sites drop or the distance between them increases, the concentration of IBE adsorbed on the catalyst would decrease to a certain extent that could be beneficial to the dimerization process. In addition, the selectivity to a particular C$_8$ could increase (mainly TP2) while that towards other C$_{8s}$ decreases.

The principal cause for the catalyst deactivation seems to be that the surface of the solid acids is not capable of releasing C$_8$ products formed at the catalyst surface rapidly enough and at the same time, they favor the augment of side reactions. In this sense, the acid sites are especially active for oligomerization causing side reactions. These side species serve as starting points for the growth of carbonaceous species that ultimately masks the active sites making them inaccessible and leading to deactivation [19]. Corma and Ortega [45] also highlight that reactants' and products' adsorptions on the active sites could play a vital role in catalyst deactivation. In order to avert these limitations, most industrial processes for IBE dimerization function at conversion values between 20% and 60% [43]. The low thermal stability of sulfonic groups and their decomposition under the conditions utilized in this study could be another reason for the catalyst deactivation. However, TG profiles indicate that the sulfonic group stability is quite high under the conditions elected this study. In addition, some reaction tests involving the incorporation of steam in the reactor feed show that deactivation is not related with the loss of structural water from the A15 material, since the catalyst also ended up deactivated.

Catalytic properties will be compared with those of Amberlyst "A15". In the first place, with the purpose of clarifying the effects of surface acidity, we have compared Amberlyst (A15) and H$_4$[SiW$_{12}$O$_{40}$] (STA) supported over a high surface area graphite HSAG$_{100}$ (STA/HSAG$_{100}$) (Figure 4). Pure heteropolyacids usually display very low catalytic reactivity due to their reduced surface area. As aforementioned we had verified that HSAG did not exhibit significant activity for IBE conversion. Therefore, supported heteropoly acids, commonly prepared by their impregnation over classical high surface area materials, are adequate for catalytic reactions. Moreover, their strong acidity makes them attractive candidates to tackle the current challenges found in alkene dimerization.

In ours experiments of IBE conversion and product yield (Figures 4 and 5), the evolution with the reaction time of catalytic parameters can be compared for both catalysts (A15 and STA/HSAG$_{100}$). Both catalysts presented an initial high activity in the IBE dimerization gas phase reaction. The acidity of these materials was enough to activate this reaction, observing TP1 and TP2 as the main reaction products. However, one main aspect can be featured: A15 is initially (during the first 50 min) more active than STA/HSAG$_{100}$, the latter becoming less deactivated than A15 upon the first 50 min into the reaction. Deconvolution of TPD-NH$_3$ portrayed similar profiles at low temperature (LT) and high temperature (HT), being LT/HT similar in both samples. As refers to the acid sites, Guerrero et at. [46] studied the surface acidity of STA/HSAG in the decomposition reaction of isopropanol. The sole product obtained in this study was propylene, which results as a dehydration product favored by the presence of acidic surface centers. Therefore, from the estimation of specific activities for isopropanol, they could achieve a direct and quantitative estimate of acid sites, this being higher for STA. From these results, it was obvious that the density of acid surface sites exposed on A15 material is somewhat lower than on the STA supported sample. This subtle difference cannot justify the lower initial catalytic activity of STA/HSAG$_{100}$. Hence, this commercial material (A15) probably possesses acid surface sites not similar in nature or of different distances between neighbor surrounding acid sites, when compared with the acid sites of STA/HSAG$_{100}$.

Concerning the STA/HSAG$_{100}$ catalyst, the graphite support presents a basic nature even though it is not functionalized with any surface specie. This is due to the delocalized p-electrons present on the basal planes of carbon supports [47]. On the other hand, the heteropolyacid solids anchored could be accommodated on defective carbon surface sites or at the graphite crystallite edges, which will be impeded by the oxygen groups (of acidic nature), whose exit creates defective sites. A good dispersion of the STA on the graphite could take place by an effective interaction of STA with the basal graphitic surfaces, i.e., small crystallites sizes. Furthermore, carbon supports can facilitate the products' diffusion from the catalyst surface improving the IBE isomerization. Guerrero et al. [45] studied how for the graphite supported samples the heteropolyacid crystallites are practically undistinguished from the pristine support. It is worth emphasizing that the heteropolyacid also retains its acidity despite being supported as it was determined by NH_3 adsorption. These compounds are highly acidic when the giver is a proton (Brønsted acidic sites) and they are sometimes considered as superacidic. It could be suggested that the combination of the particularly small STA supported crystallites, nature of the acid sites (Brønsted) and the surrounding atomic layout defines the activity of the catalysts. This cooperative effect leads to C8$_s$ (mainly TP2) production.

Finally, it should be noted that the STA/HSAG$_{100}$ also suffers deactivation as a result of carbonaceous deposition. Furthermore, the comparison between both catalysts evidenced that the deactivation balance is worse in the case of A15 due to higher carbon deposits accumulation during process of the reaction on this material. In addition, STA/HSAG$_{100}$ displays higher selectivity to C$_8$ and higher catalytic durability than that of A15 in IBE dimerization. This can be attributed to the fact that the surface-appropriate acidity of former may be more effective in decreasing the interaction between the catalyst surface and coke precursors and this favors the readily desorption of coke deposits. This behavior also gives away some differences in the chemical and physical characteristics of the active sites displayed in STA/HSAG$_{100}$ with respect to those in A15. Acid supported STA and A15 have proven to be active although they deactivate rapidly. This behavior is caused by formation of polyolefins as a result of the polymerization of IBE, leading to strongly adsorbed molecules on the acid sites, especially resin material. Therefore, it is necessary to prevent this poisoning of acid sites with an appropriate design of the catalyst.

Subsequently, the above mention catalysts were compared with the sulfated zirconia (ZS) sample. Activity versus time on reaction is reported in Figure 4, which reveals a rapid decline in the activity at the beginning of the catalytic run ascribed to a partial process of deactivation. This process of deactivation has already been described for A15 and for STA/HSAG$_{100}$. After about 50 min, the activity stabilizes at values close to 15–18% remaining practically constant for at least 300 min. Note that the activity of ZS is lower than that of A15 and STA/HSAG$_{100}$. However, its deactivation is less pronounced in comparison with that of A15 or STA/HSAG$_{100}$.

Despite the fact that no a general agreement on the structure of ZS has not been achieved, it has been well established that acidic SO_3 with an adjacent Brønsted site is the acidic active site [48]. Therefore, sulfated metal oxides such as ZrO_2, have both Brønsted acid sites arising from the existing sulfate groups at the surface of the catalyst and Lewis acid sites derived from the metal oxides [49]. In our particular case of the ZS material, Lewis acidity could be inherent to Zr^{4+} sites. Both Brønsted and Lewis acid sites seem to be present on the ZS surface (Scheme 2).

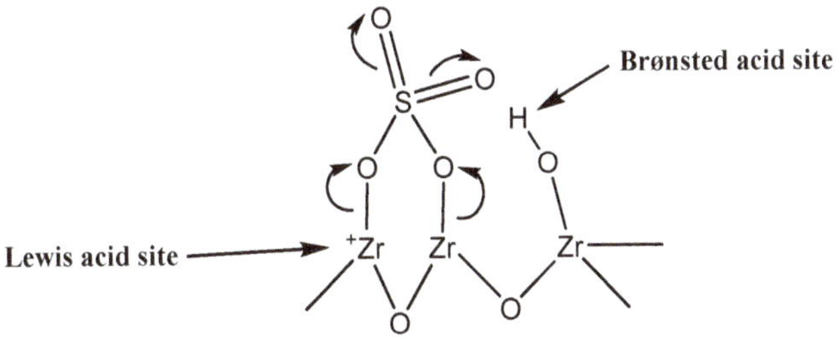

Scheme 2. Brønsted and Lewis sites on the sulfated zirconia.

Xuan Menga et at. [50] studied the surface of sulfated zirconia synthetized by sulfated zirconium hydroxides. Here, Lewis acid are the main active sites on the surface, dependent upon the calcination temperature. Protonic (Brønsted) acid sites play a pivotal role, though the existence of plenty Lewis sites (made superacidic by the surface sulfates´ existence and inductive effect therefrom) could be crucial for the development of the IBE dimerization under the limitations applied in this study. Certainly, according to the results obtained with A15, high density and strength of Brønsted sites leads to a rapid deactivation by blockage of these sites. Therefore, there must be a correlation between strong acid sites and activity [51]. ZS catalyst present both Brønsted and Lewis sites on surface but the total acid centers of the pair decreased, which is reflected in the lower activity and yet a low deactivation rate in comparison with A15 or STA/HSAG$_{100}$. Hence, it could be assumed that the acid sites (Brønsted and Lewis) must be in the accurate density and formation in order to carry out the reaction. The IBE isomerization could be thought to proceed over a Brønsted acid site with the Lewis acid site playing a supporting role. On the other hand, since the activity of ZS is low, its deactivation is also low. Now, the surface of the solid is capable of eliminating C$_8$ products from the catalyst surface. As regards the deactivation, TGA and TPD-MS experiments have shown that low thermal stability of sulfonic groups or their decomposition under the limitations imposed in this study can be discarded.

In contrast, NiO/Al$_2$O$_3$ catalysts showed a remarkably low catalytic activity, compared with the all the other catalysts (Figure 4), which reflected the ineffective strong acidity of this material. It should be recalled that blank experiments involving IBE and the bare supports (Al$_2$O$_3$) showed no catalytic activity. The γ-alumina acidic properties are well-known and usually ascribed to incompletely coordinated aluminum ions of Lewis acid nature [52]. Thermally activated alumina did not display strong Brønsted acid sites on its surface. Regarding the acidity of our catalyst (NiO/Al$_2$O$_3$), which was prepared by impregnation employing nickel (II) sulfate, it could exert certain influence on this property. The sulfate ions at the surface led to the occurrence of strong Lewis acid sites, if during calcination there is no decomposition of sulfate anion (SO$_4^{2-}$). The introduction of a certain amount of sulfur compounds to γ-alumina (verified by TPD-MS) is recognized to improve its acidity. The SO$_4^{2-}$ ions exist as a chelating bidentate ligand and enrich the Lewis metal centers' acidity (Al and/or Ni) and the catalyst would be highly active for the oligomerization of olefins [53]. On the other hand, the Ni (II) sulfate impregnation on γ-Al$_2$O$_3$ and the subsequent calcination at 500 °C could provoke to the formation of either nickel aluminate or nickel oxide. The existence of only one type of surface nickel species should not be discarded in the calcined catalyst. In terms of activity, the catalyst treatment prior to its use under hydrogen flow has indicated that metallic Ni is not active under this reaction conditions.

Normally, unsaturated Ni species interact with IBE in the course of the oligomerization reaction [53]. The surface species interact with the Ni close by, producing a strong acidity, which allows the activation of the dimerization of IBE with demand for strong acid sites.

However, NiO/Al$_2$O$_3$ shows low catalytic activity most likely due to the insufficient acidity at the catalyst surface. The addition of NiSO$_4$ as an acid precursor and the strength derived owing to this group has not induced the presence of strong Lewis acid sites and it has not permitted the activation of this demanding reaction. It should be emphasized that NiO/Al$_2$O$_3$ presents weak Lewis acid sites and that all the above discussed experimental evidences confirmed that the Brønsted-type acidity seem to be a more effective criterion in the IBE dimerization.

On the basis of the results obtained, it can be hinted that a high acid site concentration, type of acidity (Brønsted or Lewis) or distance among acidic sites must be the factors which allow the activation in the gas phase reaction. The thermal properties of these acid sites, as well as the textural properties also seem to be important. Moreover, the active adsorbed intermediated should be closely linked to the surface Brønsted/Lewis groups. The higher the acidity of the Brønsted sites, the higher is the reaction rate, which is in agreement with the following order of the total Brønsted acidity of the catalysts: A15 > STA/HSAG$_{100}$ > ZS > NiO/Al$_2$O$_3$. However, it is worth noting that the correlation between the catalyst activity and the acid loading was not linear. This may suggest that in the process of IBE dimerization, not only surface chemistry (acid loading), but also textural properties of catalysts play a significant role.

It should also be mentioned that the selectivity (or products distribution) obtained with A15 was not the same as when using STA/HSAG$_{100}$, ZS or NiO/Al$_2$O$_3$ where the main product was TP2. The results obtained clearly suggest that C$_8$ olefins formation in the reaction must be rapidly and easily detached from the active sites, which would avoid the formation of C$_{12}$ and C$_{16+}$ olefin fractions and therefore the deactivating process.

5. Conclusions

In this paper, the study of the IBE dimerization, under gas phase at moderately low temperatures, over different solid catalysts has led to outstanding results. The acidic sites present and their strengths have been associated with the catalytic activities and the respective selectivities to C$_8$. These solids have proven to be highly selectivity to C8 (TP2 and TP1), disclosing the interest of this work. Moreover, the formation of minor amounts of others C$_{8s}$ has been related with the existence of strong Brønsted sites. As a tentative explication, the solid acid catalysts release proton species that interact with close by active sites, producing a strong acidity which allows the activation of a gas phase reaction with demand for strong acid sites. In the case of Lewis sites, these seem to be less active than Brønsted sites. The catalytic performance of our catalysts might be assigned to a combination of their acidic and textural properties. Moreover, other factors such as high concentration of acid sites, thermal properties of these sites and ratio between Brønsted/Lewis sites or distance among acidic sites stimulate this reaction. On the other hand, if the acid sites are very active for the IBE dimerization, they would also favor the side polymerization reactions, modifying the surface acid capacity of the catalyst, which leads to rapid deactivation. In order to avoid limitations caused by deactivation for IBE dimerization, it is necessary to operate at relatively low conversions (20–60%) which will extend the lifespan of the catalysts. It is beneficial to prolong the durability of the catalyst. A positive effect on the catalyst stability could also be the use of an inert diluent in the IBE feed (in our case helium). This gas could be responsible for the "washing-out" of olefins stored at the active sites of the catalysts.

The high selectivity efficiency of the samples in the dimerization of IBE described in this study is of great importance as very high values (close to 100%) towards C$_8$ fraction have been reported in this work. The dimers selectivity could be significantly enhanced by effectively controlling the acid capacity, among others. However, further work is still needed in order to know the ambiguities of the surface of the acid solids.

Supplementary Materials: The following are available online at http://www.mdpi.com/2079-4991/10/6/1235/s1, Figure S1: (a) Catalytic activity at 50, 100, 150 and 180 °C for the Amberlyst; (b) Selectivity to TP2 and (c) to TP1 olefins formed in the IBE dimerization reaction.

Author Contributions: Conceptualization, methodology, writing—original draft preparation, E.C.; validation, formal analysis, investigation, writing—review and editing, E.A.-N., J.M.F.-M., A.G.-R. and A.B.D.; project administration, funding acquisition, A.G.-R. and I.R.-R. All authors have read and agreed to the published version of the manuscript.

Funding: This research received external funding from CTQ2017-89443-C3-1-R and -3-R.

Acknowledgments: We acknowledge financial support from the Spanish Government (CTQ2017-89443-C3-1-R and -3-R).

Conflicts of Interest: The authors declare no conflict of interest.

References

1. Oil and Petroleum Products—A Statistical Overview—Statistics Explained. Available online: https://ec.europa.eu/eurostat/statistics-explained/index.php?title=Oil_and_petroleum_products_-_a_statistical_overview (accessed on 2 February 2020).

2. Archive:Consumption of Energy—Statistics Explained. Available online: https://ec.europa.eu/eurostat/statistics-explained/index.php?title=Archive:Consumption_of_energy (accessed on 2 February 2020).

3. Sarathy, S.M.; Farooq, A.; Kalghatgi, G.T. Recent progress in gasoline surrogate fuels. *Prog. Energy Combust. Sci.* **2018**, *65*, 67–108. [CrossRef]

4. Mortensen, P.M.; Grunwaldt, J.D.; Jensen, P.A.; Knudsen, K.G.; Jensen, A.D. A review of catalytic upgrading of bio-oil to engine fuels. *Appl. Catal. A Gen.* **2011**, *407*, 1–19. [CrossRef]

5. Dagle, V.L.; Smith, C.; Flake, M.; Albrecht, K.O.; Gray, M.J.; Ramasamy, K.K.; Dagle, R.A. Integrated process for the catalytic conversion of biomass-derived syngas into transportation fuels. *Green Chem.* **2016**, *18*, 1880–1891. [CrossRef]

6. Antunes, B.M.; Rodrigues, A.E.; Lin, Z.; Portugal, I.; Silva, C.M. Alkenes oligomerization with resin catalysts. *Fuel Process. Technol.* **2015**, *138*, 86–99. [CrossRef]

7. Marchionna, M.; Di Girolamo, M.; Patrini, R. Light olefins dimerization to high quality gasoline components. *Catal. Today* **2001**, *65*, 397–403. [CrossRef]

8. Rorrer, J.E.; Toste, F.D.; Bell, A.T. Mechanism and Kinetics of Isobutene Formation from Ethanol and Acetone over Zn xZr yO z. *ACS Catal.* **2019**, *9*, 10588–10604. [CrossRef]

9. Hidalgo, J.M.; Zbuzek, M.; Černý, R.; Jíša, P. Current uses and trends in catalytic isomerization, alkylation and etherification processes to improve gasoline quality. *Cent. Eur. J. Chem.* **2014**, *12*, 1–13. [CrossRef]

10. Tang, S.; Scurto, A.M.; Subramaniam, B. Improved 1-butene/isobutane alkylation with acidic ionic liquids and tunable acid/ionic liquid mixtures. *J. Catal.* **2009**, *268*, 243–250. [CrossRef]

11. Di Girolamo, M.; Marchionna, M. Acidic and basic ion exchange resins for industrial applications. *J. Mol. Catal.* **2001**, *177*, 33–40. [CrossRef]

12. Liu, J.; Ge, Y.; Song, Y.; Du, M.; Zhou, X.; Wang, J.A. Dimerization of iso-butene on sodium exchanged Amberlyst-15 resins. *Catal. Commun.* **2019**, *119*, 57–61. [CrossRef]

13. Alcántara, R.; Alcántara, E.; Canoira, L.; Franco, M.J.; Herrera, M.; Navarro, A. Trimerization of isobutene over Amberlyst-15 catalyst. *React. Funct. Polym.* **2000**, *45*, 19–27. [CrossRef]

14. Mantilla, A.; Tzompantzi, F.; Ferrat, G.; López-Ortega, A.; Alfaro, S.; Gómez, R.; Torres, M. Oligomerization of isobutene on sulfated titania: Effect of reaction conditions on selectivity. *Catal. Today* **2005**, *107–108*, 707–712. [CrossRef]

15. Padilla, J.M.; Del Angel, G.; Tzompantzi, F.J.; Manrı, M.E. One pot preparation of NiO/ZrO2 sulfated catalysts and its evaluation for the isobutene oligomerization. *Catal. Today* **2008**, *135*, 154–159. [CrossRef]

16. Mantilla, A.; Ferrat, G.; López-Ortega, A.; Romero, E.; Tzompantzi, F.; Torres, M.; Ortíz-Islas, E.; Gómez, R. Catalytic behavior of sulfated TiO2in light olefins oligomerization. *J. Mol. Catal. Chem.* **2005**, *228*, 333–338. [CrossRef]

17. Davis, R.J. The Role of Transition Metal Promoters on Sulfated Zirconia Catalysts for Low-Temperature Butane Isomerization. *J. Catal.* **1996**, *133*, 125–133.

18. Querini, C.A.; Roa, E. Deactivation of solid acid catalysts during isobutane alkylation with C4 olefins. *Appl. Catal. A Gen.* **1997**, *163*, 199–215. [CrossRef]

19. Chen, G.; Li, J.; Yang, X.; Wu, Y. Surface-appropriate lipophobicity-Application in isobutene oligomerization over Teflon-modified silica-supported 12-silicotungstic acid. *Appl. Catal. A Gen.* **2006**, *310*, 16–23. [CrossRef]

20. Corma, A. Solid acid catalysts. *Curr. Opin. Solid State Mater. Sci.* **1997**, *2*, 63–75. [CrossRef]

21. Lefebvre, F. Acid Catalysis by Heteropolyacids: Transformations of Alkanes. *Curr. Catal.* **2017**, *6*, 77–89. [CrossRef]

22. Yoon, J.W.; Jhung, S.H.; Choo, D.H.; Lee, S.J.; Lee, K.Y.; Chang, J.S. Oligomerization of isobutene over dealuminated Y zeolite catalysts. *Appl. Catal. A Gen.* **2008**, *337*, 73–77. [CrossRef]

23. Yoon, J.W.; Chang, J.S.; Lee, H.D.; Kim, T.J.; Jhung, S.H. Trimerization of isobutene over a zeolite beta catalyst. *J. Catal.* **2007**, *245*, 253–256. [CrossRef]

24. Dalla Costa, B.O.; Querini, C.A. Isobutane alkylation with solid catalysts based on beta zeolite. *Appl. Catal. A Gen.* **2010**, *385*, 144–152. [CrossRef]

25. Corma, A.; Martinez, A.; Martinez, C. Isobutane/2-butene alkylation on MCM-22 catalyst. Influence of zeolite structure and acidity on activity and selectivity. *Catal. Lett.* **1994**, *28*, 187–201. [CrossRef]

26. Corma, A.; Martínez, A. Chemistry, catalysts, and processes for isoparaffin-olefin alkylation: Actual situation and future trends. *Catal. Rev.* **1993**, *35*, 483–570. [CrossRef]

27. Fernández-Morales, J.M.; Lozano, L.A.; Castillejos-López, E.; Rodríguez-Ramos, I.; Guerrero-Ruiz, A.; Zamaro, J.M. Direct sulfation of a Zr-based metal-organic framework to attain strong acid catalysts. *Microporous Mesoporous Mater.* **2019**, *290*, 109686. [CrossRef]

28. Trickett, C.A.; Osborn Popp, T.M.; Su, J.; Yan, C.; Weisberg, J.; Huq, A.; Urban, P.; Jiang, J.; Kalmutzki, M.J.; Liu, Q.; et al. Identification of the strong Brønsted acid site in a metal–organic framework solid acid catalyst. *Nat. Chem.* **2019**, *11*, 170–176. [CrossRef] [PubMed]

29. Ehrmaier, A.; Liu, Y.; Peitz, S.; Jentys, A.; Chin, Y.H.C.; Sanchez-Sanchez, M.; Bermejo-Deval, R.; Lercher, J. Dimerization of Linear Butenes on Zeolite-Supported Ni 2+. *ACS Catal.* **2019**, *9*, 315–324. [CrossRef]

30. Ehrmaier, A.; Peitz, S.; Sanchez-Sanchez, M.; Bermejo-Deval, R.; Lercher, J. On the role of co-cations in nickel exchanged LTA zeolite for butene dimerization. *Microporous Mesoporous Mater.* **2019**, *284*, 241–246. [CrossRef]

31. Comito, R.J.; Metzger, E.D.; Wu, Z.; Zhang, G.; Hendon, C.H.; Miller, J.T.; Dincă, M. Selective Dimerization of Propylene with Ni-MFU-4l. *Organometallics* **2017**, *36*, 1681–1683. [CrossRef]

32. Almohalla, M.; Rodríguez-Ramos, I.; Guerrero-Ruiz, A. Comparative study of three heteropolyacids supported on carbon materials as catalysts for ethylene production from bioethanol. *Catal. Sci. Technol.* **2017**, *7*, 1892–1901. [CrossRef]

33. Álvarez-Rodríguez, J.; Rodríguez-Ramos, I.; Guerrero-Ruiz, A.; Gallegos-Suarez, E.; Arcoya, A. Influence of the nature of support on Ru-supported catalysts for selective hydrogenation of citral. *Chem. Eng. J.* **2012**, *204–205*, 169–178. [CrossRef]

34. Lónyi, F.; Valyon, J. On the interpretation of the NH 3 -TPD patterns of. *Microporous Mesoporous Mater.* **2001**, *47*, 293–301. [CrossRef]

35. Wijaya, Y.P.; Winoto, H.P.; Park, Y.K.; Suh, D.J.; Lee, H.; Ha, J.M.; Jae, J. Heteropolyacid catalysts for Diels-Alder cycloaddition of 2,5-dimethylfuran and ethylene to renewable p-xylene. *Catal. Today* **2017**, *293–294*, 167–175. [CrossRef]

36. Li, N.; Wang, A.; Zheng, M.; Wang, X.; Cheng, R.; Zhang, T. Probing into the catalytic nature of Co/sulfated zirconia for selective reduction of NO with methane. *J. Catal.* **2004**, *225*, 307–315. [CrossRef]

37. Charisiou, N.D.; Iordanidis, A.; Polychronopoulou, K.; Yentekakis, I.V.; Goula, M.A. Studying the stability of Ni supported on modified with CeO_2 alumina catalysts for the biogas dry reforming reaction. *Mater. Today Proc.* **2018**, *5*, 27607–27616. [CrossRef]

38. Wang, A.; Arora, P.; Bernin, D.; Kumar, A.; Kamasamudram, K.; Olsson, L. Investigation of the robust hydrothermal stability of Cu/LTA for NH3-SCR reaction. *Appl. Catal. B Environ.* **2019**, *246*, 242–253. [CrossRef]

39. Katada, N.; Tsubaki, T.; Niwa, M. Measurements of number and strength distribution of Brønsted and Lewis acid sites on sulfated zirconia by ammonia IRMS-TPD method. *Appl. Catal. A Gen.* **2008**, *340*, 76–86. [CrossRef]

40. Saboya, R.M.A.; Cecilia, J.A.; García-Sancho, C.; Sales, A.V.; de Luna, F.M.T.; Rodríguez-Castellón, E.; Cavalcante, C.L. Assessment of commercial resins in the biolubricants production from free fatty acids of castor oil. *Catal. Today* **2017**, *279*, 274–285. [CrossRef]

41. Fan, G.; Liao, C.; Fang, T.; Luo, S.; Song, G. Amberlyst 15 as a new and reusable catalyst for the conversion of cellulose into cellulose acetate. *Carbohydr. Polym.* **2014**, *112*, 203–209. [CrossRef]

42. Ouni, T.; Honkela, M.; Kolah, A.; Aittamaa, J. Isobutene dimerisation in a miniplant-scale reactor. *Chem. Eng. Process. Process Intensif.* **2006**, *45*, 329–339. [CrossRef]

43. Goortani, B.M.; Gaurav, A.; Deshpande, A.; Ng, F.T.T.; Rempel, G.L. Production of isooctane from isobutene: Energy integration and carbon dioxide abatement via catalytic distillation. *Ind. Eng. Chem. Res.* **2015**, *54*, 3570–3581. [CrossRef]

44. Saus, A.; Schmidl, E. Benzyl sulfonic acid siloxane as a catalyst: Oligomerization of isobutene. *J. Catal.* **1985**, *94*, 187–194. [CrossRef]

45. Corma, A.; Ortega, F.J. Influence of adsorption parameters on catalytic cracking and catalyst decay. *J. Catal.* **2005**, *233*, 257–265. [CrossRef]

46. García Bosch, N.; Bachiller-Baeza, B.; Rodriguez-Ramos, I.; Guerrero-Ruiz, A. Fructose transformations in ethanol using carbon supported polyoxometalate acidic solids for 5-Etoxymethylfurfural production. *ChemCatChem* **2018**. [CrossRef]

47. Bringué, R.; Ramírez, E.; Iborra, M.; Tejero, J.; Cunill, F. Influence of acid ion-exchange resins morphology in a swollen state on the synthesis of ethyl octyl ether from ethanol and 1-octanol. *J. Catal.* **2013**, *304*, 7–21. [CrossRef]

48. Yan, G.X.; Wang, A.; Wachs, I.E.; Baltrusaitis, J. Critical review on the active site structure of sulfated zirconia catalysts and prospects in fuel production. *Appl. Catal. A Gen.* **2019**, *572*, 210–225. [CrossRef]

49. El-Dafrawy, S.M.; Hassan, S.M.; Farag, M. Kinetics and mechanism of Pechmann condensation reaction over sulphated zirconia-supported zinc oxide. *J. Mater. Res. Technol.* **2020**, *9*, 13–21. [CrossRef]
50. Liu, N.; Ma, Z.; Wang, S.; Shi, L.; Hu, X.; Meng, X. Palladium-doped sulfated zirconia: Deactivation behavior in isomerization of n-hexane. *Fuel* **2020**, *262*, 116566. [CrossRef]
51. Ecormier, M.A.; Wilson, K.; Lee, A.F. Structure-reactivity correlations in sulphated-zirconia catalysts for the isomerisation of α-pinene. *J. Catal.* **2003**, *215*, 57–65. [CrossRef]
52. Knözinger, H.; Ratnasamy, P. Catalysis Reviews: Science and Engineering Catalytic Aluminas: Surface Models and Characterization of surface sites. *Catal. Rev.* **1978**, *17*, 31–70. [CrossRef]
53. Sarkar, A.; Seth, D. Active Sites of a NiSO4/γ-Al2O3 Catalyst for the Oligomerization of Active Sites of a NiSO 4 / c -Al 2 O 3 Catalyst for the Oligomerization of Isobutene. *Top. Catal.* **2014**. [CrossRef]

Article

Upgrading the Properties of Reduced Graphene Oxide and Nitrogen-Doped Reduced Graphene Oxide Produced by Thermal Reduction toward Efficient ORR Electrocatalysts

Carolina S. Ramirez-Barria [1,2], Diana M. Fernandes [3,*], Cristina Freire [3], Elvira Villaro-Abalos [4], Antonio Guerrero-Ruiz [2,*] and Inmaculada Rodríguez-Ramos [1,*]

[1] Instituto de Catálisis y Petroleoquímica, CSIC, Marie Curie 2, 28049 Madrid, Spain; ramirez.carolina@outlook.com
[2] Dpto. Química Inorgánica y Técnica, Facultad de Ciencias UNED, Senda del Rey 9, 28040 Madrid, Spain
[3] REQUIMTE/LAQV, Departamento de Química e Bioquímica, Faculdade de Ciências, Universidade do Porto, 4169-007 Porto, Portugal; acfreire@fc.up.pt
[4] Interquimica, Antonio de Nebrija 8, 26006 Logroño, Spain; evillaro@interquimica.org
* Correspondence: diana.fernandes@fc.up.pt (D.M.F.); aguerrero@ccia.uned.es (A.G.-R.); irodriguez@icp.csic.es (I.R.-R.); Tel.: +34-915854765 (I.R.-R.)

Received: 21 November 2019; Accepted: 9 December 2019; Published: 11 December 2019

Abstract: N-doped (NrGO) and non-doped (rGO) graphenic materials are prepared by oxidation and further thermal treatment under ammonia and inert atmospheres, respectively, of natural graphites of different particle sizes. An extensive characterization of graphene materials points out that the physical properties of synthesized materials, as well as the nitrogen species introduced, depend on the particle size of the starting graphite, the reduction atmospheres, and the temperature conditions used during the exfoliation treatment. These findings indicate that it is possible to tailor properties of non-doped and N-doped reduced graphene oxide, such as the number of layers, surface area, and nitrogen content, by using a simple strategy based on selecting adequate graphite sizes and convenient experimental conditions during thermal exfoliation. Additionally, the graphenic materials are successfully applied as electrocatalysts for the demanding oxygen reduction reaction (ORR). Nitrogen doping together with the starting graphite of smaller particle size (NrGO$_{325}$-4) resulted in a more efficient ORR electrocatalyst with more positive onset potentials (E_{onset} = 0.82 V versus RHE), superior diffusion-limiting current density ($j_{L, 0.26V, 1600rpm}$ = −4.05 mA cm^{-2}), and selectivity to the direct four-electron pathway. Moreover, all NrGO$_m$-4 show high tolerance to methanol poisoning in comparison with the state-of-the-art ORR electrocatalyst Pt/C and good stability.

Keywords: graphite; reduced graphene oxide; nitrogen-doped reduced graphene oxide; exfoliation; oxygen reduction reaction; electrocatalysis

1. Introduction

In recent years, the interest in graphene has grown due to its many outstanding electronic, thermal, chemical, and mechanical properties [1]. These properties give graphene an enormous versatility. It can be extensively used in diverse applications including energy conversion and storage technologies [2–5], electronics [6], and sensing device applications [7]. Graphenic materials have also been widely employed as solid (electro)catalytic materials, either as active phases or as supports [8]. These (electro)catalytic applications strongly depend on its surface chemical properties. It is well known that the surface chemical properties of graphenic materials can be tuned by covalent added adatoms [9]. Thus, the presence of nitrogen or boron atoms in the basal planes of graphene layers produce changes to many of their chemical properties. It was proven that the doping of nitrogen into a graphenic nanostructure modifies its chemical and electrical properties, [10]. Thus, the N atoms inserted in the graphene structure increase the electronic density of the graphenic layer, since one more electron is added to the graphenic layer, making it more basic [11,12]. In addition, the easily tunable surface chemical properties and specific surface area of the graphenic materials make them very versatile materials for application in this field.

An extensive number of works have been devoted to the preparation of graphene and N-doped graphene. The synthesis of graphene can be performed using different methods, among them the micromechanical cleavage method [13], chemical vapor deposition (CVD) [14], and epitaxial growth on silicon carbide surfaces [15]. However, they show low productivity and lack of properties' selectivity. One of the common approaches used for a large-scale graphene production is based on the chemical oxidation of graphite (G) flakes to produce graphite oxide (GO) [16,17] with strong oxidizing agents. GO can be later exfoliated and converted to reduced graphene oxide (rGO) through a reduction procedure. The most used routes for the reduction of GO are chemical [18,19], electrochemical [20,21], solvothermal [22,23], and thermal treatments [24]. The use of thermal reduction is preferred over the rest of the methods due to its simplicity and industrial scalability [25].

The incorporation of nitrogen in graphenic materials to produce N-doped reduced graphene oxide (NrGO) has been described using several methods as well. For example, CVD using NH_3 [26], acetonitrile [27], or pyridine [28] as the N source; the arc discharge of graphite [29] in the presence of pyridine or NH_3; the nitrogen plasma treatment of graphene [30]; and the thermal treatment of GO with melamine [31,32], urea [33], or NH_3 [9,34].

The improvement of some of the properties of rGO and NrGO has been undertaken using different synthetic strategies. Wu et al. [35] studied a chemical exfoliation of GO to produce reduced graphene oxide with a selective number of layers based on different starting materials such as pyrolytic graphite, natural flake graphite, Kish graphite, flake graphite powder, and artificial graphite. They reported the effect of the lateral size and crystallinity of these starting graphite materials on the number of graphene layers presented in the obtained graphenic material. They found that graphite samples with a small lateral size and low crystallinity produce a higher proportion of single layer graphene. Li et al. [34] obtained NrGO through the thermal annealing of GO in NH_3 sweeping temperatures between 300 and 1100 °C. Annealing at 500 °C afforded the highest N-doping level of ~5%, showing the strong influence of the temperature on the nitrogen content of the obtained materials. They claim that the N-doping degree depends on the amount of oxygen functional groups of graphene as they are responsible for the formation of C–N bonds. The higher the annealing temperature is, the lower the content of oxygen, leading to a lower reactivity between the graphene layer and NH_3. The production of NrGO based on the annealing of GO in the presence of melamine at high temperature (700–1000 °C) has been reported by Sheng et al. [32]. They noted that nitrogen content depends on the mass ratio between GO and melamine as well as on the temperature, reaching values of 10.1 at % using a 1:5 ratio of GO to melamine at 700 °C. A similar approach was used by Canty et al. [33] working with urea and GO as precursor materials of NrGO. They used the ratio of GO to urea as a way to control the amount of nitrogen inserted and the surface area values obtained. Nonetheless, neither Li et al. [34] nor Canty et al. [33] reported a systematic study of the synthesis conditions to optimize NrGO properties. Menendez's group [25] analyzed the effect of the experimental conditions of the thermal transformation of GO onto rGO, finding that the treatment temperature strongly affects the type and amount of functional groups obtained. Following this line of research, they proposed that the temperature of the initial flash thermal treatment allows the control of the surface area obtained [36]. However, the surface area values achieved were lower than 500 m^2g^{-1}. Zhang et al. [37] proposed a vacuum-promoted thermal exfoliation method for different samples of GO, which were obtained from natural flake graphites with particle sizes ranging from 100 to 5000 mesh. Nevertheless, neither the morphology nor the structures of different graphenic samples were affected by the parent graphite particle size. The surface areas observed were around 490 m^2g^{-1} and the C/O ratio determined by XPS analysis revealed that all the samples had almost the same oxygen content. Thus, the vacuum-promoted exfoliation method minimizes the differences originated from the raw graphite particle size. The effect of the raw graphite size on the rGO properties has been also described by Dao et al. [38]. They prepared rGO by the rapid heating of dry GO using three graphite particle sizes obtained from grinding a large graphite sample. An increase in the surface area was observed as the particle size of the samples was reduced, reaching a value of 739 m^2g^{-1} in their best sample. This improvement was explained as being due to the better oxidation degree achieved with the decrease of the graphite particle size, which also favors a better exfoliation of the GO.

On the other hand, a previous study of our group [10] reported the synthesis of NrGO from GO obtained from three different graphite particle sizes. We found that the quantity of nitrogen and the surface properties obtained were dependent on the particle size of the graphite used. However, the surface areas obtained were not optimized enough.

Due to the lack of simultaneous and systematic reports regarding the combined effect of raw graphite size and the conditions of the thermal treatment (heating rate and temperature) on the properties of non-doped and N-doped reduced graphene oxide, we have investigated the preparation of graphenic materials using three different particle sizes of starting graphite and using various thermal treatments. Moreover, a deep characterization of the graphenic materials was performed before applying those as electrocatalysts (ECs) for the oxygen reduction reaction, in order to optimize the properties of the obtained materials—particularly, the surface area and content of nitrogen—due to the importance of these parameters in their electrocatalytic performance. Finally, the synthesized materials have been tested as ECs in the oxygen reduction reaction (ORR).

2. Materials and Methods

2.1. Preparation and Characterization of Graphenic Materials

Graphenic materials were prepared by the treatment of graphite oxide (GO) at high temperature. Natural graphite powders of various grain sizes—10 mesh, 100 mesh, and 325 mesh (supplied by Alfa Aesar, Thermo Fisher Scientific, UK, purity 99.8%) were used for the synthesis of GO through a modified Brodie's method [17]. This procedure consists of the addition of 10 g of graphite (G) over 200 mL of fuming HNO_3, keeping the mixture at 0 °C. First, 80 g of $KClO_3$ was gradually added over 2 h. Afterward, the mixture was stirred for 21 h, maintaining the 0 °C temperature. The GO obtained was filtered and washed systematically with water until neutral pH and dried under vacuum at room temperature. The resultant samples were labeled GO_m, where m indicates the mesh size used.

The GO_m were exfoliated in a vertical quartz reactor under two different atmospheres. The first atmosphere consists of nitrogen (87 mL/min) producing rGO, while the second was a mixture of NH_3, H_2, and N_2 (10, 3, and 87 mL/min, respectively) generating NrGO.

In order to study the effect of temperature and heating rate on the properties of rGO and NrGO, for each atmosphere described, five different exfoliation ramps were applied over GO_{325}. In a first ramp, 0.3 g of GO was introduced in the furnace and heated at 5 °C min^{-1} to 250 °C; then, the sample was kept at this temperature for 30 min. The temperature was increased from 250 to 500 °C with a heating rate of 5 °C min^{-1} and then kept at this temperature for 30 minutes. In a second and third ramp, GO was heated at 5 °C min^{-1} and 10 °C/min respectively to 250 °C; then, the sample was kept at this temperature for 30 min. The temperature was increased from 250 to 700 °C using the same heating rates, and then this temperature was maintained for 30 min. In a fourth ramp, GO was heated at 5 °C min^{-1} to 100 °C; then, the sample was kept at this temperature for 1 h. The temperature was increased from 100 to 700 °C with a heating rate of 10 °C min^{-1}; then, the sample was kept at this temperature for 5 min. Finally, in the fifth ramp, the GO was heated up to 250 °C at 20 °C min^{-1}; then, the sample was kept at this temperature for 30 min. The temperature was subsequently increased from 250 to 500 °C with a heating rate of 20 °C min^{-1}; then, the sample was kept at this temperature for 30 min.

For the exfoliation of the samples GO_{10} and GO_{100}, the temperature and heating rate used were selected according to the ramp that gave the higher value of surface area among the exfoliated samples of GO_{325}. The samples obtained were labeled rGO_m-r and $NrGO_m$-r, respectively, where m indicates the mesh size used and r indicates the ramp used.

The characterization of the as-prepared materials was carried out by different techniques: elemental analysis, nitrogen adsorption isotherms, X-ray diffraction (XRD), scanning electron microscopy (SEM), transmission electron microscopy (TEM), and X-ray photoelectron spectroscopy (XPS). The detailed equipment and methods used are described in the electronic supporting information (ESI) file.

2.2. ORR Electrocatalytic Activities

For the cyclic voltammetry (CV) and linear sweep voltammetry (LSV) experiments, an Autolab PGSTAT 302N potentiostat/galvanostat (EcoChimie B.V. Netherlands) was used. A modified glassy carbon rotating disk electrode, RDE ($d = 3$ mm, Metrohm, Switzerland) was used as the working electrode, a carbon rod ($d = 2$ mm, Metrohm, Switzerland) was used as the counter, and an Ag/AgCl (3 mol dm^{-3} KCl, Metrohm, Switzerland) was used as the reference electrode.

Before any type of modification, the electrode was cleaned (detailed information in ESI). The electrode modification was achieved by dropping 2×2.5 µL of electrocatalyst dispersion onto the RDE surface followed by solvent evaporation under a flux of hot air. To prepare the electrocatalyst dispersion, 1 mg of material was mixed with 20 µL of Nafion, 125 µL of ultrapure water, and 125 µL of isopropanol. Then, the mixture was sonicated for 15 min.

The CV and LSV tests were performed in KOH electrolyte (0.1 mol dm^{-3}) saturated in O_2 and N_2 (30 min purge for each gas). The potential range used for CV and LSV tests was between 0.26 and 1.46 V versus reversible hydrogen electrode (RHE) at a scan rate of 0.005 V s^{-1}. In addition, for the LSV tests, rotation speeds between 400 and 3000 rpm were used. The effective ORR current was determined through the subtraction of the current obtained in KOH saturated with N_2 by that saturated with O_2. The stability was evaluated by chronoamperometry (CA) for 20,000 s at 0.5 V versus RHE and 1600 rpm. CA was also used to determine the tolerance to methanol crossover by applying a $E = 0.5$ V versus RHE for 2000 s and a 1600 rpm rotation speed.

All relevant ORR parameters (effective currents, diffusion-limiting current densities (j_L), onset potential (E_{onset}), Tafel slope, and the number of electrons transferred for each O_2 molecule ($\tilde{n}_{O2}$)) were calculated as described in the ESI file.

Experiments with the rotating ring disk electrode (RRDE) were also conducted in KOH electrolyte saturated with O_2 to determine the amount of H_2O_2 produced. This was achieved using Equation (1) where i_R and i_D correspond to the ring and disk currents, respectively, and N is the current collection efficiency of the Pt ring ($N = 0.25$) [39].

$$\% \ H_2O_2 \ = \ 200 \times \frac{i_R / N}{i_D + i_R/N},\tag{1}$$

3. Results and Discussion

3.1. Characteristics of the Graphenic Materials

XRD is a powerful tool to evaluate the interlayer changes of graphene-based materials. The XRD patterns of G_{10}, G_{100}, G_{325}, GO_{10}, GO_{100}, and GO_{325} samples are shown in Figure 1. The G samples presented the characteristic diffraction peak corresponding to pristine graphite (002) reflection at $2\theta{\sim}26°$ [10].

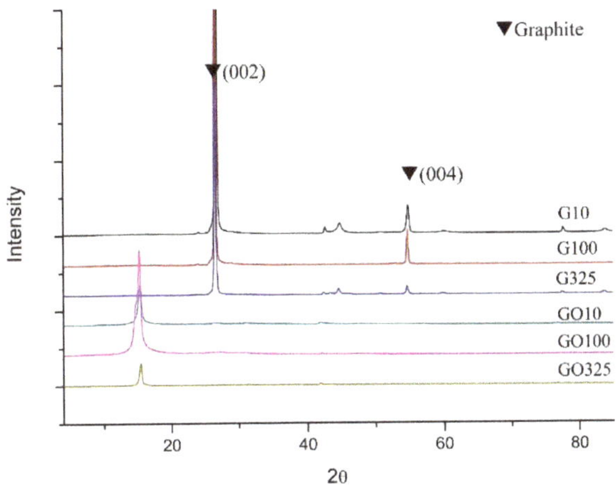

Figure 1. XRD patterns of graphene (G$_m$) and graphite oxide (GO$_m$), where m indicates the mesh samples used.

A downshift for (002) reflection peak to 2θ~15° after the oxidation treatment was observed for all the GO samples. It indicates that a successful oxidization of all G samples was achieved [8]. The distance between layers (d$_{(002)}$) increased from 0.33 nm (for G) to 0.57 nm (for GO samples), which is due to the presence of interlayered species incorporated during the oxidation of graphite [25].

The morphology of the GOs was observed by SEM. Figure 2 shows representative images of the GO samples. The analysis of the particle size distribution based on a minimum of 200 particles shows that the average particle sizes of GO$_{10}$, GO$_{100}$, and GO$_{325}$ samples are 86, 38, and 25 μm, respectively.

Slow heating rates were selected to be evaluated during the exfoliation process of GOs, since fast heating rates produce more wrinkled sheets [25]. Thus, small heating rates are fast enough to produce an effective expansion allowing the exfoliation and minimizing the distortion of the graphene sheets. It is known that oxygen groups decompose at high temperatures, reducing the number of reactive sites for N doping [34], and that high annealing temperatures (>700 °C) could break C–N bonds in the NrGO materials leading to a low doping level [40]. So, temperatures up to 700 °C were used in the thermal treatments applied to the GO samples.

Table 1 compiles the main characteristics of the non-doped and N-doped reduced graphite oxides obtained. Firstly, the GO$_{325}$ sample was submitted in both inert atmosphere and atmosphere with the presence of ammonia to five different thermal ramps described in the experimental section. Application of the BET method to N$_2$ adsorption isotherms (these type IV isotherms were displayed in the Figure S1 of the Electronic Supplementary Information, ESI) measured over the GO$_{325}$-derived non-doped and N-doped reduced graphene oxides gave surface area (S$_{BET}$) values ranging from 667 to 867 m^2 g^{-1} for the different rGO$_{325}$ samples and from 427 to 492 m^2 g^{-1} for the NrGO$_{325}$ samples. These obtained S$_{BET}$ values are significantly lower than the theoretical value calculated for a single layer of graphene (2630 m^2 g^{-1}) [8]. This finding indicates the piling up of graphene layers and the formation of a few-layer graphene structure for both rGO and NrGO. However, these values are much higher than the values previously reported using thermal exfoliation to produce rGO [36,38] and NrGO [10]. From the results shown in Table 1, it can be seen that ramp 3 for non-doped reduced graphene oxide (rGO$_{325}$) and ramp 4 for N-doped reduced graphene oxide (NrGO$_{325}$) led to reduced materials having an enhanced S$_{BET}$. The ramps used for the exfoliation of GO$_{10}$ and GO$_{100}$ were selected based on these results. Thus, the ramp used for the preparation of rGO$_{10}$ and rGO$_{100}$ was ramp 3, and for NrGO$_{10}$ and NrGO$_{100}$ was ramp 4. A significant and gradual decrease of the surface area values was observed for rGO and NrGO sample series as the size of starting graphite increases. These findings are coherent with the tendency previously reported for non-doped graphene by Dao et al. [38] where the lower the starting G size, the higher the oxidation degree of the obtained GO, which could lead to a better exfoliation of rGO.

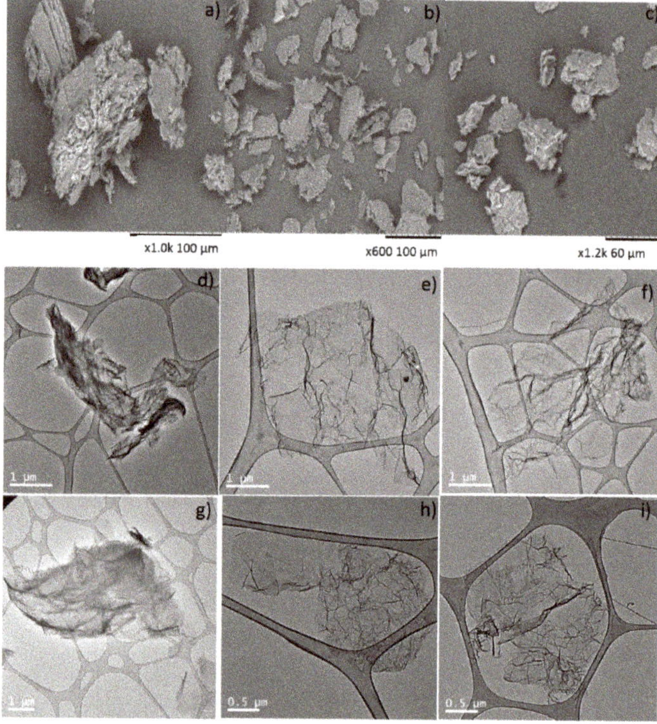

Figure 2. Representative SEM images of (**a**) GO_{10}, (**b**) GO_{100} and (**c**) GO_{325} samples. TEM micrographs for (**d**) rGO_{10}-3, (**e**) rGO_{100}-3, (**f**) rGO_{325}-3, (**g**) $NrGO_{10}$-4, (**h**) $NrGO_{100}$-4 and (**i**) $NrGO_{325}$-4 samples. NrGO: N-doped reduced graphene oxide, rGO: reduced graphene oxide.

Table 1. Interlayer distance $d_{(002)}$, estimated number of layers (N_L), S_{BET}, and N content (%) for N doped and non-doped reduced graphene oxide samples.

GO_m	Ramp	Atmosphere	Sample	$d_{(002)}$ (nm)	N_L*	S_{BET} (m^2 g^{-1})	N (%)
GO_{325}	1	Inert	rGO_{325}-1	0.35	13	767	-
	2		rGO_{325}-2	0.35	10	804	-
	3		rGO_{325}-3	0.34	12	867	-
	4		rGO_{325}-4	0.35	17	667	-
	5		rGO_{325}-5	0.36	26	866	-
GO_{100}	3	Inert	rGO_{100}-3	0.35	10	778	-
GO_{10}	3	Inert	rGO_{10}-3	0.34	18	505	-
GO_{325}	1	Ammonia	$NrGO_{325}$-1	0.34	14	428	4.8
	2		$NrGO_{325}$-2	0.35	9	427	4.4
	3		$NrGO_{325}$-3	0.34	14	460	5.0
	4		$NrGO_{325}$-4	0.34	14	492	5.0
	5		$NrGO_{325}$-5	0.34	13	476	4.1
GO_{100}	4	Ammonia	$NrGO_{100}$-4	0.35	10	420	3.8
GO_{10}	4	Ammonia	$NrGO_{10}$-4	0.34	40	236	1.8

* $N_L = (L_{002} + d_{002})/d_{002}$.

The completion of the exfoliation process for the prepared graphene materials was investigated by XRD. Examination of the rGO_{325}-r and $NrGO_{325}$-r patterns (Figure S2 ESI) indicates that the selected exfoliation ramps, 3 (inert) and 4 (ammonia), conduce to a successful exfoliation of the GO325 sample since the characteristic diffraction peak at 2θ~15° of GO (see Figure 1) disappeared, suggesting a reduction of GO325 to NrGO325 and rGO325 samples [10]. In addition, the absence of reflections corresponding to crystalline graphite for GO325-3 and NrGO325-4 samples supports the adequacy of ramps 3 (inert) and 4 (ammonia). The greater or lesser success in the GO exfoliation process can be evaluated by the appearance or not of crystalline graphite reflections. Figure 3 shows the XRD patterns of $NrGO_m$-4 and rGO_m-3. It can be seen that the exfoliation process was fully successful in inert atmosphere for all the GO and also in ammonia atmosphere, especially for the GO_{100} and GO_{325} samples, because only a small and broad peak appeared at 2θ slightly lower than 26°, corresponding to the graphite (002) reflection. This fact indicates that the sample contains some small restacking of graphene layers and the formation of a few-layer graphene structures. The intensity of this peak displays a progressive increase with the particle size of the GO starting material, showing a higher number of restacked layers ($rGO_{10} > rGO_{100} \geq rGO_{325}$). Particularly, the progressive association of the graphene sheets produced after blasting is favored under ammonia reactive conditions but without reaching the level typical for crystalline graphite. This tendency is coherent with the values of S_{BET}. The average stacking number of graphene layers (N_L) in the exfoliated samples was estimated by using the layer-to-layer distance ($d_{(002)}$) and the size of the crystallite measured from the width of the diffraction peak of the (002) reflection, using the Scherrer equation [12]. The N_L calculated show higher values than the ones expected according to the S_{BET} obtained. This may be due to the fact that the XRD signal is strongly influenced by the thicker particles, because although these constitute a minor proportion in the total of the sample, they are the ones with the highest diffraction signal. The X-ray crystalline parameters are shown in Table 1.

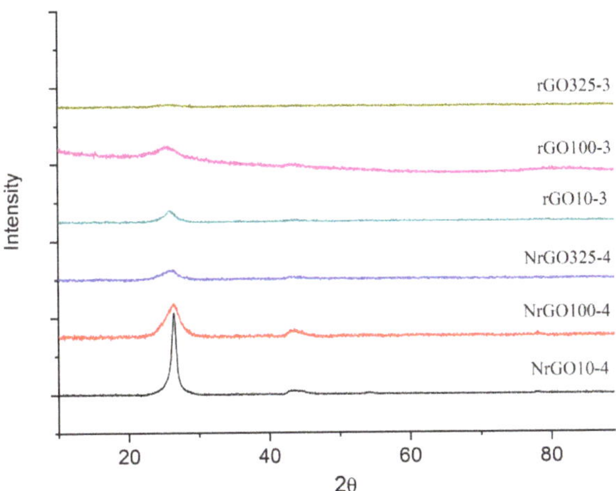

Figure 3. XRD patterns of reduced graphene materials.

Some morphological evidences of the differences between rGO and NrGO were determined by TEM (Figure 2). The rGO TEM images exhibited the presence of winkled structures of graphene. The introduction of nitrogen in the graphitic structure—the NrGO samples—did not produce noticeable difference in the morphology of the graphene sheets. High resolution transmission electron microscopy (HRTEM) characterization further showed that rGO_m-3 and $NrGO_m$-4 samples consist of 5–12 graphene layers (Figure S3).

The elemental analysis technique was used to determine the amount of nitrogen atoms in the graphene materials (see Table 1 for N wt %). For the rGO sample, nitrogen was not found. The

NrGO$_{325}$-4 sample presents the higher N content among the NrGO samples with 5.0 wt %, and a gradual decrease of the N content values was observed as the size of starting graphite increases. There are two explanations for this phenomenon. First, this tendency can be explained in terms of the higher degree of oxidation obtained in GO samples with smaller graphite particle sizes [38]. It is known that the oxygen functional groups in GO including carbonyl, carboxylic, lactone, and quinone groups are responsible for reacting with NH$_3$ to form C–N bonds, allowing the incorporation of N in the structure [34]. Thus, a smaller graphite particle size favors the formation of GO with a higher degree of oxidation, which could lead to higher N contents. Besides, the higher degree of exfoliation of small crystals also facilitates the contact of the ammonia with the oxygen groups in the basal plane of the sheets, leading to a better incorporation of N atoms.

XPS is a powerful technique to identify the chemical states of the surface species. It was used to analyze the surface of the different graphene materials from the characteristic XPS peaks corresponding to C, N, and O regions (Figure S4 ESI shows general XPS spectra). The results obtained from the analysis of the C1s, O1s, and N1s individual high-resolution spectra are shown in Table 2. The assignment of the components of the N1s, O1s, and C1s region is not straightforward. The value of binding energy observed for the different functional groups of these elements varies in the literature. It may be due to the specific environments of the atoms and the redistribution of electrons after the ionization of the sample [30,41–43]. Nitrogen peak deconvolution for the NrGO$_m$-4 samples (Figure 4) indicated the presence of four elementary peaks: pyridinic nitrogen (399.5–398.5 eV), pyrrolic nitrogen (400.8–399.8 eV), quaternary nitrogen (403.0–401.0 eV), and NOx groups (404.9–405.6 eV) [10,44–48]. XPS analysis indicated that about 3.2–3.4 at % N was found in the surface of the graphene sheets after ammonia treatment. These values are lower than those corresponding to the bulk N% content obtained by elemental analysis. This difference could be attributed to inhomogeneous nitrogen doping of the graphenic materials, since it was inferior at the surface analyzed by XPS. For rGO samples, the nitrogen peak was undetected. The percentage of pyridinic N species was slightly higher for the samples obtained from smaller sizes and quaternary nitrogen was higher for the samples obtained from bigger ones. The pyridinic nitrogen is the most basic among the different N species. Therefore, the basicity of carbon catalysts is related to the content of pyridinic groups [48,49].

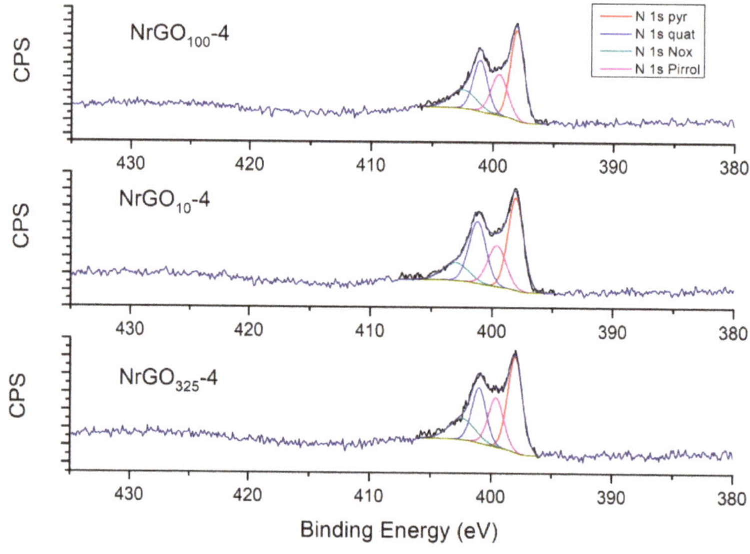

Figure 4. XPS spectra of the N 1s region for NrGO100-4, NrGO10-4, and NrGO325-4 samples.

Table 2. XPS deconvolution results and RAMAN I_D/I_G ratio for rGO$_m$-3 and NrGO$_m$-4 samples.

Sample	O (at %)	N (at %)	O 1s				N1s				I_D/I_G
			C–O	C=O	COOH	H$_2$O	Pyr	Pyrr	Quat	Nox	
rGO$_{325}$-3	7.0	-	57.0	21.1	15.7	6.2	-	-	-	-	0.63
rGO$_{100}$-3	7.0	-	63.1	17.0	13.7	6.3	-	-	-	-	0.39
rGO$_{10}$-3	6.8	-	54.6	20.7	18.7	6.0	-	-	-	-	0.53
NrGO$_{325}$-4	4.1	3.4	54.7	23.5	17.6	4.2	40.0	21.7	22.7	15.6	0.75
NrGO$_{100}$-4	3.6	3.3	54.2	21.2	18.9	5.7	39.5	21.7	23.6	15.3	0.63
NrGO$_{10}$-4	3.6	3.2	56.2	17.4	18.7	7.6	38.5	20.5	28.5	12.4	0.58

The C1s spectra was solved considering five components, which can be assigned to graphitic sp^2 carbon atoms (284.6–285.1 eV), the C–O bonds present in alcohol or ether groups (286.3–287.0 eV), C=O functional groups (287.5–288.1 eV), carboxyl or ester groups (289.3–290 eV), and a fifth wide shake-up satellite peak representing the π–π* transitions of aromatic rings (291.2–292.1 eV) [46,50].

Concerning O1s, the curve was fitted considering four contributions corresponding to carbonyl groups (531.1–531.8 eV), epoxide and hydroxyl groups (532.3–533.3 eV), carboxylic groups (534.0–534.4 eV), and chemisorbed H$_2$O or oxygen (535.5–536.1 eV) [43,46,50–52]. The spectra analysis revealed that surface oxygen content varies from 3.6–4.1 at % to 6.8–7.0 at % for N-doped and non-doped samples, respectively. The relatively small amount of oxygen with respect to that of carbon may be attributed to the process temperature. It is known that higher temperatures during the thermal treatment produce a decrease of the amount of oxygen functional groups [25]. The oxygen/carbon atomic ratios also confirm this, as can be seen in Table 2. The differences in these ratios for the various samples can be ascribed to an enhancement in oxidation degree from the original GO$_m$. For samples from raw graphite with smaller particle sizes, the oxidation degree of GO is higher. After the thermal treatment, the final content of oxygen is slightly higher for samples from graphite with a smaller particle size.

The point of zero charge (PZC) was used to assess the surface chemistry of the graphene materials [9], because the surface charge of carbon materials is directed by the type and population of functional groups and the pH. The NrGO samples (Figure 5) show a PZC of 8.5–8.7, while rGO samples exhibit lower PZC values of 7.2–7.4. Therefore, the rGO samples have a practically neutral surface, while the NrGO surfaces are more basic. The incorporation of nitrogen atoms into the graphene structure means more electrons in comparison with carbon atoms and this fact favors the delocalization of p electrons in N-doped samples, leading to changes in the hydrophobicity of the surface. This excess of electrons produces a higher basic strength of NrGO surfaces [47].

Raman spectroscopy is a very useful tool to evaluate the degree of disorder in the structure of graphene [8]. Figure 6 shows Raman spectra for rGO$_m$-3 and NrGO$_m$-4 samples. As Raman spectra at discrete spots cannot provide an overall picture in the case of non-uniform defects distributions in the sample, spectral mapping was used to acquire 25 points over a 50 × 50 μm^2 area. Two main peaks corresponding to vibrations with E2g symmetry in the graphitic lattice (G band) at 1580 cm^{-1} and to graphite edges or structural defects (D bands) at 1345 cm^{-1} were observed. Another two featured peaks have been reported in previous studies. A band D′ peak appears at 1625 cm^{-1} as a shoulder of the G band. It arises from alterations in the tension of sp^2 carbon atoms in the lattice caused by the arrangement of the electronic cloud [50]. A second peak assigned as 2D (historically called G′) is always present at 2700 cm^{-1} in the spectra of graphene materials [40].

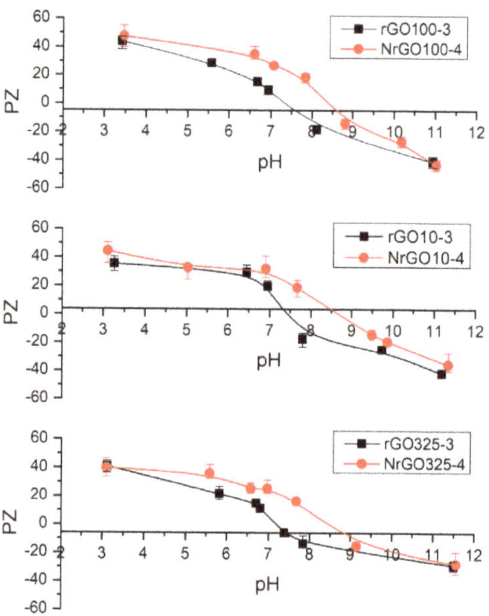

Figure 5. PZC for rGOm-3 and N-doped reduced graphene oxide (NrGOm-4) samples.

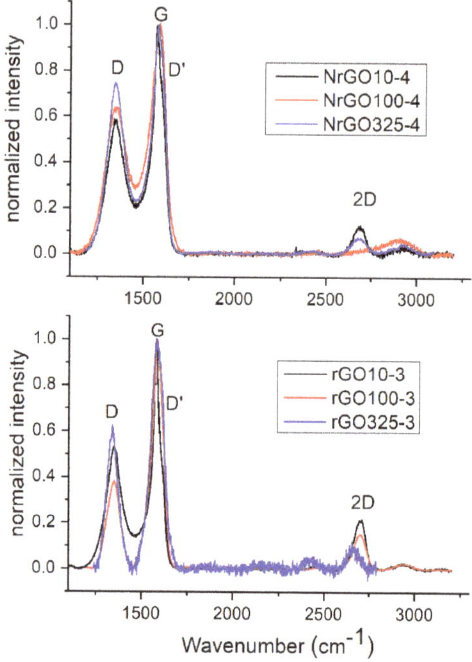

Figure 6. Raman spectra for rGO_m-3 and $NrGO_m$-4 samples.

The intensity ratio of the D and G bands (I_D/I_G) can be used as a quantitative indicator of the amount of disorder or edges within the structure of the samples. I_D/I_G ratios were calculated (Table 2). A comparison of rGO_m samples with their $NrGO_m$ counterparts shows that the ratio increases when N is introduced in the reduced graphite oxide. In addition, an increase with N content in the $NrGO_m$ series is observed. This is due to a disruption of the symmetry of the lattice produced by the incorporation of heteroatoms into the graphitic structure. This effect has been described previously by Chen et al. [53]. They claimed that the introduction of N into the carbon lattice could produce distortions, transforming the graphitic region into an sp^3 domain. From the Raman spectra, it can be observed that an increment in the N content leads to a shift of the D band to lower frequencies.

For a bulk graphite sample, the 2D band consists of two contributions. For single-layer graphene, the 2D band appears as a single sharp peak at the lower frequency (around 2690 cm^{-1}). As the number of layers increases, the 2D band changes its shape, width, and position, and the G peak position shifts to lower frequencies [8]. A systematic study of the in-plane crystallite size was carried out in 1970 by Tuinstra and Koenig [54]. They found that the ratio of the D and G band intensities (I_D/I_G) is inversely proportional to the in-plane crystallite sizes (L_a). The crystallite sizes (L_a) can be calculated from $L_a(nm)$ = $(2.4 \times 10^{-10}) \lambda^4 (I_D/I_G)^{-1}$ (λ being the Raman excitation wavelength) [55]. L_a were 25.6–33.1 nm for NrGO samples and 30.5–49.3 nm for rGO samples; it is concluded that the crystallite size decreases with the presence of defects, and therefore the doping level. It is consistent with the bibliography [40], which point outs that since L_a is the average interdefect distance, the introduction of nitrogen atoms accompanied by defects implies a smaller L_a.

The stability of the prepared graphenic materials in air atmosphere with the temperature was studied by thermal gravimetric analysis (Figure S5, ESI). Differential thermogravimetric profiles for rGO_m-3 samples show pronounced peaks near 598–642 °C that could be attributed to a weight loss due to the oxidation of a graphitized carbon structure. In comparison, for $NrGO_m$-4 samples, this peak moved toward higher temperatures, showing oxidation temperatures between 663 and 671 °C. These findings reveal that the presence of nitrogen in the NrGO samples rise the stability in air of the graphenic material, which is in agreement with previous results [33].

3.2. ORR Electrocatalytic Activities

To assess the electrocatalyst's properties toward ORR of the N-doped ($NrGO_{325}$-4, $NrGO_{100}$-4, and $NrGO_{10}$-4) and non-doped (rGO_{325}-3, rGO_{100}-3, and rGO_{10}-3) graphenic materials, a cyclic voltammetry study was initially performed in a KOH electrolyte (0.1 mol dm^{-3}) saturated with N_2 and O_2. Figure S6 (ESI) shows the CVs in the N_2 and O_2 electrolytes for the six samples tested. In the presence of N_2, no redox processes are detected, while in the electrolyte purged with O_2, the six graphene materials studied showed an irreversible ORR peak with $0.74 \geq E_{pc} \geq 0.72$ V versus RHE. For comparison, the commercial Pt/C (20 wt %) was also measured in the same experimental conditions toward ORR (Figure S7a, ESI), presenting analogous behavior to the graphene materials with E_{pc} = 0.88 V versus RHE.

Then, the electrocatalytic properties toward ORR were further studied by LSV with an RDE at several rotation speeds in KOH purged with O_2. The LSVs of all graphene materials are presented in Figure 7, while those corresponding to Pt/C can be seen in Figure S7b. All graphene materials presented a linear relationship between j_L and rotation speed, suggesting that the electron transfer reaction is diffusion limited. The LSVs show three different regions: for E higher than ≈ 0.80 V, the process is kinetically controlled; potential values between ≈ 0.80 V and ≈ 0.50 V indicate the mixed kinetic-diffusion region, and for $E \leq 0.50$ V, the process is controlled by O_2 diffusion.

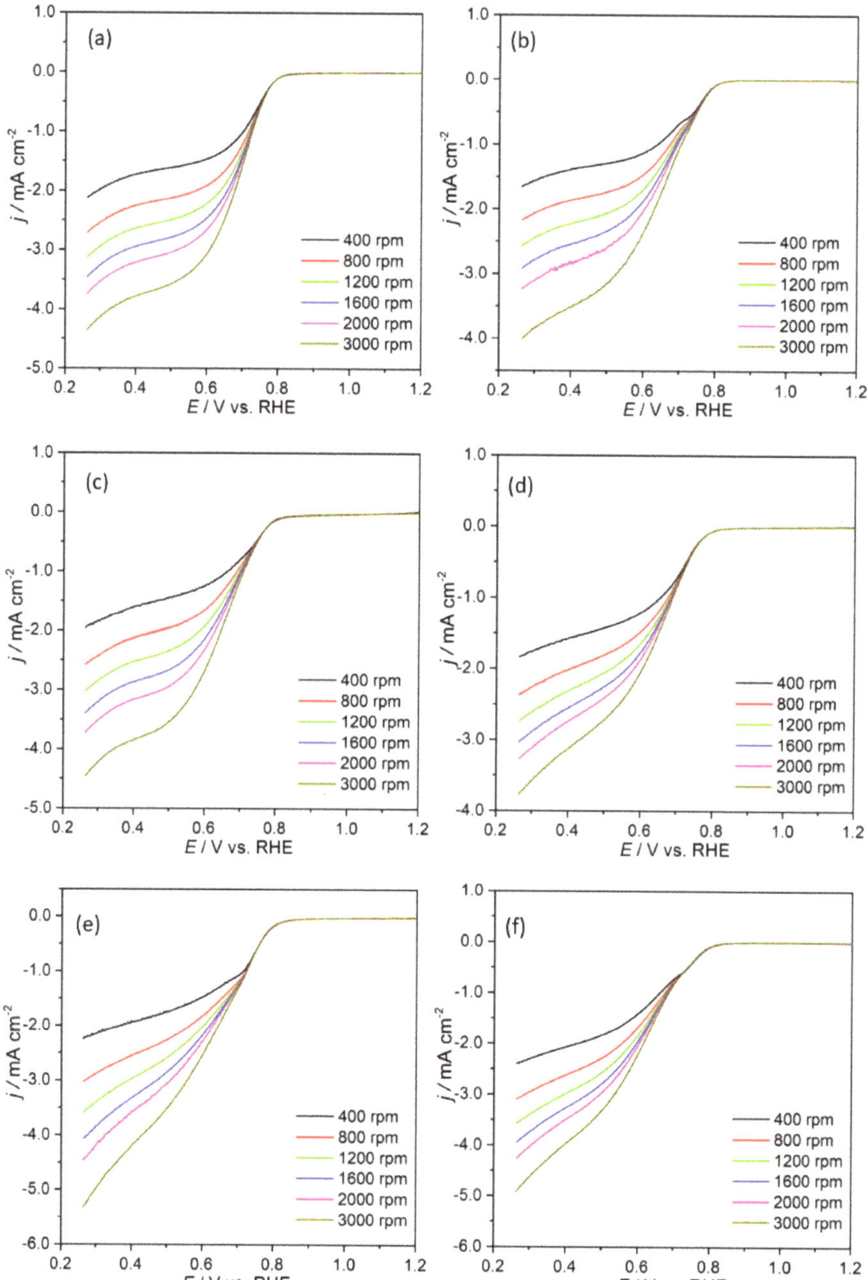

Figure 7. Oxygen reduction reaction (ORR) polarization curves of rGO$_{10}$-3 (**a**), rGO$_{100}$-3 (**b**), rGO$_{325}$-3 (**c**), NrGO$_{10}$-4 (**d**), NrGO$_{100}$-4 (**e**), and NrGO$_{325}$-4 (**f**) modified electrodes at different rotation rates in KOH purged with O$_2$ at 0.005 V s^{-1}.

Table 3 shows comparisons of the obtained E_{onset} and $j_{L,\,0.26V,\,1600rpm}$ values for the N-doped and non-doped graphene materials at 1600 rpm. All the ECs showed similar onset potentials ($0.82 \geq E_{onset}$

≥ 0.79 V versus RHE), but higher $j_{L,\,0.26V,\,1600rpm}$ values were obtained for the N-doped materials. For the non-doped graphene materials, these values varied between -2.92 and -3.46 mA cm^{-2}, while for the N-doped materials, they varied between -3.03 and -4.05 mA cm^{-2}. The NrGO$_{325}$-4 electrocatalyst showed the most promising result of all the prepared graphenic materials with the more positive E_{onset} of 0.82 V versus RHE (0.12 V more negative than Pt/C) and the higher $j_{L,\,0.26V,\,1600rpm}$ of -4.05 mA cm^{-2}. The introduction of nitrogen species led to an improvement in the ORR properties, which is in agreement with several works that have reported the enhancement of ORR electrocatalytic activity after the N-doping of carbon materials [56,57]. Furthermore, comparing the results for the non-doped graphenic materials, it seems that the particle size of the starting graphite does not have a clear direct relationship with their ORR properties, unlike is observed for the N-doped materials. For the latter, as the particle size of the starting graphite decreased, there was an improvement of the ORR features (more positive E_{onset} values and higher j_L), and this is a consequence of the higher degree of oxidation that favored the incorporation of increased amounts of nitrogen, leading to higher possible active sites for oxygen reduction. This is supported by the elemental analysis results where the percentages of nitrogen obtained were 5.0%, 3.8%, and 1.8% for NrGO$_{325}$-4, NrGO$_{100}$-4, and NrGO$_{10}$-4, respectively.

Table 3. Relevant ORR parameters for commercial Pt/C and activated carbons prepared.

Sample	E_{onset} **vs. RHE**	j_L **(mA cm^{-2})**	$\tilde{n}_{O2}$	**Tafel Slopes (mV dec^{-1})** [*]
Pt/C (20 wt %)	0.94	-4.70	4.0	89
rGO$_{325}$-3	0.81	-3.40	2.6	98
rGO$_{100}$-3	0.80	-2.92	2.2	48
rGO$_{10}$-3	0.80	-3.46	3.1	65
NrGO$_{325}$-4	0.82	-4.05	3.9	62
NrGO$_{100}$-4	0.80	-3.94	3.3	99
NrGO$_{10}$-4	0.79	-3.03	3.0	78

[*] normalized by the mass of catalysts

Then, the ORR kinetics at different potentials were assessed with the Koutecky-Levich (K-L) plots obtained with the application of the K-L equation to the LSVs (in Figure 7) at rotation speeds in the range of 400 to 3000 rpm. The K-L plots of the graphene materials are presented in Figure S8, and those corresponding to Pt/C are shown in Figure S7c. For the graphene materials, the j^{-1} increases with increasing $\omega^{-1/2}$, which suggests a first-order electrocatalytic oxygen reduction with respect to the concentration of oxygen dissolved. Additionally, all the K-L plots of the non-doped materials show different slopes, suggesting that $\tilde{n}_{O2}$ depends on the applied potential, while for the N-doped ones, the differences are less significant. This suggests that in this case, n$_{O2}$ is less dependent on the potential. The ORR process in alkaline medium can proceed by two pathways: a direct one involving one step (Equation 2) or an indirect one involving two steps (Equation 3 and 4) [reference 2 in ESI].

$$O_2 + H_2O + 4e^- \rightarrow HO^- \tag{2}$$

$$O_2 + H_2O + 2e^- \rightarrow HO_2^- + HO^- \tag{3}$$

$$HO_2^- + H_2O + 2e^- \rightarrow 3HO^- \tag{4}$$

The mean n_{O2} ($\tilde{n}_{O2}$) values estimated from Equation 3 in ESI are depicted in Table 3, and Figure 8a shows the changes of n_{O2} with the applied potential. For the non-doped materials, as the potential decreases from 0.46 to 0.26 V versus RHE, there is an increase in n$_{O2}$ values with ΔE values of 1.0, 0.53, and 0.71 V for rGO$_{10}$-3, rGO$_{100}$-3, and rGO$_{325}$-3, respectively. The variation for the N-doped materials is less significant with $0.40 \geq \Delta E \geq 0.22$ V versus RHE. The results of mean n$_{O2}$ values obtained suggest that rGO$_{100}$-3 and rGO$_{325}$-3 proceed through the indirect pathway, while rGO$_{10}$-3, NrGO$_{10}$-4, and NrGO$_{100}$-4 seem to proceed through a mixed mechanism, similar to $\tilde{n}_{O2} \approx 3$. This behavior has already been reported for other N-doped structures [58]. On the other hand, for the NrGO$_{325}$-4, a one-step four-electron transfer mechanism seems to be the leading process, since $\tilde{n}_{O2} = 3.9$. This value is equal to the one obtained for Pt/C. Additionally, this result is better than that obtained for the other N-doped carbon nanotubes and graphene [32,59,60].

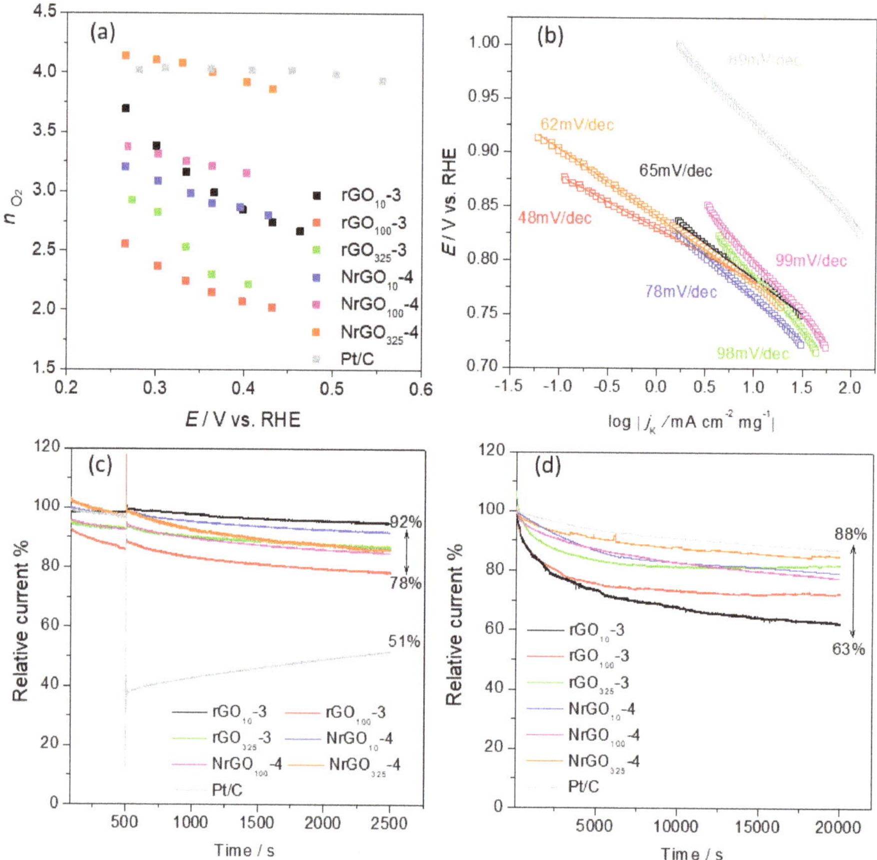

Figure 8. n_{O2} at several potential values (**a**); ORR Tafel plots (**b**); chronoamperograms in O_2-saturated KOH electrolyte at 1600 rpm and $E = 0.50$ V vs. reversible hydrogen electrode (RHE), with the addition of methanol (0.5 mol dm^{-3}) at $t = 500$ s (**c**) and without for 20000 s (**d**).

The better ORR performances (j_L and $\tilde{n}_{O2}$) of the N-doped materials in comparison with the non-doped may be related to the presence of nitrogen as explained before, while the differences between the three N-doped materials to the percentages and types of N species found could also be related to the particle size of the starting graphite. As discussed above in Section 3.1, usually a higher degree of oxidation is obtained in GO samples with smaller graphite particle size, and on the other hand, oxygen functional groups are responsible for reacting with NH_3, leading to the incorporation of nitrogen in the structure, which will favor ORR. Elemental analysis showed that NrGO$_{325}$-4, NrGO$_{100}$-4, and NrGO$_{10}$-4 presented 5.0%, 3.8%, and 1.8% of N, respectively, whereas the XPS results indicated percentages of 3.4%, 3.3%, and 3.2%. The NrGO$_{325}$-4 sample presents a higher N% at the surface and has also a higher percentage of pyridinic nitrogen. The latest have been reported to work as ORR electrocatalytic sites due to the electron donated to the conjugated π-bond of graphene and to the lone pair of electrons they have [61,62].

To confirm the excellent result obtained for $NrGO_{325}$-4, RRDE tests were performed. The j values obtained at the disk and ring are depicted in Figure S9a. For comparison, the Pt/C data were also included. Equation (1) was used to estimate the percentage of H_2O_2 produced, and the results are presented in Figure S9b at several potential values. The $NrGO_{325}$-4 electrocatalyst presented a low percentage of H_2O_2 ($\approx 15\%$), which is in accordance with the high $\bar{n}_{O2}$ value ($\bar{n}_{O2} = 3.9$), while a value of 3% was obtained for Pt/C.

Tafel plots (Figure 8b) were obtained from LSV data in Figures S8 and S7b in KOH electrolyte purged with O_2 at 1600 rpm. The Tafel slopes obtained between $E = 1.00$ to 0.70 V versus RHE were 65, 48, 98, 78, 99, 62, and 89 mv dec^{-1} for rGO_{10}-3, rGO_{100}-3, rGO_{325}-3, $NrGO_{10}$-4, $NrGO_{100}$-4, $NrGO_{325}$-4, and Pt/C, respectively. All electrocatalysts except for rGO_{325}-3 and $NrGO_{100}$-4 presented Tafel slopes lower than those obtained for Pt/C, which suggests that the graphene materials can easily adsorb O_2 onto their surface and activate it, promoting a robust electrocatalytic activity toward ORR. The different Tafel slopes obtained for rGO_{325}-3 and $NrGO_{100}$-4 are most likely associated to fluctuations in the oxygenated intermediates adsorption strength or to an incomplete electrocatalyst, which is used as a consequence of mass transport losses [63,64]. It is recognized that the intermediates adsorption strength is dependent on the physical properties and on the chemical nature of the selected electrocatalyst, which rules the determining step rate. For these two electrocatalysts, the first discharge step upon the consumption of the MOOH with a high coverage of MOO^- is the rate-determining step, whereas for the remaining electrocatalysts, the conversion of MOO^- to MOOH rules the overall reaction rate (where M stands for an empty site on the electrocatalyst surface) [63].

Tolerance to methanol crossover is another parameter that is usually evaluated because in fuels cells run on methanol, this can be a problem, as the performance of the cathode can be drastically reduced if the catalyst in sensitive to methanol. Therefore, to evaluate this parameter, chronoamperometric tests were performed in a KOH electrolyte purged with oxygen for 2500 s, to which 0.5 mol dm^{-3} of methanol were added at $t = 500$ s (Figure 8c). The presence of methanol led to an impressive decrease ($\approx 49\%$) of Pt/C current. Oppositely, the graphene materials are much more stable and less sensible to methanol with current retentions between 78% and 92% suggesting higher selectivity toward oxygen reduction.

The EC stability is also of extreme importance. So, this was evaluated, for all graphene materials, using chronoamperometric runs in KOH electrolyte (in O_2) for 20,000 s applying a potential of $E = 0.50$ V versus RHE at 1600 rpm. The results obtained for Pt/C and all graphenic materials are presented in Figure 8d. The best result was observed for Pt/C with a current retention of 88%. The $NrGO_{325}$-4 electrocatalyst showed the best result from all the graphene materials with a current retention of 85%, which is only 3% less than Pt/C. For the other ECs, the current retention percentages were 63%, 72%, 82%, 80%, and 78% for rGO_{10}-3, rGO_{100}-3, rGO_{325}-3, $NrGO_{10}$-4, and $NrGO_{100}$-4, respectively. The best performance of the $NrGO_{325}$-4 electrocatalyst can be ascribed to the combination of two features, the N-doping and the smaller particle size of the starting graphite.

4. Conclusions

In this paper, we report a viable method for the synthesis of graphene materials and nitrogen-doped reduced graphene oxide derivatives with enhanced properties. These are based on the chemical oxidation of graphite flakes with different particle sizes and selecting the experimental conditions used during the subsequent thermal exfoliation process.

The characterization results suggest that the starting particle size and thermal conditions applied during the exfoliation treatment remarkably affect the final surface properties of the prepared materials. The results point out that smaller particle sizes lead to higher surface areas. We achieved surface areas of 867 m^2g^{-1} for rGO. For NrGO samples obtained from graphite 10, 100, and 325, we observed BET surface areas of 236, 420, and 492 m^2g^{-1} respectively, which may provide a new way to control the surface area of N-doped graphene.

The use of different reduction atmosphere (NH_3 versus inert) allow to successfully introduce nitrogen within the graphenic structure, enhancing the electronic properties and basicity of the doped materials. For the N-doped samples, the amount of nitrogen introduced and surface areas could be also tailored.

As a result of their tailored enhanced properties, these optimized NrGO and rGO samples were successfully applied as ORR electrocatalysts. The NrGO samples showed better ORR performance in alkaline medium with onset potentials ranging from 0.79 and 0.82 V versus RHE, low Tafel slopes (62–99 mV dec^{-1}), and good j_L values (-3.03–-4.05 mA cm^{-2}). The better performance of NrGO$_{325}$-4 was attributed not only to the higher content of nitrogen but also to the smaller particle size of the starting material. Moreover, all the graphene materials presented good durability/stability and very low sensitivity to methanol. This work has led to a new class of metal-free ORR electrocatalysts with good efficiency and stability.

Supplementary Materials: The following are available online at http://www.mdpi.com/2079-4991/9/12/1761/s1, Experimental details; Figure S1 displays N_2 adsorption–desorption isotherm for all samples; Figure S2 contains XRD spectra for rGO and NrGO samples; Figure S3 contains TEM images of the different graphene materials; Figure S4 shows survey XPS for rGO$_m$-2 and NrGO$_m$-5 samples; Figure S5 contains TGA for graphene materials; Figure S6 contains the CVs of rGO$_m$-3 and NrGO$_m$-4 modified electrodes in N_2-saturated and O_2-saturated 0.1 mol dm^{-3} KOH solution; Figure S7 contains CVs, LSVs, and K-L plots for Pt/C; Figure S8 contains the ORR polarization curves of rGO$_m$-3 and NrGO$_m$-4 modified electrodes at different rotation rates in O_2-saturated solution; Figure S9 contains LSVs recorded with RRDE and the estimated percentage of H_2O_2 formed for Pt/C and NrGO$_{325}$-4.

Author Contributions: A.G.-R. and I.R.-R. conceived and designed the experiments; C.S.R.-B. carried out the synthetic experiments; D.M.F. and C.F. conducted the electrocatalytic measurements; E.V.-A. performed the Raman characterization; D.M.F. and I.R.-R. wrote the original draft manuscript; and all the authors discussed the results and contributed to the manuscript.

Funding: This research was supported by the Spanish Agencia Estatal de Investigación (AEI) under projects CTQ-2017-89443-C3-1-R and CTQ-2017-89443-C3-3-R. C.S.R.B. gratefully acknowledges financial support from Spanish Ministerio de Educacion, Cultura y Deporte, Grant N° FPU15/01838. D.M.F. also thanks Project UNIRCELL - POCI-01-0145-FEDER-016422 funded by European Structural and Investment Funds (FEEI) through - Programa operacional Competitividade e Internacionalização - COMPETE2020 and by national funds through FCT - Fundação para a Ciência e a Tecnologia, I.P.

Conflicts of Interest: The authors declare no conflict of interest.

References

1. Geim, A.K.; Novoselov, K.S. The Rise of Graphene. *Nat. Mater.* **2007**, *6*, 183–191. [CrossRef] [PubMed]
2. Stoller, M.D.; Park, S.; Zhu, Y.; An, J.; Ruoff, R.S. Graphene-Based Ultracapacitors. *Nano Lett.* **2008**, *8*, 3498–3502. [CrossRef] [PubMed]
3. Sun, Y.; Wu, Q.; Shi, G. Graphene based new energy materials. *Energy Environ. Sci.* **2011**, *4*, 1113–1132. [CrossRef]
4. Raccichini, R.; Varzi, A.; Passerini, S.; Scrosati, B. The role of graphene for electrochemical energy storage. *Nat. Mater.* **2015**, *14*, 271–279. [CrossRef] [PubMed]
5. Bonaccorso, F.; Colombo, L.; Yu, G.; Stoller, M.; Tozzini, V.; Ferrari, A.C.; Ruoff, R.S.; Pellegrini, V. Graphene, Related Two-Dimensional Crystals, and Hybrid Systems for Energy Conversion and Storage. *Science* **2015**, *347*, 1246501. [CrossRef]
6. Weiss, N.O.; Zhou, H.; Liao, L.; Liu, Y.; Jiang, S.; Huang, Y.; Duan, X. Graphene: An Emerging Electronic Material. *Adv. Mater.* **2012**, *24*, 5782–5825. [CrossRef]
7. Soldano, C.; Mahmood, A.; Dujardin, E. Production, Properties and Potential of Graphene. *Carbon* **2010**, *48*, 2127–2150. [CrossRef]
8. Machado, B.F.; Serp, P. Graphene-Based Materials for Catalysis. *Catal. Sci. Technol.* **2012**, *2*, 54–75. [CrossRef]
9. Serp, P.; Figueiredo, J.L. *Carbon Materials for Catalysis*; John Wiley & Sons, Inc: Hoboken, NJ, USA, 2008.
10. Asedegbega-Nieto, E.; Perez-Cadenas, M.; Morales, M.V.; Bachiller-Baeza, B.; Gallegos-Suarez, E.; Rodriguez-Ramos, I.; Guerrero-Ruiz, A. High Nitrogen Doped Graphenes and their Applicability as Basic Catalysts. *Diam. Relat. Mater.* **2014**, *44*, 26–32. [CrossRef]
11. Castillejos-Lopez, E.; Bachiller-Baeza, B.; Asedegbega-Nieto, E.; Guerrero-Ruiz, A.; Rodriguez-Ramos, I. Selective 1,3-butadiene Hydrogenation by Gold Nanoparticles Deposited & Precipitated onto Nano-Carbon Materials. *RSC Adv.* **2015**, *5*, 81583–81598.

12. Geng, D.; Yang, S.; Zhang, Y.; Yang, J.; Liu, J.; Li, R.; Sham, T.-K.; Sun, X.; Ye, S.; Knights, S. Nitrogen doping effects on the structure of graphene. *Appl. Surf. Sci.* **2011**, *257*, 9193–9198. [CrossRef]

13. Novoselov, K.S.; Geim, A.K.; Morozov, S.V.; Jiang, D.; Zhang, Y.; Dubonos, S.V.; Grigorieva, I.V.; Firsov, A.A. Electric Field Effect in Atomically Thin Carbon Films. *Science* **2004**, *306*, 666–669. [CrossRef] [PubMed]

14. Kim, K.S.; Zhao, Y.; Jang, H.; Lee, S.Y.; Kim, J.M.; Kim, K.S.; Ahn, J.-H.; Kim, P.; Choi, J.-Y.; Hong, B.H. Large-Scale Pattern Growth of Graphene Films for Stretchable Transparent Electrodes. *Nature* **2009**, *457*, 706–710. [CrossRef] [PubMed]

15. Berger, C.; Song, Z.; Li, X.; Wu, X.; Brown, N.; Naud, C.; Mayou, D.; Li, T.; Hass, J.; Marchenkov, A.N.; et al. Electronic Confinement and Coherence in Patterned Epitaxial Graphene. *Science* **2006**, *312*, 1191–1196. [CrossRef] [PubMed]

16. Hummers, W.S.; Offeman, R.E. Preparation of Graphitic Oxide. *J. Am. Chem. Soc.* **1958**, *80*, 1339. [CrossRef]

17. Brodie, B.C. On the Atomic Weight of Graphite. *Philos. Trans. R. Soc.* **1859**, *149*, 249–259.

18. Stankovich, S.; Dikin, D.A.; Piner, R.D.; Kohlhaas, K.A.; Kleinhammes, A.; Jia, Y.; Wu, Y.; Nguyen, S.T.; Ruoff, R.S. Synthesis of graphene-based nanosheets via chemical reduction of exfoliated graphite oxide. *Carbon* **2007**, *45*, 1558–1565. [CrossRef]

19. Shin, H.-J.; Kim, K.K.; Benayad, A.; Yoon, S.-M.; Park, H.K.; Jung, I.-S.; Jin, M.H.; Jeong, H.-K.; Kim, J.M.; Choi, J.-Y.; et al. Efficient Reduction of Graphite Oxide by Sodium Borohydride and Its Effect on Electrical Conductance. *Adv. Funct. Mater.* **2009**, *19*, 1987–1992. [CrossRef]

20. Zhou, M.; Wang, Y.; Zhai, Y.; Zhai, J.; Ren, W.; Wang, F.; Dong, S. Controlled Synthesis of Large-Area and Patterned Electrochemically Reduced Graphene Oxide Films. *Chem. A Eur. J.* **2009**, *15*, 6116–6120. [CrossRef]

21. Sundaram, R.S.; Gómez-Navarro, C.; Balasubramanian, K.; Burghard, M.; Kern, K. Electrochemical Modification of Graphene. *Adv. Mater.* **2008**, *20*, 3050–3053. [CrossRef]

22. Choucair, M.; Thordarson, P.; Stride, J.A. Gram-scale Production of Graphene Based on Solvothermal Synthesis and Sonication. *Nat. Nanotechnol.* **2009**, *4*, 30–33. [CrossRef] [PubMed]

23. Wang, H.; Robinson, J.T.; Li, X.; Dai, H. Solvothermal Reduction of Chemically Exfoliated Graphene Sheets. *J. Am. Chem. Soc.* **2009**, *131*, 9910–9911. [CrossRef] [PubMed]

24. McAllister, M.J.; Li, J.-L.; Adamson, D.H.; Schniepp, H.C.; Abdala, A.A.; Liu, J.; Herrera-Alonso, M.; Milius, D.L.; Car, R.; Prud'homme, R.K.; et al. Single Sheet Functionalized Graphene by Oxidation and Thermal Expansion of Graphite. *Chem. Mater.* **2007**, *19*, 4396–4404. [CrossRef]

25. Botas, C.; Álvarez, P.; Blanco, C.; Santamaría, R.; Granda, M.; Gutiérrez, M.D.; Rodriguez-Reinoso, F.; Menéndez, R. Critical Temperatures in the Synthesis of Graphene-like Materials by Thermal Exfoliation-Reduction of Graphite Oxide. *Carbon* **2013**, *52*, 476–485. [CrossRef]

26. Luo, Z.; Lim, S.; Tian, Z.; Shang, J.; Lai, L.; MacDonald, B.; Fu, C.; Shen, Z.; Yu, T.; Lin, J. Pyridinic N Doped Graphene: Synthesis, Electronic Structure, and Electrocatalytic Property. *J. Mater. Chem.* **2011**, *21*, 8038–8044. [CrossRef]

27. Reddy, A.L.M.; Srivastava, A.; Gowda, S.R.; Gullapalli, H.; Dubey, M.; Ajayan, P.M. Synthesis of Nitrogen-Doped Graphene Films for Lithium Battery Application. *ACS Nano* **2010**, *4*, 6337–6342. [CrossRef] [PubMed]

28. Jin, Z.; Yao, J.; Kittrell, C.; Tour, J.M. Large-Scale Growth and Characterizations of Nitrogen-Doped Monolayer Graphene Sheets. *ACS Nano* **2011**, *5*, 4112–4117. [CrossRef] [PubMed]

29. Panchakarla, L.S.; Subrahmanyam, K.S.; Saha, S.K.; Govindaraj, A.; Krishnamurthy, H.R.; Waghmare, U.V.; Rao, C.N.R. Synthesis, Structure and Properties of Boron and Nitrogen Doped Graphene. *Adv. Mater.* **2009**, *21*, 4726–4738. [CrossRef]

30. Wang, Y.; Shao, Y.; Matson, D.W.; Li, J.; Lin, Y. Nitrogen-Doped Graphene and Its Application in Electrochemical Biosensing. *ACS Nano* **2010**, *4*, 1790–1798. [CrossRef]

31. Lin, Z.; Song, M.; Ding, Y.; Liu, Y.; Wong, C. Facile Preparation of Nitrogen-Doped Graphene as a Metal-Free Catalyst for Oxygen Reduction Reaction. *Phys. Chem. Chem. Phys.* **2012**, *14*, 3381–3387. [CrossRef]

32. Sheng, Z.H.; Shao, L.; Chen, J.J.; Bao, W.J.; Wang, F.B.; Xia, X.H. Catalyst-Free Synthesis of Nitrogen-Doped Graphene Via Thermal Annealing Graphite Oxide with Melamine and its Excellent Electrocatalysis. *ACS Nano* **2011**, *5*, 4350–4358. [CrossRef] [PubMed]

33. Canty, R.; Gonzalez, E.; MacDonald, C.; Osswald, S.; Zea, H.; Luhrs, C.C. Reduction Expansion Synthesis as Strategy to Control Nitrogen Doping Level and Surface Area in Graphene. *Materials* **2015**, *8*, 7048–7058. [CrossRef] [PubMed]

34. Li, X.; Wang, H.; Robinson, J.T.; Sanchez, H.; Diankov, G.; Dai, H. Simultaneous Nitrogen Doping and Reduction of Graphene Oxide. *J. Am. Chem. Soc.* **2009**, *131*, 15939–15944. [CrossRef] [PubMed]

35. Wu, Z.S.; Ren, W.; Gao, L.; Liu, B.; Jiang, C.; Cheng, H.M. Synthesis of High-Quality Graphene with a Pre-Determined Number of Layers. *Carbon* **2009**, *47*, 493–499. [CrossRef]

36. Paez, A.; Jesús, G.; Alvarez, P.; Granda, M.; Blanco, C.; Santamaria, P.; Blanco, R.; Fernandez, L.; Menendez, R.M.; Calle, F. Methods for Producing Graphene with Tunable Properties by a Multi-Step Thermal Reduction Process. WO 2016042099 A1, 18 September 2014.

37. Zhang, C.; Lv, W.; Xie, X.; Tang, D.; Liu, C.; Yang, Q.-H. Towards Low Temperature Thermal Exfoliation of Graphite Oxide for Graphene Production. *Carbon* **2013**, *62*, 11–24. [CrossRef]

38. Dao, T.D.; Jeong, H.M. Graphene Prepared by Thermal Reduction—Exfoliation of Graphite Oxide: Effect of Raw Graphite Particle Size on the Properties of Graphite Oxide and Graphene. *Mater. Res. Bull.* **2015**, *70*, 651–657. [CrossRef]

39. Muthuswamy, N.; Buan, M.E.M.; Walmsley, J.C.; Rønning, M. Evaluation of ORR Active Sites in Nitrogen-Doped Carbon Nanofibers by KOH Post Treatment. *Catal. Today* **2018**, *301*, 11–16. [CrossRef]

40. Wang, H.; Maiyalagan, T.; Wang, X. Review on Recent Progress in Nitrogen-Doped Graphene: Synthesis, Characterization, and Its Potential Applications. *ACS Catal.* **2012**, *2*, 781–794. [CrossRef]

41. Serp, P.; Machado, B.F. *Nanostructured Carbon Materials for Catalysis, Royal Society of Chemistry*; Royal Society of Chemistry: Cambridge, UK, 2015.

42. Wepasnick, K.A.; Smith, B.A.; Bitter, J.L.; Fairbrother, D.H. Chemical and Structural Characterization of Carbon Nanotube Surfaces. *Anal. Bioanal. Chem.* **2010**, *396*, 1003–1014. [CrossRef]

43. Desimoni, E.; Casella, G.I.; Morone, A.; Salvi, A.M. XPS Determination of Oxygen-Containing Functional Groups on Carbon-Fibre Surfaces and the Cleaning of these Surfaces. *Surf. Interface Anal.* **1990**, *15*, 627–634. [CrossRef]

44. Matter, P.H.; Zhang, L.; Ozkan, U.S. The Role of Nanostructure in Nitrogen-Containing Carbon Catalysts for the Oxygen Reduction Reaction. *J. Catal.* **2006**, *239*, 83–96. [CrossRef]

45. Pels, J.R.; Kapteijn, F.; Moulijn, J.A.; Zhu, Q.; Thomas, K.M. Evolution of Nitrogen Functionalities in Carbonaceous Materials during Pyrolysis. *Carbon* **1995**, *33*, 1641–1653. [CrossRef]

46. Biniak, S.; Szymański, G.; Siedlewski, J.; Światkoski, A. The Characterization of Activated Carbons with Oxygen and Nitrogen Surface Groups. *Carbon* **1997**, *35*, 1799–1810. [CrossRef]

47. Faba, L.; Criado, Y.A.; Gallegos-Suarez, E.; Pérez-Cadenas, M.; Díaz, E.; Rodriguez-Ramos, I.; Guerrero-Ruiz, A.; Ordóñez, S. Preparation of Nitrogen-Containing Carbon Nanotubes and Study of their Performance as Basic Catalysts. *Appl. Catal. A Gen.* **2013**, *458*, 155–161. [CrossRef]

48. García-García, F.R.; Álvarez-Rodríguez, J.; Rodriguez-Ramos, I.; Guerrero-Ruiz, A. The Use of Carbon Nanotubes with and without Nitrogen Doping as Support for Ruthenium Catalysts in the Ammonia Decomposition Reaction. *Carbon* **2010**, *48*, 267–276. [CrossRef]

49. van Dommele, S.; de Jong, K.P.; Bitter, J.H. Nitrogen-Containing Carbon Nanotubes as Solid Base Catalysts. *Chem. Commun.* **2006**, *76*, 4859–4861. [CrossRef]

50. Dongil, A.B.; Bachiller-Baeza, B.; Guerrero-Ruiz, A.; Rodríguez-Ramos, I.; Martínez-Alonso, A.; Tascón, J.M.D. Surface Chemical Modifications Induced on High Surface Area Graphite and Carbon Nanofibers Using Different Oxidation and Functionalization Treatments. *J. Colloid Interface Sci.* **2011**, *355*, 179–189. [CrossRef]

51. Figueiredo, J.L.; Pereira, M.F.R.; Freitas, M.M.A.; Órfão, J.J.M. Modification of the Surface Chemistry of Activated Carbons. *Carbon* **1999**, *37*, 1379–1389. [CrossRef]

52. Zielke, U.; Hüttinger, K.J.; Hoffman, W.P. Surface-Oxidized Carbon Fibers: I. Surface Structure and Chemistry. *Carbon* **1996**, *34*, 983–998. [CrossRef]

53. Chen, C.-M.; Zhang, Q.; Zhao, X.-C.; Zhang, B.; Kong, Q.-Q.; Yang, M.-G.; Yang, Q.-H.; Wang, M.-Z.; Yang, Y.-G.; Schlögl, R.; et al. Hierarchically Aminated Graphene Honeycombs for Electrochemical Capacitive Energy Storage. *J. Mater. Chem.* **2012**, *22*, 14076–14084. [CrossRef]

54. Tuinstra, F.; Koenig, J.L. Raman Spectrum of Graphite. *J. Chem. Phys.* **1970**, *53*, 1126–1130. [CrossRef]

55. Pimenta, M.A.; Dresselhaus, G.; Dresselhaus, M.S.; Cançado, L.G.; Jorio, A.; Saito, R. Studying Disorder in Graphite-Based Systems by Raman Spectroscopy. *Phys. Chem. Chem. Phys.* **2007**, *9*, 1276–1290. [CrossRef] [PubMed]

56. Qu, L.T.; Liu, Y.; Baek, J.B.; Dai, L.M. Nitrogen-Doped Graphene as Efficient Metal-Free Electrocatalyst for Oxygen Reduction in Fuel Cells. *ACS Nano* **2010**, *4*, 1321–1326. [CrossRef] [PubMed]

57. Rocha, I.M.; Soares, O.G.P.; Fernandes, D.M.; Freire, C.; Figueiredo, J.L.; Pereira, M.F.R. N-Doped Carbon Nanotubes for the Oxygen Reduction Reaction in Alkaline Medium: Synergistic Relationship between Pyridinic and Quaternary Nitrogen. *Chem. Sel.* **2016**, *1*, 2522–2530. [CrossRef]

58. Kim, D.-W.; Lia, O.L.; Saito, N. Enhancement of ORR catalytic activity by multiple heteroatom-doped carbon materials. *Phys. Chem. Chem. Phys.* **2015**, *17*, 407–413. [CrossRef]

59. Fernandes, D.M.; Novais, H.C.; Bacsa, R.; Serp, P.; Bachiller-Baeza, B.; Rodríguez-Ramos, I.; Guerrero-Ruiz, A.; Freire, C. Polyoxotungstate@Carbon Nanocomposites as Oxygen Reduction Reaction (ORR) Electrocatalysts. *Langmuir* **2018**, *34*, 6376–6387. [CrossRef]

60. Li, H.; Liu, H.; Jong, Z.; Qu, W.; Geng, D.S.; Sun, X.; Wang, H. Nitrogen-doped carbon nanotubes with high activity for oxygen reduction in alkaline media. *Int. J. Hydrogen Energy* **2011**, *36*, 2258–2265. [CrossRef]

61. O'Hayre, R.; Cha, S.; Colella, W.; Prinz, F. *Fuel Cell Fundamentals*; John Wiley & Sons: New York, NY, USA, 2005.

62. Subramanian, N.P.; Li, X.G.; Nallathambi, V.; Kumaraguru, S.P.; Colon-Mercado, H.; Wu, G.; Lee, J.W.; Popov, B.N. Nitrogen-Modified Carbon-based Catalysts for Oxygen Reduction Reaction in Polymer Electrolyte Membrane Fuel Cells. *J. Power Sources* **2009**, *188*, 38–44. [CrossRef]

63. Shinagawa, T.; Garcia-Escudero, A.T.; Takanabe, K. Insight on Tafel Slopes from a Microkinetic Analysis of Aqueous Electrocatalysis for Energy Conversion. *Sci. Rep.* **2015**, *5*, 13801. [CrossRef]

64. Chlistunoff, J. RRDE and Voltammetric Study of ORR on Pyrolyzed Fe/Polyaniline Catalyst. On the Origins of Variable Tafel Slopes. *J. Phys. Chem. C* **2011**, *115*, 6496–6507.

Article

Hydrogen Production by Formic Acid Decomposition over Ca Promoted Ni/SiO₂ Catalysts: Effect of the Calcium Content

B. Faroldi [1,2,*], M. A. Paviotti [2], M. Camino-Manjarrés [3], S. González-Carrazán [3], C. López-Olmos [1] and I. Rodríguez-Ramos [1,*]

1   Instituto de Catálisis y Petroleoquímica, CSIC, C/Marie Curie 2, 28049 Madrid, Spain; cristina.lopez.olmos@csic.es
2   Instituto de Investigaciones en Catálisis y Petroquímica (INCAPE-CONICET), Facultad de Ingeniería Química, Universidad Nacional del Litoral, Santiago del Estero 2829, Santa Fe 3000, Argentina; apaviotti@fiq.unl.edu.ar
3   Departamento de Química Inorgánica, Facultad de Ciencias Químicas, Universidad de Salamanca, 37008 Salamanca, Spain; mcamino@usal.es (M.C.-M.); silviag@usal.es (S.G.-C.)
*   Correspondence: bfaroldi@fiq.unl.edu.ar (B.F.); irodriguez@icp.csic.es (I.R.-R.); Tel.: +345854765 (I.R.-R.)

Received: 3 October 2019; Accepted: 23 October 2019; Published: 25 October 2019

Abstract: Formic acid, a major product of biomass processing, is regarded as a potential liquid carrier for hydrogen storage and delivery. The catalytic dehydrogenation of FA to generate hydrogen using heterogeneous catalysts is of great interest. Ni based catalysts supported on silica were synthesized by incipient wet impregnation. The effect of doping with an alkaline earth metal (calcium) was studied, and the solids were tested in the formic acid decomposition reaction to produce hydrogen. The catalysts were characterized by X-ray diffraction (XRD), X-ray photoelectron spectroscopy (XPS), temperature-programmed reduction (TPR), Fourier transform infrared spectroscopy (FTIR), transmission electron microscopy (TEM), and programmed temperature surface reaction (TPSR). The catalyst doped with 19.3 wt.% of Ca showed 100% conversion of formic acid at 160 °C, with a 92% of selectivity to hydrogen. In addition, all the tested materials were promising for their application, since they showed catalytic behaviors (conversion and selectivity to hydrogen) comparable to those of noble metals reported in the literature.

Keywords: hydrogen production; formic acid decomposition; nickel catalyst; calcium oxide promoter; silica support

---

1. Introduction

Hydrogen (H_2) has significant advantages as an energy vector compared to petroleum or other conventional fossil fuels, although currently there are problems associated with its production, storage, and transportation that must be solved [1]. One possible solution for mobile applications, such as fuel cells, is to produce H_2 in situ by a reaction such as the reforming of methanol, although these reactions have been associated with the generation of CO_2, a greenhouse gas. There would be a significant advantage if H_2 was produced from a chemical product derived from biomass since, in it, the CO_2 formed in parallel must be considered a product of a carbon-neutral balance process. Formic acid is a chemical substance of relatively low specific volume and has limited uses, including its application as an antibacterial and antifungal agent. It can be produced by chemical methods such as the hydrolysis of methyl formate, but it is also obtained in equimolar proportions, together with levulinic acid, by hydrolysis of cellulose raw materials derived from biomass. Currently, with the increased interest in the production of levulinic acid and other valuable chemicals from biomass, it is important to develop processes to use the derived formic acid, since, otherwise, it constitutes a waste material [1]. In this direction, the interest in the use of the decomposition reaction of formic acid to produce H_2 has increased remarkably. Therefore, the challenge is to produce pure H_2, with minimal CO content, at the lowest possible temperature. This demand can be achieved through the careful choice of the catalyst and the reaction conditions, so a lot of research is currently being done in this direction.

The production of H_2 from formic acid using heterogeneous catalysts has been studied in liquid [2] and vapor phases [3], but, in most cases, formulations based on noble metals, such as Rh, Pt, Ru, Au, Ag, and Pd that are supported on C, Al_2O_3, and SiO_2 have been investigated [1–3]. For the vapor-phase reaction, Solymosi et al. [3] found the following order of activity on a set of carbon-supported noble metals: Ir > Pt > Rh > Pd > Ru. They reported that the decomposition of formic acid started at and above 77 °C on all catalysts, and that decomposition was complete at 200–250 °C. On the other hand, in the case of non-noble metals, the activities of supported Ni catalysts have been measured, proving to be active at relatively higher temperatures (>220 °C) than noble metal catalysts [4].

Ni catalysts with different support matrices show adequate activity and selectivity for H_2 production in various processes [5–8]. SiO_2 stands out among the supports used due to its high surface area and low acidity [5,6,8]. Alkaline-earth metal oxides act as structural promoters by increasing the dispersion of the active phase and stabilizing the dispersed metallic phase against sintering [9–12]. Also, these additives act as chemical promoters by influencing the acid–base properties of support [13–15] or the electron density of dispersed metal crystallites [16,17]. In particular, the use of CaO as a promoter has exerted a positive effect on the increase in the interaction of the Ni with the support and the resistance of the Ni catalysts to the sintering [18–20].

In this work, Ni catalysts supported on a SiO_2 matrix are used in the decomposition reaction of formic acid in the vapor phase. The effect of doping the support with different loadings of calcium is studied. The catalysts were characterized by X-ray diffraction (XRD), X-ray photoelectron spectroscopy (XPS), temperature-programmed reduction (TPR), Fourier transform infrared spectroscopy (FTIR), transmission electron microscopy (TEM), and programmed temperature surface reaction (TPSR).

2. Materials and Methods

2.1. Catalysts Preparation

Ni catalysts were supported on commercial SiO_2 AEROSIL 200 (S_{BET} = 200 m^2/g, Degussa, previously calcined at 900 °C) or on Ca-SiO_2 solids. The binary supports were prepared by incipient wetness impregnation of SiO_2 with $Ca(NO_3)_2·6H_2O$ (Panreac Química SLU, Castellar del Vallès, Spain). Different Ca loadings were used (3.4, 6.8, and 19.3 wt.%). The Ca–SiO_2 supports were maintained at room temperature for 12 h and then dried in an oven at 90 °C overnight. The solids thus obtained were finally calcined in flowing air, at 550 °C for 6 h. Samples are denoted as Ca(X)–SiO_2, where X stands for the nominal Ca content in wt.%.

The Ni metal was incorporated by incipient wetness impregnation with a concentration of 5 wt.%, using $Ni(NO_3)_2·6H_2O$ (Alfa Aesar, Thermo Fisher Scientific, UK) as the precursor. These samples were subjected to a drying process in equal conditions to those of the binary support.

2.2. Sample Characterization

The surface area of the material was measured by BET analysis of the N_2 adsorption isotherms collected at −196 °C (ASAP 2020 Micromeritics Instrument Corp., Norcross, GA, USA), with pretreatment at 200 °C for 2 h. The crystalline phases of the samples were examined by X-ray diffraction (XRD), using an X'Pert Pro PANalytical B.V., Almelo, The Netherlands. The TPR experiments were carried out in a conventional fixed-bed flow reactor, and the effluent gases were continuously monitored by online mass spectrometry (Pfeiffer/Balzers Quadstar GmbH, Asslar, Germany, QMI422 QME125); the samples were heated up in a 5% H_2/Ar stream with a rate of 10 °C/min up to 800 °C. The XPS measurements were carried out using a multi-technique system (SPECS GmbH, Berlin, Germany) equipped with a dual Mg/Al X-ray source and a hemispherical PHOIBOS 150 analyzer. The catalysts were analyzed after two reduction treatments under H_2 atmosphere, first at 400 °C for 1 h in a tubular quartz reactor and then at 400 °C for 15 min in the load-lock XPS chamber. Transmission electron microscopy (TEM) images of the reduced catalysts were acquired using a JEOL, Ltd., Tokyo, Japan, JEM 2100F field emission gun electron microscope equipped with an energy dispersive X-ray (EDX) detector. The fresh samples were reduced at 400 °C for 1 h in a pure H_2 stream, while used samples were measured without any treatment. The particle size was determined by counting 300 particles. The temperature-programmed surface reaction (TPSR) measurements were carried out in conventional dynamic vacuum equipment coupled to a quadrupole mass spectrometer (RGA-200, SRS Inc., Sunnyvale, CA, USA). The catalysts were reduced before experiments in hydrogen flow at 400 °C and were degassed in a high vacuum at the same temperature. The adsorption was then carried out using a 40 Torr pulse of HCOOH at 40 °C. Once the gas phase was evacuated, the desorption step was carried out at a programmed temperature, and the gases released were analyzed with a mass spectrometer.

2.3. Catalytic Test

The catalytic activity measurements for the formic acid decomposition in the vapor phase were carried out in a conventional fixed-bed flow reactor. The catalysts were pretreated in H_2 flux at 400 °C for 1 h and then cooled in N_2 flux at the reaction temperature. A mixture of formic acid diluted with N_2 was fed to the reactor using a saturator–condenser at 15 °C (HCOOH concentration equal to 6%, with a flow of 25 mL·min^{-1}). The reactants and products were analyzed by gas chromatography with a Carboxen 1000 column and a TCD detector.

3. Results

3.1. Catalysts Characterization

Figure 1 shows the diffractograms obtained for the Ni/SiO$_2$ and Ni over the binary supports after reduction in hydrogen at 400 °C for 1 h. From the XRD patterns, the characteristic diffraction broad peak centered on 2θ = 23° confirmed the amorphous nature of silica in Ni/SiO$_2$ sample. No reflections from CaO species were observed in the diffraction patterns obtained for the Ca(X)-SiO$_2$ supported catalysts.

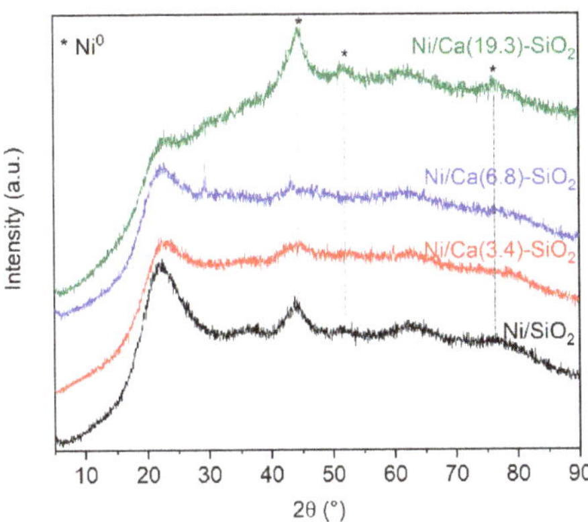

Figure 1. X-ray diffractograms of reduced Ni/SiO$_2$ and Ni/Ca(X)-SiO$_2$ catalysts.

The XRD patterns of Ni/SiO$_2$ and Ni/Ca(19.3)-SiO$_2$ catalysts exhibit broad peaks which could be assigned to the Ni species. The XRD peaks at 2θ = 44.5°, 51.9°, and 76.4° indicate the presence of metallic nickel (JCPDS 04-0850), although small broad peaks at 2θ = 37.2°, 43.3° and 62.9° reveal that there is also oxidized nickel phase (JCPDS 47–1049) [21].

The FTIR spectra of the reduced Ni/SiO_2 and $Ni/Ca(19.3)$-SiO_2 catalysts were analyzed (Figure 2). FTIR spectra show a broad band at 3528–3596 cm^{-1}, which corresponds to the stretching vibration mode of the O-H bond from the silanol group (Si-OH). The band at 1050–1080 cm^{-1} is assigned to the asymmetric stretching vibration of the siloxane bonds (Si-O-Si). The network Si-O-Si symmetric bond stretching vibrations are found at 620–900 cm^{-1}, whereas the network O-Si-O bending vibration modes are observed at 469–481 cm^{-1}. It is noted that the bands decrease in intensity as the content of Ca wt.% increases [20]. For the Ca promoted sample, with 19.3 wt.% of Ca, the broad band at 1458 cm^{-1}, associated with the band at 876 cm^{-1}, is assigned to asymmetric C-O stretch and out-of-plane deformation, respectively, of monodentate carbonate species on the CaO phase [22].

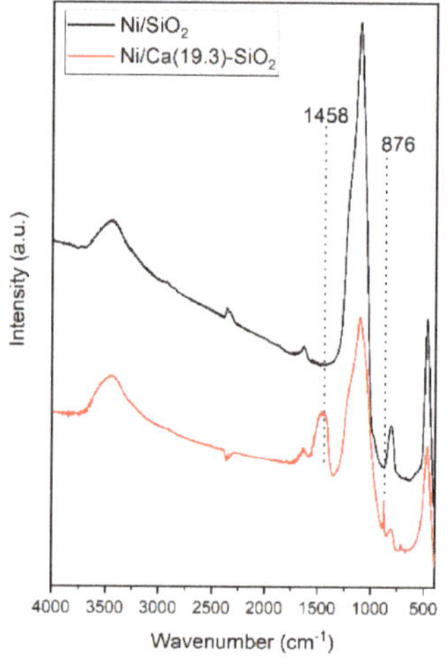

Figure 2. Fourier transform infrared spectroscopy (FTIR) spectra of reduced Ni/SiO_2 and $Ni/Ca(19.3)$-SiO_2 catalysts.

The XPS analysis of the reduced catalysts was carried out to study the presence of different surface Ni species (Figure 3). Both samples presented the Si 2s peak at 154.8 eV and the O 1s simple signal centered at 533 eV, corresponding mainly to the photoemission of the oxygen atoms presented on the siliceous support [23]. The Ca $2p_{3/2}$ core level spectrum from the reduced sample shows binding energy of 348.2 eV [24].

Figure 3 shows a difference in the Ni $2p_{3/2}$ spectra; the Ni catalysts exhibited a peak at 853.2 eV associated with reduced Ni species and another contribution related with octahedral Ni^{2+} clusters at 856.3 and 857.1 eV for Ni/SiO_2 and $Ni/Ca(19.3)$-SiO_2, respectively [25]. This difference in the binding energy values could be related to a different interaction between the Ni and the support. In both catalysts, a reduced nickel fraction was observed under the treatment conditions carried out prior to the catalytic tests. This Ni^0/Ni^{2+} surface ratio was higher for the Ni/SiO_2 catalyst. From the deconvolution of the spectra, the surface concentration of Ni^0 was estimated with respect to the total of surface Ni species, resulting in 70% and 13% for Ni/SiO_2 and $Ni/Ca(19.3)$-SiO_2, respectively.

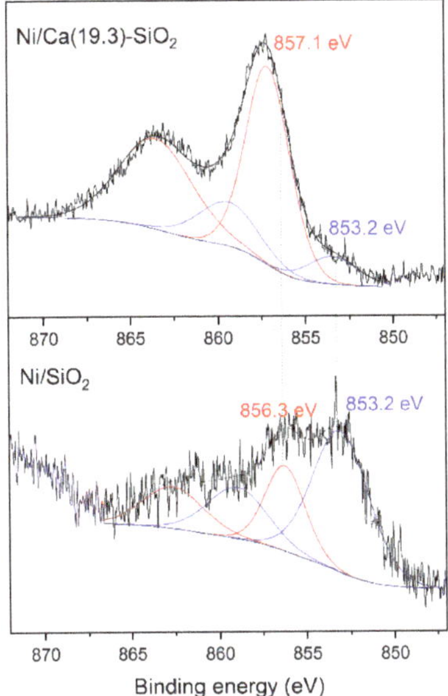

Figure 3. X-ray photoelectron spectroscopy (XPS) spectra of Ni 2p$_{3/2}$ region of reduced Ni/SiO$_2$ and Ni/Ca(19.3)-SiO$_2$ catalysts.

The reducibility of the supported nickel catalysts was studied by temperature-programmed reduction (TPR). TPR is a powerful tool for the study of the reduction behavior of oxidized phase, as NiO, and obtainment of the strength of the oxide-support interaction. Figure 4 shows the TPR profiles of the catalysts, where two main reduction peaks can be observed. Peaks in the 200–300 °C range that are attributed to the reduction of superficial oxygen [11] are not detected in these solids. Peaks above 300 °C represent reductions in Ni(II) species with different interactions with the support. Peaks between 300 °C and 600 °C can be attributed to Ni(II) species with low support interaction, as NiO. Due to high mobility, this Ni(II) phase can be easily reduced and shows a low reduction temperature. Peaks above 600 °C refer to Ni(II) with moderate/strong support interaction [26]. Reduction peaks at higher temperatures appear as Ca is added, which suggests this addition increases the interaction of Ni(II) with the support. The reduction temperature increased with the Ca loading. These results are consistent with those observed by XPS experiments.

The TEM images of the undoped and doped Ca(19.3) materials are shown in Figure 5. It can be seen that the nickel particles are evenly distributed over the support. To estimate the average size, 300 particles were measured. Values of 5.1 and 4.8 nm for the undoped and doped catalyst, respectively, were obtained. Thus, the doping with Ca did not modify the average size of the particles, but did slightly modify the particle size distribution (see histograms in Figure 5).

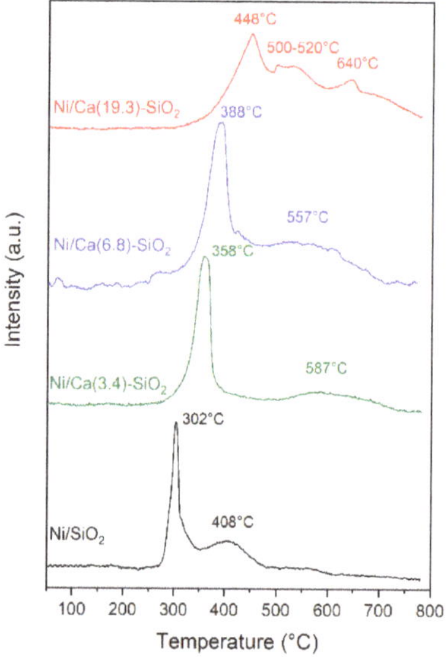

Figure 4. Temperature-programmed reduction (TPR) profiles of Ni/SiO$_2$ and Ni/Ca(X)-SiO$_2$.

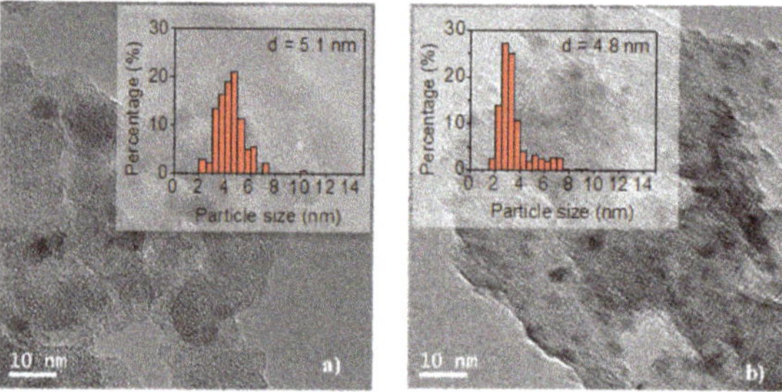

Figure 5. Transmission electron microscopy (TEM) images of reduced samples: (**a**) Ni/SiO$_2$ and (**b**) Ni/Ca(19.3)-SiO$_2$; the histograms were included.

Figure 6 shows the images obtained in the STEM mode and the EDX mapping of nickel (green), calcium (red), and silicon (white) revealed that Ni and CaO particles are evenly distributed on SiO$_2$. In addition, the EDX images showed that Ni particles coincided in space with CaO phase supporting the metal-support interaction revealed by the XPS and TPR experiments.

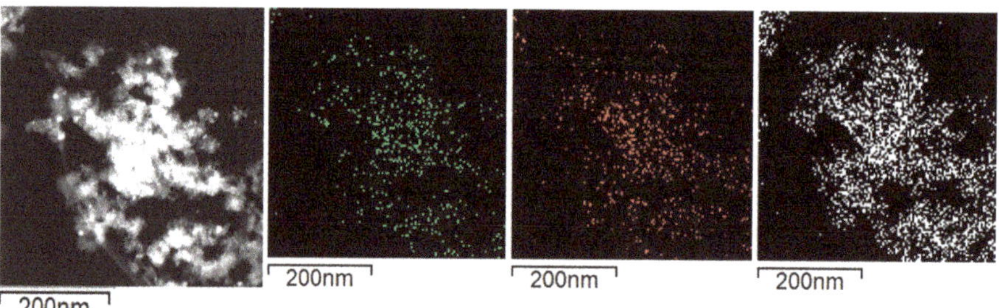

Figure 6. Selected area for the EDX mapping in reduced Ni/Ca(19.3)-SiO$_2$; mapping of nickel (green), calcium (red), and silicon (white).

3.2. Catalytic Performance in Fixed-Bed Reactor

Ni catalysts supported on SiO_2 and on Ca-SiO_2 were used in the decomposition reaction of formic acid to produce hydrogen. The catalytic activity of all materials was evaluated by operating the reactor in the experiments with a mass/flow ratio (W/F) equal to 5×10^{-5} g·h·mL^{-1}. The decomposition of formic acid can give the following as products:

$$HCOOH(g) \rightarrow H_2 + CO_2 \tag{1}$$

$$HCOOH(g) \rightarrow H_2O + CO \tag{2}$$

The values of reaction temperature for which the materials reach a 50% or 100% conversion of formic acid and its H_2 selectivity are compared in Table 1.

Table 1. Catalytic activity of Ni solids: temperature reaction and H_2 selectivity for 50% and 100% of HCOOH conversions (W/F = 5×10^{-5} g·h·mL^{-1}; feed composition: 6% HCOOH/N_2).

Catalyst	Ca/Ni Ratio	$T_{50\%}$	$S_{50\%}$	$T_{100\%}$	$S_{100\%}$	H_2/CO_2 Ratio
Ni/SiO$_2$	-	148	91	180	87	1.01
Ni/Ca(3.4)-SiO$_2$	1	153	93	180	91	1.08
Ni/Ca(6.8)-SiO$_2$	2	153	93	180	90	1.04
Ni/Ca(19.3)-SiO$_2$	5.6	145	92	160	92	0.95

It can be observed that the Ni/SiO$_2$ catalyst reached the 50% conversion at a temperature of 148 °C and 100% at 180 °C, while the selectivity was 91% and 87%, respectively (Table 1). In the catalysts supported on binary systems, the selectivity was higher in all cases. It is important to note that the catalyst with the highest Ca content (19.3 wt.%) reached 100% conversion at 160 °C, this being 20 °C lower than the undoped one.

Liu et al. [27] reported the study of the effect of different temperature pretreatments and atmospheres on the catalytic behavior of Ni catalysts for the dry reforming of methane. They observed that materials treated in He compared with those treated in H_2 achieved better yields. During the pretreatment with He, a small fraction of Ni particles was reduced. However, a short period of exposure to reactants was sufficient to achieve the formation of metallic Ni nanoparticles that are particularly active under reaction conditions [27]. This phenomenon could explain the high activity of the Ni/Ca(19.3)-SiO$_2$ catalyst, even though a low proportion of surface metallic species was observed after the reduction treatment.

Figure 7 shows the conversion of formic acid as a function of the reaction temperature for the series of Ni catalysts. The light-off curves were made following the same procedure in all the samples. After the reduction of the catalytic material, it was cooled in N_2 flux to 60 °C, and then the reaction mixture was fed with a concentration of HCOOH of 6% in N_2. After the curve measured from 0% to 100% (1st evaluation—Figure 7), the temperature was lowered to leave it in isothermal conditions and to measure the stability of the samples (Figure 8). After 16 h of reaction, the temperature was lowered and the complete curve was again measured from 0% to 100% (2nd evaluation in Figure 7) of the light off curve. The behavior throughout the conversion range shows the same tendency observed in Table 1. The catalysts were relatively stable under the conditions tested, although it can be observed that the points corresponding to the 2nd evaluation are below those obtained in the 1st, probably due to a restructuring of the material at the temperature reached (160–180 °C) and with conversion values close to 100%. In the Ca(19.3)-SiO$_2$ catalyst, this behavior is less marked.

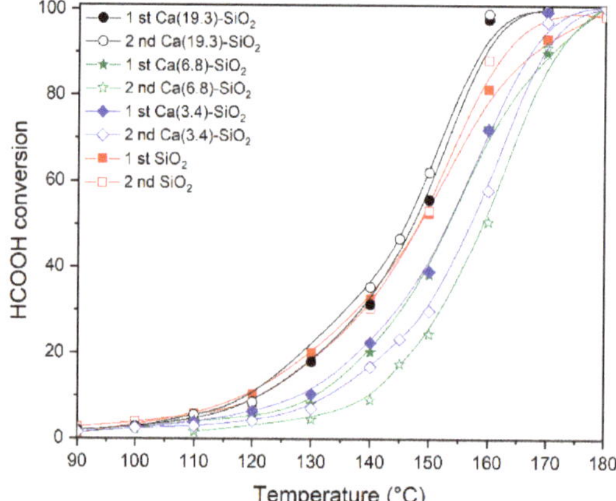

Figure 7. Catalytic activity of Ni solids after reduction at 400 °C at different reaction temperatures. The HCOOH conversions are plotted as function of the reaction temperature (W/F = 5×10^{-5} g·h·mL^{-1}; feed composition: 6% HCOOH/N$_2$).

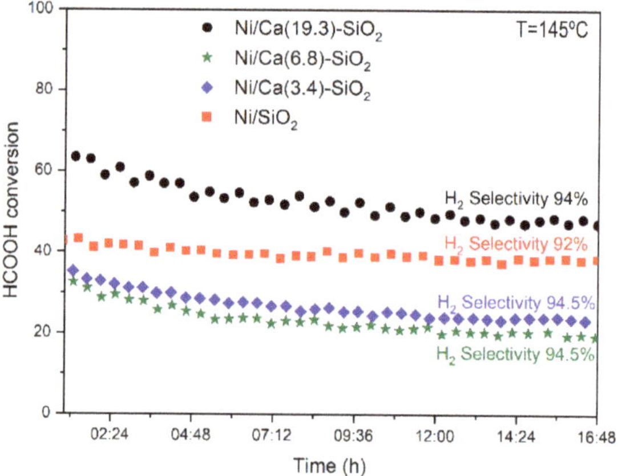

Figure 8. Stability test of the catalysts after reduction at 400 °C in the fixed-bed reactor. (Reaction temperature = 145 °C, W/F = 5×10^{-5} g·h·mL^{-1}, feed composition: 6% HCOOH/N$_2$.)

The doping of K in Pd catalysts supported over SiO$_2$, Al$_2$O$_3$, and activated carbon was previously reported [28]. These authors observed a significant effect of improvement in the catalytic behavior of noble metal for the formic acid decomposition. As a reaction mechanism, they proposed, as a first step, the formation of a phase containing liquid formic acid condensed in the pores of the catalyst; this phase provides a reservoir for the formation of formate ions with the participation of K$^+$ ions that later decompose to form CO$_2$ and H$_2$. In our materials, since the support is a nonporous material, condensation of formic acid is not likely to occur in pores; however, formates could form in the alkaline earth oxyhydroxide phase in the doped catalysts, with these species being the reaction intermediates.

3.3. Study of the Adsorbed Species Under Reaction Conditions: Temperature Programmed Surface Reaction and FTIR Experiments

Temperature-programmed surface reaction (TPSR) experiments were carried out to try to understand the differences in the catalytic performance. The catalysts were reduced before experiments in hydrogen flow at 400 °C and were degassed in high vacuum at the same temperature. The adsorption was then performed using a pulse of 40 Torr of HCOOH at 40 °C. Once the gas phase was evacuated, the evolution of the masses desorbed as a function of temperature was followed by mass spectroscopy. The TPSR experiments for the Ni catalysts are shown in Figure 9. Among the detected gases are the evolution of H_2, CO_2, CO, H_2O, and HCOOH (m/z = 2, 44, 28, 18, and 29, respectively). At lower temperature (<100 °C) the desorption of the unreacted HCOOH is observed, and, above 80 °C, the decomposition process begins to produce H_2 and CO_2 and minority CO and H_2O. It can be observed that the undoped catalyst exhibits lower adsorption of HCOOH and, subsequently, lower production of H_2 and CO_2. In all the samples, the molar ratio CO_2/H_2 produced was equimolar, as corresponds to the decomposition of formic acid (Table 1). These values were calculated by integrating the H_2 and CO_2 signals and taking into account the relative calibration of these gases.

There are two regions marked on the profiles: the first part corresponds to the decomposition of formic acid (up to 180 °C), and the second part corresponds to the decomposition of surface or mass species (formates, bicarbonates, and carbonates). The second part is closely related to the basic component of the catalyst. These samples doped with Ca present the H_2 and CO_2 desorption at a higher temperature. This could indicate greater stability of the species, for example, formate or bicarbonate species, which store hydrogen.

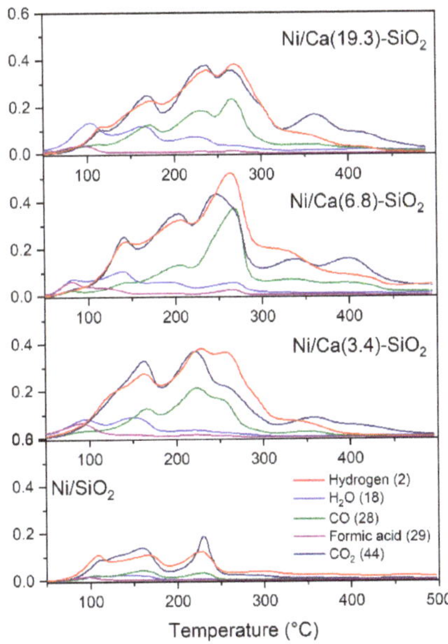

Figure 9. Temperature-programmed surface-reaction profiles of Ni catalysts after 40 Torr pulse of HCOOH at 40 °C.

3.4. Characterization of Used Catalysts

The FTIR spectra of the used Ni catalysts were analyzed (Figure 10). Figure 10a shows FTIR spectrum of used Ni/SiO$_2$ catalysts; the reduced one was included for comparison. It can be clearly observed that the spectra are identical before and after the catalytic test, and the signals described above are present. Figure 10b shows the spectrum of used Ni/Ca(19.3)-SiO$_2$ catalysts; the reduced one was included for comparison.

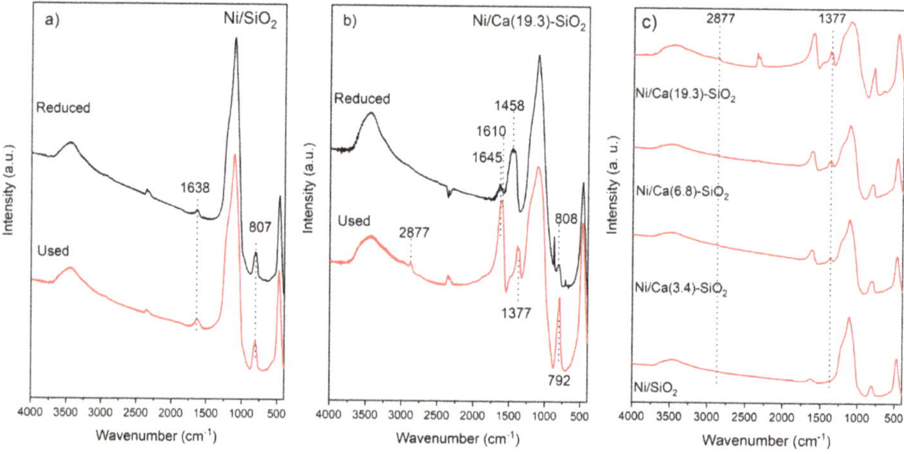

Figure 10. Fourier transform infrared spectroscopy (FTIR) spectra of (**a**) reduced and used Ni/SiO$_2$; (**b**) reduced and used Ni/Ca(19.3)-SiO$_2$; and (**c**) used Ni/Ca(X)-SiO$_2$ catalysts.

For all the samples, the fingerprints of SiO$_2$ at 475, 805, and 1115 cm^{-1} related to the Si-O-Si, Si-OH, and Si-O bonds, respectively, are observed. The characteristic bands of surface monodentate carbonate species at 1458 and 876 cm^{-1} are not present in the used Ni/Ca(19.3)-SiO$_2$ catalyst. In addition to SiO$_2$ bands, characteristic bands of formate species at 2877 and 1377 cm^{-1} are observed in the calcium doped solids used in reaction (Figure 10b,c). These bands are very weak in the spectrum of Ca(3.4)-SiO$_2$, but they are more intense at high X values (Ca(19.3)-SiO$_2$). The signal centered at 1645 cm^{-1} and the shoulder at 1240 cm^{-1} revealed the presence of asymmetric C-O stretching a C-O-H bending modes, respectively, of bicarbonate species [22,29]. For all the Ca-SiO$_2$ supported catalysts, the presence of formate and bicarbonate species in the catalysts after the reaction was confirmed by FTIR (Figure 10c) and were consistent with TPSR experiments.

Figure 11 shows the diffractograms obtained for the used Ni/SiO$_2$ and Ni/Ca-SiO$_2$ catalysts. The characteristic diffraction broad peak centered on 2θ = 23° confirmed the amorphous nature of the SiO$_2$ support. After reaction experiments, the diffraction patterns obtained for the Ca(X)-SiO$_2$ supported catalysts exhibit reflections from CaO species centered at 15.8°, 26.5°, and 30.7° [11]. The used Ni/Ca(19.3)-SiO$_2$ catalyst, as well as the reduced one, present broad peaks assigned to Ni species, indicating the presence of both metallic and oxidizes Ni particles. This result is consistent with those observed through TPR and XPS experiments. For Ni/SiO$_2$ catalyst, the peaks at 44.3°, 51.7°, and 76.2° are assigned to metallic Ni particles. Comparison with the fresh reduced samples (Figure 1) reveals that, in the case of the Ni/SiO$_2$ sample, the reflections corresponding to Ni^0 became sharper, which suggest that, in the absence of the Ca promoter, nickel is affected by the reaction conditions.

The TEM images of the undoped and doped Ca materials are shown in Figure 12. It can be observed that the nickel particles are evenly distributed over the support.

The estimated particle size using around 300 particles was 8.9 and 4.8 nm for the undoped and doped catalyst, respectively. The histogram of the Ni/SiO$_2$ particles was modified during the catalytic test, and the distribution and average particle size are doubled with respect to those of the reduced sample. However, the doping with Ca modified the interaction of the metal with the support and the Ni particles remained stable during the catalytic test (see histograms in Figure 12). These findings are in agreement with the XRD results discussed above.

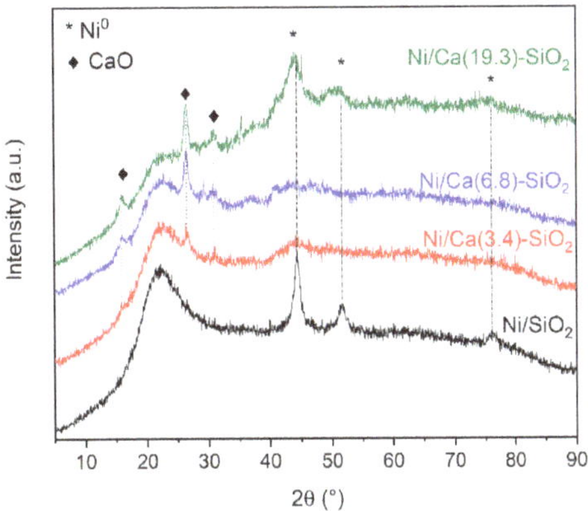

Figure 11. X-ray diffractograms of used Ni/SiO_2 and $Ni/Ca(X)-SiO_2$ catalysts.

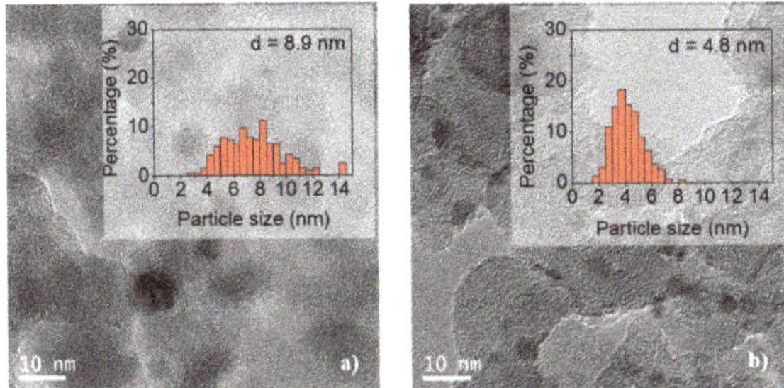

Figure 12. TEM images of used samples: (**a**) Ni/SiO_2 and (**b**) $Ni/Ca(19.3)-SiO_2$; the histograms were included.

The stability in the average size of the nickel particles in the used $Ni/Ca(19.3)-SiO_2$ catalyst could explain the difference in the behavior of this material with respect to the others presenting an equal performance in both catalytic tests (1st and 2nd evaluation in Figure 7). In addition, it can justify the highest stability of this catalyst during the long-term experiments of this reaction (Figure 8).

Figure 13 shows the images obtained in the STEM mode and the EDX mapping of nickel (green), calcium (red), and silicon (white) revealed that Ni and CaO particles are evenly distributed on SiO_2 and coincide in occupying the same space on the support.

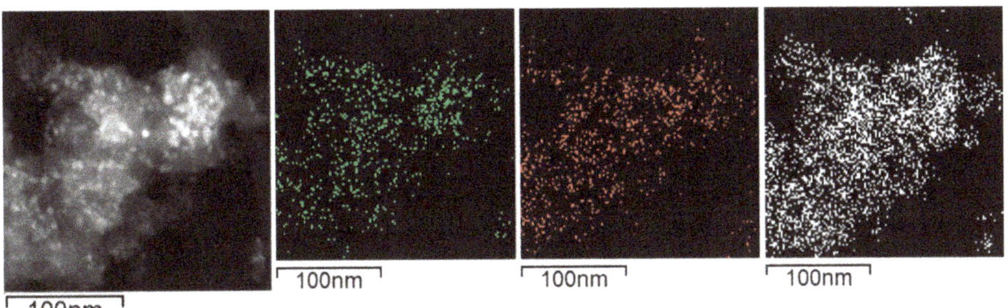

Figure 13. Selected area for the EDX mapping for used $Ni/Ca(19.3)-SiO_2$ catalyst; mapping of nickel (green), calcium (red), and silicon (white).

4. Conclusions

Ni catalysts supported on SiO_2 and on Ca-SiO_2 were synthesized. These materials were employed in the formic acid decomposition reaction to produce hydrogen. On the catalytic system, calcium species act as structural promoters by stabilizing the dispersed metallic phase against sintering. In addition, this additive also acts as a chemical promoter by influencing the acid–base properties of support and the interaction metal-support.

The XRD patterns for Ni/SiO_2 and $Ni/Ca(19.3)$-SiO_2 catalysts indicated the presence of metallic and oxidized nickel particles after the reduction step at 400 °C. The reduction temperature increased with the Ca loading. These results are consistent with those of the XPS and XRD experiments. The average size of Ni particles was measured for the two samples, undoped and doped catalyst, being 5.1 and 4.8 nm, respectively. The incorporation of Ca did not modify the average size of the particles, but did slightly modify the particle size distribution.

The Ni/SiO_2 catalyst reached 50% of formic acid conversion at a temperature of 148 °C and 100% at 180 °C, while the selectivity was 91% and 87%, respectively. For the catalysts supported on binary systems, the selectivity was higher in all cases. It is important to note that the catalyst with the highest Ca content (19.3 wt.%) reached 100% conversion at 160 °C, this being 20 °C lower than that of the undoped one. The doping with Ca modified the interaction of the metal with the support and the Ni particles remained stable during the catalytic test. However, for Ni/SiO_2 catalyst the distribution and average particle size are doubled during reaction with respect to those of the reduced sample. The stability in the average size of the nickel particles in the used $Ni/Ca(19.3)$-SiO_2 catalyst could explain the difference in the behavior of this material to the others. Moreover, this catalyst was relatively stable under the reaction conditions used, presenting an equal performance in two sequential catalytic tests.

The TPRS experiments reveal that, at lower temperatures (<100 °C), the desorption of the unreacted HCOOH is observed, and above 80 °C, the decomposition process begins to produce H_2 and CO_2 and minority CO and H_2O. It can be observed that the undoped catalyst exhibits lower adsorption of HCOOH and subsequent lower production of H_2 and CO_2. In the Ca-SiO_2 supported catalysts, the presence of formate and bicarbonate species in the catalysts after the reaction was confirmed by FTIR and were consistent with TPSR experiments.

Author Contributions: B.F. and I.R.-R. conceived of and designed the experiments; M.C.-M. and S.G.-C. carried out the synthetic experiments; M.A.P. conducted the FTIR measurements; C.L.-O. performed the TEM characterization; B.F. wrote the original draft manuscript; and all authors discussed the results and contributed to the manuscript.

Funding: This research was supported by the Spanish Agencia Estatal de Investigación (AEI) under project CTQ-2017-89443-C3-3-R. CONICET Postdoctoral External Scholarship awarded to B. Faroldi is acknowledged.

References

1. Jia, L.; Bulusheva, D.A.; Beloshapkin, S.; Ross, J.R.H. Hydrogen production from formic acid vapour over a Pd/C catalyst promoted by potassium salts: Evidence for participation of buffer-like solution in the pores of the catalyst. *Appl. Catal. B Environ.* **2014**, *160–161*, 35–43. [CrossRef]

2. Wang, Z.L.; Yan, J.M.; Wang, H.L.; Ping, Y.; Jiang, Q. Pd/C synthesized with citric acid: An efficient catalyst for hydrogen generation from formic acid/sodium formate. *Sci. Rep.* **2012**, *2*, 598. [CrossRef] [PubMed]

3. Solymosi, F.; Koós, Á.; Liliom, N.; Ugrai, I. Production of CO-free H_2 from Formic Acid. A Comparative Study of the Catalytic Behaviour of Pt Metals on a Carbon Support. *J. Catal.* **2011**, *279*, 213–219. [CrossRef]

4. Iglesia, E.; Boudart, M. Decomposition of formic acid on copper, nickel, and copper-nickel alloys: III. Catalytic decomposition on nickel and copper-nickel alloys. *J. Catal.* **1983**, *81*, 224–238. [CrossRef]

5. Thyssen, V.V.; Maia, T.A.; Assaf, E.M. Ni supported on La2O3-SiO2 used to catalyse glycerol steam reforming. *Fuel* **2013**, *105*, 358–363. [CrossRef]

6. Thyssen, V.V.; Assaf, E.M. Ni/La_2O_3-SiO_2 catalysts applied to glycerol steam reforming reaction: Effect of the preparation method and reaction temperature. *J. Braz. Chem. Soc.* **2014**, *25*, 2455–2465. [CrossRef]

7. Thyssen, V.V.; Maia, T.A.; Assaf, E.M. Cu and Ni catalysts supported on g-Al$_2$O$_3$ and SiO$_2$ assessed in glycerol steam reforming reaction. *J. Braz. Chem. Soc.* **2015**, *26*, 22–31. [CrossRef]

8. Thyssen, V.V.; Georgetti, F.; Assaf, E.M. Influence of MgO content as an additive on the performance of Ni/MgO-SiO$_2$ catalysts for the steam reforming of glycerol. *Int. J. Hydrogen Energy* **2017**, *42*, 16979–16990. [CrossRef]

9. Meng, G.; Lu, G. The effect of impregnation strategy on structural characters and CO$_2$ methanation properties over MgO modified Ni/SiO$_2$ catalysts. *Catal. Commun.* **2014**, *54*, 55–60. [CrossRef]

10. Gluhoi, A.C.; Bogdanchikova, N.; Nieuwenhuys, B.E. Alkali (earth)-doped Au/Al$_2$O$_3$ catalysts for the total oxidation of propene. *J. Catal.* **2005**, *232*, 96–101. [CrossRef]

11. Yang, D.; Li, J.; Wen, M.; Song, C. Enhanced activity of Ca-doped Cu/ZrO$_2$ for nitrogen oxides reduction with propylene in the presence of excess oxygen. *Catal. Today* **2008**, *139*, 2–7. [CrossRef]

12. Xu, L.; Song, H.; Chou, L. One-Pot Synthesis of Ordered Mesoporous NiO–CaO–Al$_2$O$_3$ Composite Oxides for Catalyzing CO$_2$ Reforming of CH$_4$. *ACS Catal.* **2012**, *2*, 1331–1342. [CrossRef]

13. Park, J.; McFarland, E. A highly dispersed Pd–Mg/SiO$_2$ catalyst active for methanation of CO$_2$. *J. Catal.* **2009**, *266*, 92–97. [CrossRef]

14. Kim, H.Y.; Lee, H.M.; Park, J. Bifunctional Mechanism of CO2 Methanation on Pd-MgO/SiO$_2$ Catalyst: Independent Roles of MgO and Pd on CO$_2$ Methanation. *J. Phys. Chem. C* **2010**, *114*, 7128–7131. [CrossRef]

15. Xu, L.; Song, H.; Chou, L. Ordered mesoporous MgO–Al$_2$O$_3$ composite oxides supported Ni based catalysts for CO$_2$ reforming of CH$_4$: Effects of basic modifier and mesopore structure. *Int. J. Hydrogen Energy* **2013**, *38*, 7307–7325. [CrossRef]

16. Larsen, G.; Haller, G. Metal-support effects in Pt/L-zeolite catalysts. *Catal. Lett.* **1989**, *3*, 103–110. [CrossRef]

17. Panagiotopoulou, P.; Kondarides, D.I. Effects of promotion of TiO$_2$ with alkaline earth metals on the chemisorptive properties and water–gas shift activity of supported platinum catalysts. *Appl. Catal. B Environ.* **2011**, *101*, 738–746. [CrossRef]

18. Borowiecki, T.; Denis, A.; Rawski, M.; Gołębiowski, A.; Stołecki, K.; Dmytrzyk, J.; Kotarba, A. Studies of potassium-promoted nickel catalysts for methane steam reforming: Effect of surface potassium location. *Appl. Surf. Sci.* **2014**, *300*, 191–200. [CrossRef]

19. Alipour, Z.; Rezaei, M.; Meshkani, F. Effect of alkaline earth promoters (MgO, CaO, and BaO) on the activity and coke formation of Ni catalysts supported on nanocrystalline Al2O3 in dry reforming of methane. *J. Ind. Eng. Chem.* **2014**, *20*, 2858–2863. [CrossRef]

20. Múnera, J.; Faroldi, B.; Frutis, E.; Lombardo, E.; Cornaglia, L.; Carrazán, S.G. Supported Rh nanoparticles on CaO-SiO$_2$ binary systems for the reforming of methane by carbon dioxide in membrane reactors. *Appl. Catal. A Gen.* **2014**, *474*, 114–124. [CrossRef]

21. Le, T.A.; Kang, J.K.; Park, E.D. Active Ni/SiO$_2$ catalysts with high Ni content for benzene hydrogenation and CO methanation. *Appl. Catal. A Gen.* **2019**, *581*, 67–73. [CrossRef]

22. Mutch, G.; Anderson, J.; Vega-Maza, D. Surface and bulk carbonate formation in calcium oxide during CO$_2$ capture. *Appl. Energy* **2017**, *202*, 365–376. [CrossRef]

23. Faroldi, B.M.; Lombardo, E.A.; Cornaglia, L.M. Surface properties and catalytic behavior of Ru supported on composite La$_2$O$_3$–SiO$_2$ oxides. *Appl. Catal. A Gen.* **2009**, *369*, 15–26. [CrossRef]

24. Faroldi, B.; Múnera, J.; Falivene, J.; Rodriguez-Ramos, I.; Gutiérrez García, A.; Tejedor Fernández, L.; Carrazán, S.G.; Cornaglia, L. Well-dispersed Rh nanoparticles with high activity for the dry reforming of methane. *Int. J. Hydrogen Energy* **2017**, *42*, 16127–16138. [CrossRef]

25. Biesinger, M.C.; Payne, B.P.; Grosvenor, A.P.; Lau, L.W.M.; Gerson, A.R.; Smart, R.S.C. Resolving surface chemical states in XPS analysis of first row transition metals, oxides and hydroxides: Cr, Mn, Fe, Co and Ni. *Appl. Surf. Sci.* **2011**, *257*, 2717–2730. [CrossRef]

26. Thyssen, V.V.; Nogueira, F.G.E.; Assaf, E.M. Study of the influence of nickel content and reaction temperature on glycerol steam reforming with Ni/La$_2$O$_3$-SiO$_2$ catalysts. *Int. J. Res. Eng. Sci.* **2017**, *5*, 54–62.

27. Liu, D.; Lau, R.; Borgna, A.; Yang, Y. Carbon dioxide reforming of methane to synthesis gas over Ni-MCM-41 catalysts. *Appl. Catal. A* **2009**, *358*, 110–118. [CrossRef]

28. Jia, L.; Bulusheva, D.A.; Ross, J.R.H. Formic Acid Decomposition over Palladium Based Catalysts Doped by Potassium Carbonate. *Catal. Today* **2016**, *259*, 453–459. [CrossRef]
29. Anderson, J.; Rochester, C. Infrared Studies of Probe Molecules adsorbed on Calcium Oxide. *J. Chem. Soc. Faraday Trans. I* **1986**, *82*, 1911–1922. [CrossRef]

 *nanomaterials*

Article

(Ag)Pd-Fe₃O₄ Nanocomposites as Novel Catalysts for Methane Partial Oxidation at Low Temperature

Blanca Martínez-Navarro [1,2], **Ruth Sanchis** [3], **Esther Asedegbega-Nieto** [1], **Benjamín Solsona** [3] and **Francisco Ivars-Barceló** [1,2,*]

[1] Departmento de Química Inorgánica y Química Técnica, Facultad de Ciencias, UNED, Paseo Senda del Rey, 9, 28040 Madrid, Spain; blancamartinez@ccia.uned.es (B.M.-N.); easedegbega@ccia.uned.es (E.A.-N.)
[2] Instituto de Catálisis y Petroleoquímica (ICP-CSIC), C/Marie Curie, 2, Cantoblanco, 28049 Madrid, Spain
[3] Departmento de Ingeniería Química, Universitat de València, C/Dr. Moliner 50, Burjassot, 46100 Valencia, Spain; rut.sanchis@uv.es (R.S.); benjamin.solsona@uv.es (B.S.)
* Correspondence: franciscoivars@ccia.uned.es

Received: 18 April 2020; Accepted: 18 May 2020; Published: 21 May 2020

Abstract: Nanostructured composite materials based on noble mono-(Pd) or bi-metallic (Ag/Pd) particles supported on mixed iron oxides (II/III) with bulk magnetite structure (Fe_3O_4) have been developed in order to assess their potential for heterogeneous catalysis applications in methane partial oxidation. Advancing the direct transformation of methane into value-added chemicals is consensually accepted as the key to ensuring sustainable development in the forthcoming future. On the one hand, nanosized Fe_3O_4 particles with spherical morphology were synthesized by an aqueous-based reflux method employing different Fe (II)/Fe (III) molar ratios (2 or 4) and reflux temperatures (80, 95 or 110 °C). The solids obtained from a Fe (II)/Fe (III) nominal molar ratio of 4 showed higher specific surface areas which were also found to increase on lowering the reflux temperature. The starting 80 m^2 g^{-1} was enhanced up to 140 m^2 g^{-1} for the resulting optimized Fe_3O_4-based solid consisting of nanoparticles with a 15 nm average diameter. On the other hand, Pd or Pd-Ag were incorporated post-synthesis, by impregnation on the highest surface Fe_3O_4 nanostructured substrate, using 1–3 wt.% metal load range and maintaining a constant Pd:Ag ratio of 8:2 in the bimetallic sample. The prepared nanocomposite materials were investigated by different physicochemical techniques, such as X-ray diffraction, thermogravimetry (TG) in air or H_2, as well as several compositions and structural aspects using field emission scanning and scanning transmission electron microscopy techniques coupled to energy-dispersive X-ray spectroscopy (EDS). Finally, the catalytic results from a preliminary reactivity study confirmed the potential of magnetite-supported (Ag)Pd catalysts for CH_4 partial oxidation into formaldehyde, with low reaction rates, methane conversion starting at 200 °C, far below temperatures reported in the literature up to now; and very high selectivity to formaldehyde, above 95%, for Fe_3O_4 samples with 3 wt.% metal, either Pd or Pd-Ag.

Keywords: methane; oxidation catalysis; formaldehyde; magnetite iron oxide; Fe_3O_4; heterogeneous catalysis; palladium; Pd; silver; Ag; low-temperature activity; nanocomposite; Raman; TG in air; TG in hydrogen; XRD; electron microscopy; EDS

1. Introduction

Transformation of methane into valuable compounds is one of the hardest challenges in petrochemistry. The large abundance of methane in natural gas and the recent discoveries of new shale gas reserves around the world make methane functionalization even more interesting. If partial oxidation of methane to compounds with high value, such as methanol and especially formaldehyde, could take place with high efficiency, it could be competitive against the indirect process in several steps from synthesis gas [1]. Unfortunately, no catalytic system has shown promising values neither regarding reaction rates nor selectivity at medium and high methane conversions. The main drawback is related to the high stability of the methane molecule, which presents strong methyl C–H bonds, which are not easy to activate. Thus, it is usual that this reaction takes place at high reaction temperatures so that in those conditions the partial oxidation products are quickly oxidized into carbon oxides. As in mild reaction conditions, the reaction rates are low, and in harder conditions the carbon oxide formation is predominant; an intermediate trade-off working seems to be the best option [2]. In any case, the direct oxidation of methane into methanol, formaldehyde, and other oxygenated products is still very far from being competitive for commercial implementation [2,3]. Among the catalysts employed for direct oxidation of methane to formaldehyde using molecular oxygen, those based on vanadium, molybdenum [4–7], and especially iron [3,8–17] have shown the most promising performance. Nevertheless, the yield of formaldehyde reported with these catalysts does not exceed 5% [2,4,8,18–20], although higher yields can be found in the literature for other catalysts. A significant example of this is the 14–17% yield of formaldehyde patented for Mo-based heteropoly acid catalysts at 600–650 °C [8], never again mentioned, with CH_4 conversions and formaldehyde selectivity within the 20–23% and 65–84% ranges, respectively.

In the case of iron-based catalysts, bulk $FePO_4$ systems and mainly iron supported on a range of siliceous materials (standard silica, MCM-41, SBA-15, and others) have been studied in this reaction [10,14–17,21]. Although it is accepted that iron sites must not be highly aggregated, otherwise formaldehyde readily decomposes, there is not a wide consensus about the exact nature of the iron active and selective species. It has been proposed that isolated tetrahedral Fe^{3+} sites are the most selective sites [22] whereas other authors have suggested that Fe with higher aggregation (2D FeOx oligonuclear sites) is more effective than isolated Fe^{3+} sites [16]. In any case, the reaction temperatures employed in these articles exceed 400 °C and are often over 500 °C [3,8,12]. Interestingly, recent work by Zhao et al. showed that iron species supported on different zeolites (ZSM-5, beta, and ferrierite) activate methane at 350 °C using N_2O as an oxidant, reaching methane conversions until 3%, with the formation of different partial oxidation products (methanol, formaldehyde, dimethyl ether) but also carbon oxides, obtaining up to 10% selectivity to formaldehyde [9]. It must be mentioned that the transformation of methane at room temperature has also been studied but with no competitive results or using reaction conditions not viable for commercial applications [23,24]. In this vein, the direct transformation of methane into formaldehyde has been interestingly demonstrated to be possible at room temperature using gaseous $[Al_2O_3]^+$ clusters, although with low rates [25].

On the other hand, palladium catalysts are, together with platinum ones, usually employed in the total oxidation of methane into carbon dioxide [26–29], a very much applied reaction from an environmental viewpoint. Thus, palladium sites can activate methane at low temperatures and high conversions can be easily reached, although the methane transformation is directed towards a total combustion product of limited interest. These works mainly pay attention to the light-off curves in order to see the temperatures required for all methane to be converted and to follow the stability with the time on stream. However, due to the massive CO_2 formation, no special interest has been paid to see what takes place at low methane conversions.

Within this context, the aim of the current work is to rationally design and develop catalytic systems with potential for the partial oxidation of methane at ambient pressure and relatively mild reaction temperatures. For such a purpose, we have drawn inspiration from materials, chemical species, and results, all known from the scientific literature as well as from our own experience. Moreover, we kept the premises of using simple preparation methods and raw materials not especially expensive, in order to develop a multifunctional catalyst which was not specifically reported for partial oxidation of methane. Thus, Fe_3O_4-based nanostructures supporting, in turn, smaller Pd nanoparticles, were the catalytic system selected and developed. To complement this, a systematic study on some synthesis parameters in order to improve the surface area of the iron oxide substrate without using sophisticated methods was successfully performed. An additional purpose was also to prove the use of Ag as a dopant to tune the high reactivity of Pd toward moderate catalytic behaviors rather than its usual propensity for methane combustion [30,31].

The structural and elemental compositions, textural properties, morphology, and reactivity against reduction and oxidation for the developed (Ag)Pd-Fe_3O_4 nanocomposites were investigated by several physicochemical techniques, as well as their catalytic properties for methane partial oxidation at temperatures not higher than 250 °C. Promising results were achieved from the preliminary catalytic screening which will be discussed and contrasted with the characterization results throughout the manuscript.

2. Materials and Methods

Nanostructured magnetite iron oxides were synthesized by the coprecipitation of the corresponding iron hydroxide species formed in aqueous NH_4OH medium, starting from the mixture of $FeCl_2 \cdot 4H_2O$ and $FeCl_3 \cdot 6H_2O$ reagents dissolved in deionized water, according to the following reaction:

$$FeCl_3 + FeCl_2 + 8\,NH_4OH \rightarrow Fe_3O_4 + 8\,NH_4Cl + 4\,H_2O \tag{1}$$

Magnetite-based solids differing in their specific surface areas were obtained using a nominal Fe^{3+}/Fe^{2+} molar ratio of 2 or 4, and different coprecipitation temperatures (within the 80–110 °C range) in an inert atmosphere (N_2) under reflux conditions. NH_4OH was added at the corresponding synthesis temperature to get the basic medium for the iron hydroxides coprecipitation. In all cases the reflux temperature was maintained for 30 min. The precipitates were filtered, washed with deionized water, and dried at 100 °C/12 h.

(Ag)Pd-Fe_3O_4 nanocomposite materials were prepared by an impregnation procedure of Pd or Ag and Pd cations over the magnetite solid obtained by reflux at 80 °C with a nominal Fe^{3+}/Fe^{2+} molar ratio of 4. The impregnation was carried out using the proper amount of the corresponding metal nitrate salt/s to get a metal load of 1, 2, or 3 wt.% Pd or 3 wt.% Ag-Pd. In the latter case, no sequential method was followed but both metals were simultaneously incorporated, employing an Ag:Pd molar ratio of 2:8. The impregnation procedure consisted of dissolving the $Pd(NO_3)_2$ and $AgNO_3$ amounts needed in acetone and adding the magnetite solid into the acetone metal solution which was kept under stirring for 10 min, followed by an ultrasonication treatment for 30 min. This two-step cycle was repeated 4 times for every sample in order to favor a homogeneous dispersion of the noble metal/s over the magnetite. After the last cycle, the mixture was stirred at room temperature until the acetone was completely evaporated. The magnetite solid containing Pd or Ag/Pd was dried at 100 °C/12 h, and afterwards treated at 200 °C for 1 h, under a reducing gas stream consisting of an 80 mL min^{-1} total flow of H_2/N_2 (50/50), in order to selectively reduce the noble metal cations to their metallic state, avoiding the reduction of iron from the magnetite which starts over 350 °C.

XRD patterns were collected using an X'Pert PRO diffractometer with θ–2θ configuration, from PANalytical (Almelo, The Netherlands), equipped with an X'Celerator fast detector, operating at 45 kV and 40 mA and using a nickel-filtered Cu Kα radiation source which produces a nearly monochromatic X-ray beam (λ = 1.5406 nm). Diffractograms were obtained within the 4–90° 2θ range employing 20 s cumulative time.

The specific surface areas were determined by the Brunauer–Emmett–Teller (BET) method from N_2 adsorption isotherms at 77 K measured in a Micromeritics ASAP2020 instrument (Norcross, GA, USA). A desorption treatment at 150 °C (10 °C min^{-1} heating rate) for 5 h, at high vacuum range conditions, was conducted for every sample just prior to the BET analysis.

The scanning electron microscopy analyses were performed employing a field emission scanning electron microscope (FESEM; model GeminiSEM 500, ZEISS, Oberkochen, Germany). Two different electron detection modes were used for FESEM imaging: (i) a secondary electron In-Lens detector, located inside the electron column, which works with low-energy secondary electrons and provides images with high resolution, and (ii) a energy selective backscattered (EsB) in-lens detector, independent of the secondary in-lens detector, which provides a pure backscattered signal with no secondary electron contamination and very low acceleration potential, providing a higher Z-contrast image than any other backscattered detector. Elemental chemical microanalyses in selected areas from the scanning electron micrographs were carried out with an X-ray energy-dispersive detector (EDS) from Oxford Instruments (Abingdon, UK), coupled with the FESEM microscope (ZEISS, Oberkochen, Germany).

A 200 kV analytical electron microscope equipped with a field emission electron gun (FEG; model JEM-2100F, JEOL, Tokio, Japan) allowing ultrahigh resolution in scanning transmission mode (STEM) through the sub-nanometric highly stable and bright electron probe (0.2 nm) provided by the FEG, was employed for the high-angle annular dark-field (HAADF) imaging in STEM mode. A high-resolution CCD (charge-coupled device) camera (2048 × 2048 pixels; model SC200, GATAN, Pleasanton, CA, USA), was used for image acquisition controlled by Digital Micrograph software. Image processing was performed by ImageJ free-software (Windows 10, 64-bit, version) [32]. The JEOL JEM-2100F microscope was equipped with a detector for X-ray energy-dispersive spectroscopy (EDS; model X-Max 80, Oxford Instruments, Abingdon, UK). The EDS detector, with 127 eV resolution and 0.5–2.4 nm spot size range, was used for elemental chemical analysis of punctual sites and selected areas from the STEM imaging.

Selected samples were measured by confocal micro-Raman spectroscopy using a LabRam-IR HR-800 model instrument (HORIBA Jobin Yvon, Edison, NJ, USA) coupled to an optical microscope model Olympus BX41. Excitation was provided by a He-Ne laser operating at a wavelength of 632.8 nm through a 100× objective lens, and with the laser power set to 0.6 mW, below the limit reported to induce phase transformations in iron oxides [33–36]. The regular Raman spectra were collected within the 100–1670 cm^{-1} region with 1 cm^{-1} spectral resolution, using an integration time of 8–16 s and 8–16 accumulations. In these conditions the lateral (xy) spatial resolution on the sample was 1 μm. For a high-resolution spectral profile, the integration time and the number of accumulations were increased to 32 s and 64, respectively, within a measuring region reduced to 320–777 cm^{-1}.

Thermogravimetric analyses (TGA) were carried out in an SDT Q 600 apparatus (TA Instruments, New Castle, DE, USA) under reducing (H_2) or oxidizing (synthetic air) conditions. Prior to analysis, samples were pre-treated at 150 °C for 1 h under a He stream (30 mL min^{-1} flow) which was maintained thereafter during the cooling down to room temperature. Following this pre-treatment, and according to the desired reducing or oxidizing experiment, the gas was switched from He to H_2 (5 mL min^{-1} flow) or synthetic air (24 mL min^{-1} flow), respectively. After the adequate stabilization time (ca. 30–40 min.), the temperature was raised, at a 5 °C min^{-1} rate, to 200 °C; then until 800 °C using a different heating rate of 10 °C min^{-1}. The evolution of species (CO, CO_2, H_2O, O_2, N_2, and H_2) in the gas phase produced during the thermogravimetric experiments was followed with a quadrupole mass spectrometer (Pfeiffer Vacuum Omnistar™ GSD 301, Asslar, Germany) coupled with the TGA equipment. For this purpose, the following mass fragments (m/z) were measured: 2 (H_2), 28 and 16 (CO), 44 (CO_2), 18 and 17 (H_2O), 32 and 16 (O_2), 28 and 14 (N_2).

The heterogeneous catalysis experiments for partial oxidation of methane were carried out in a fixed-bed quartz tubular flow reactor at atmospheric pressure, within the 100–250 °C temperature range, and using 100 mg of catalyst diluted with SiC (mg catalyst/SiC = $\frac{1}{2}$ volume ratio). The feed consisted of a molar ratio $CH_4/O_2/N_2$ = 32/4/64, for a total flow of 50 mL min^{-1}. The separation and analysis of reactants and products were carried out online using a gas chromatograph equipped with two different chromatographic columns (3 m length molecular sieve 5 Å column and 30 m length RT-U-bond 0.53 mm i.d. column) and a thermal conductivity detector (TCD).

Blank reaction experiments up to 300 °C with no catalyst in the reactor tube showed no conversion at all. In all cases, carbon balances were 100 ± 4%, irrespective of the catalyst and mass used. Each data point for the assessment of catalytic performance was determined by averaging three different analyses of reactants and products. The minor differences observed among the repetitive analyses confirmed the steady-state conditions.

3. Results

3.1. Development of Nanostructured Fe_3O_4 Materials

From the XRD patterns of the solids obtained by coprecipitation of the corresponding iron hydroxides during the reflux syntheses at different temperatures and employing aqueous mixtures of Fe^{2+} and Fe^{3+} chloride salts with Fe^{3+}/Fe^{2+} molar ratio 2 or 4 (Figure 1), the presence of magnetite structure as the main phase formed in all cases can be confirmed. Thus, diffraction peaks at $2\theta =$ 30.2, 35.7, 43.4, 53.7, 57.3, and 62.9° are present, with a similar intensity ratio among all the samples, according to the standard magnetite (Fe_3O_4) powder diffraction patterns previously reported [37–39].

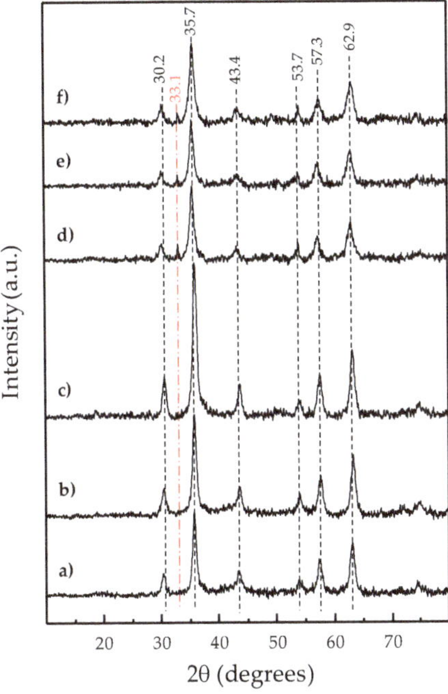

Figure 1. XRD patterns of magnetite solids prepared by coprecipitation employing a Fe^{3+}/Fe^{2+} aqueous molar ratio of 2 (**a–c**) or 4 (**d–f**), at different synthesis temperatures: 110 °C (**a,d**), 95 °C (**b,e**), or 80 °C (**c,f**). Symbols: Fe_3O_4 (- - -), α-Fe_2O_3 (- · - · -).

It is important to point out that similar 2θ positions are found in powder XRD patterns of cubic maghemite (γ-Fe_2O_3) which moreover presents a medium intensity 2θ peak at 32.17° among others of lower intensity such as the ones at 15.0 and 23.8° [37–40]. The absence of those peaks confirms the assignment of magnetite as the major phase present in every iron oxide synthesized (Figure 1). Without limiting the foregoing, the XRD patterns of the solids prepared with a nominal Fe^{3+}/Fe^{2+} molar ratio of 4 show an additional low-intensity peak at ca. 33.1°, which is not observed for those obtained from a nominal Fe^{3+}/Fe^{2+} molar ratio of 2. The 2θ peak at 33.1° can be assigned to the presence of hematite (α-Fe_2O_3) as a minority phase since it appears as the highest relative intensity diffraction for the main rhombohedral hematite ICSD (Inorganic Crystal Structure Database) references (01-079-0007, 00-024-0072, 01-072-0469, 01-085-0599). Indeed, the hematite content was found below 10% from a semiquantitative analysis of the XRD data by X′Pert Highscore Plus software.

Despite this minor structural difference dependent on the nominal molar ratio used, a significant variation was found in the specific surface areas of the prepared iron oxides (Table 1) which additionally showed a dependence with the reflux temperature employed (varied from 80 to 110 °C). Thus, the BET surface area obtained for the Fe_3O_4 precipitates from the synthesis employing a nominal Fe^{3+}/Fe^{2+} molar ratio of 4 was higher than those from a ratio of 2. For both ratios, the lower reflux temperature induced a higher specific surface area (Table 1).

Table 1. Specific surface areas of magnetite solids precipitated at different reflux temperature and nominal Fe^{3+}/Fe^{2+} molar ratio.

Fe^{3+}/Fe^{2+}	Temperature (°C)	S_{BET} (m^2 g^{-1}) [1]
2	110	77.0
2	95	91.3
2	80	96.9
4	110	125.7
4	95	135.8
4	80	137.6

[1] The standard deviation of ±0.2 m^2 g^{-1} was estimated for the BET surface areas calculated from the N_2 adsorption isotherms.

Therefore, the highest BET area (138 m^2 g^{-1}) was achieved for the magnetite solid prepared at 80 °C employing a nominal Fe^{3+}/Fe^{2+} molar ratio of 4, which was selected as the support for the following incorporation of Pd or Ag and Pd in order to prepare the target nanocomposite materials that would be catalytically tested for methane partial oxidation.

From the analysis by field emission scanning electron microscopy (FESEM), a nanometric spherical morphology was found for the crystalline Fe_3O_4 particles prepared. An example is displayed in Figure 2 for the highest surface magnetite substrate selected to incorporate the noble metals in the next step. The elemental chemical microanalyses by X-ray energy dispersive spectroscopy (EDS), from a number of FESEM images showing similar homogeneous distribution of nanosized spherical particles, including those in Figure 2, confirmed an averaged Fe:O stoichiometry of 3.0 (±0.3):4.1 (±0.3) consistent with the magnetite Fe_3O_4 structure assigned from the powder XRD results (Figure 1). A representative statistic study of the crystallite size distribution from the FESEM micrographs of the highest surface magnetite, whose resulting histogram is displayed in Figure 2, illustrates an average Fe_3O_4 particle diameter of 16.2 ± 3.2 nm.

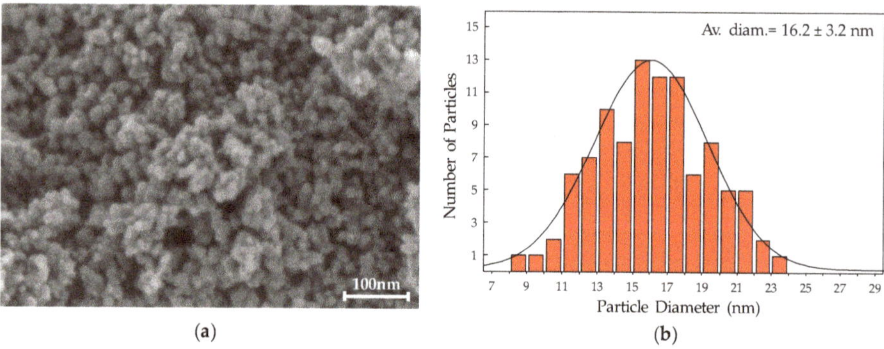

(a)　　　(b)

Figure 2. (**a**) Field emission scanning electron microscopy (FESEM) micrograph of magnetite solid precipitated from a reflux synthesis at 80 °C using a nominal Fe^{3+}/Fe^{2+} molar ratio of 4. Image conditions: in-lens mode, work distance 2.7 mm, energy selective backscattered (EsB) grid of 300 V and electron high tension (EHT) of 1 kV. (**b**) Statistic distribution of particle diameter from several FESEM micrographs of the magnetite solid shown in (**a**).

3.2. Development of Nanocomposite (Ag)Pd-Fe₃O₄ Materials

The incorporation of 1, 2, and 3 wt.% Pd, or 3 wt.% Ag-Pd (molar Ag:Pd ratio of 2:8) to the selected magnetite substrate by impregnation and subsequent reduction treatment (250 °C/1 h under 50% H_2 in N_2 stream), gave rise to the nanostructured (Ag)Pd-Fe₃O₄ composite materials that were tested as catalysts for methane partial oxidation. The XRD patterns of these nanocomposites are shown in Figure 3. All display the 2θ peaks common to those found for the (Ag)Pd-free iron oxide substrate (Figure 1f); i.e., the peaks' set of 30.2, 35.7, 43.4, 53.7, 57.3, and 62.9° assigned to magnetite [37–39], along with the one at 33.1° from traces of hematite (ICSD: 01-079-0007). However, low-intensity 2θ peaks at 71.2 and 74.3° clearly appear for all the (Ag)Pd-containing magnetite samples (Figure 3). These two peaks can also be glimpsed, to a greater or lesser extent, in some XRD patterns of the pristine iron oxides synthesized from Figure 1, especially in those obtained from Fe^{3+}/Fe^{2+} = 2 synthesis. A thorough analysis, comparing some representative references for indexed XRD patterns of Fe_3O_4 structures, allows us to identify that the set of the given main six peaks assigned above to Fe_3O_4 structure is common for both the cubic (ICSD: 01-076-0957, 01-075-0449) and the orthorhombic (ICSD: 01-075-1609) crystal systems of magnetite. Nevertheless, the low-intensity 2θ peaks at 71.2 and 74.3° appear specifically characteristic of orthorhombic magnetite.

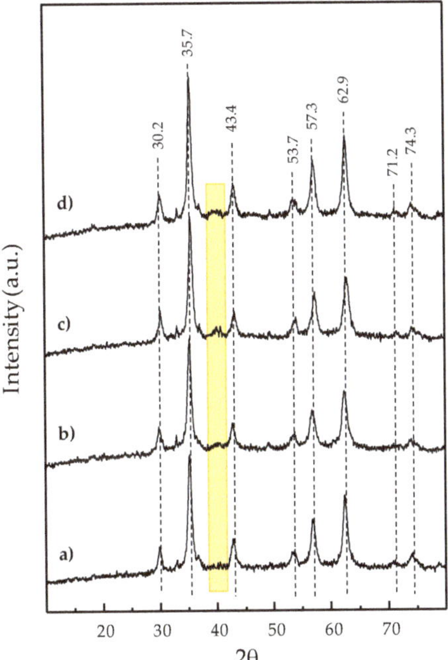

Figure 3. XRD patterns of nanocomposite catalysts based on magnetite (from 85 °C synthesis using Fe^{3+}/Fe^{2+} molar ratio of 4) doped with Pd or Pd-Ag: 1% Pd-Fe₃O₄ (**a**), 2% Pd-Fe₃O₄ (**b**), 3% Pd-Fe₃O₄ (**c**), 3% AgPd-Fe₃O₄ (**d**). Symbols: (- - -) Fe₃O₄ (ICSD: 01-075-1609); the highlighted strip defines the region where the highest intensity diffractions for pure Pd, pure Ag, and AgPd alloys are found [41–44].

Furthermore, semiquantitative analysis of the XRD data by X'Pert Highscore Plus software confirmed the major presence of an orthorhombic Fe_3O_4 structure in all (Ag)Pd-Fe₃O₄ catalysts from Figure 3. Only for 1% Pd-Fe₃O₄ and 3% AgPd-Fe₃O₄ samples did the semiquantitative phase analysis provide a minor coexistence with the cubic Fe_3O_4 structure (ca. 10% and 20%, respectively).

On the other hand, an additional low-intensity peak at 2θ = 40.1° can be observed for the 3% Pd-Fe₃O₄ catalyst (Figure 3c), which corresponds to the (111) plane of metal Pd [41,42,44].

This low-intensity peak turns into a small broadband within the 2θ range from 38.7 to 41.6° for the bimetallic 3% AgPd-Fe_3O_4 sample. This band, along with the glimpsed wide bump centered at 45.5°, is in full agreement with previously reported XRD patterns of bimetallic AgPd alloys presenting fcc crystalline structure [41–44], especially with those containing Ag:Pd molar ratio of 2:8 [42] or between 3:7 and 1:9 [41]. The main XRD diffraction of metal Pd is barely observed for the 2% Pd-Fe_3O_4 sample and the region becomes totally flat for the 1% Pd-Fe_3O_4, which is coherent to the quantitative detection limit (<2%) associated to the XRD technique for multiphase mixtures [45].

The monometallic Pd-Fe_3O_4 samples were investigated by field emission scanning electron microscopy (FESEM) using an energy selective backscattered (EsB) electron in-lens detector which can select the electrons according to their energy. Thus, the EsB detector is very sensitive to variations in the atomic number (Z) of atoms, allowing Z-contrast images (i.e., brighter image appearance for atoms with a higher Z), since the higher the atomic number the more electrons are scattered due to greater electrostatic interactions between the nucleus and the microscope electron beam at high angles. Indeed, Figure 4 shows Z-contrast micrographs for 1% Pd-, 2% Pd-, and 3% Pd-Fe_3O_4 samples, all with glaring bright spots consistent with metal Pd clusters, over a matt grey matrix assigned to Fe_3O_4. Congruently, a higher population density of these metal clusters is displayed as the nominal Pd load increases in the magnetite (Figure 4).

The average diameter of the metal Pd clusters was statistically determined for each Pd-Fe_3O_4 specimen. By direct measurement using ImageJ software [32], over more than 150 shiny clusters from several micrographs were acquired by EsB mode. Thus, the results in Table 2 show an average diameter of 3.3 ± 1.2 nm for the Pd clusters in 1% Pd-Fe_3O_4 catalyst, which remains practically unchanged for 2% Pd-Fe_3O_4, while it slightly increases to 4.4 ± 1.6 nm for the highest Pd load (i.e., 3% Pd-Fe_3O_4). It should be mentioned that the EsB detector employed gives a higher Z-contrast than any other backscattered detector, enabling differentiation between elements that are only distinguished by a few atoms.

Table 2. Characteristics of the metal nanoparticles (NPs) on the magnetite-based catalysts.

Catalyst	Metal Load [1] (wt.%)	Pd Load [1] (wt.%)	Pd:Ag [2]	Pd:Ag (SA-EDS)	Diam. Metal NPs [4] (nm)
1% Pd-Fe_3O_4	1	1	1:0	-	3.3 ± 1.2
2% Pd-Fe_3O_4	2	2	1:0	-	3.4 ± 1.1
3% Pd-Fe_3O_4	3	3	1:0	-	4.4 ± 1.6
3% AgPd-Fe_3O_4	3	2.4	0.8:0.2	0.8:0.2 [3]	4.3 ± 1.2

[1] Pd or Ag-Pd metal load nominally introduced by wet impregnation. [2] Atomic ratios employed for wet impregnation. [3] An averaged atomic ratio determined by chemical analysis (standard deviation = 0.1) of selected areas employing X-ray energy-dispersive spectroscopy (SA-EDS) coupled to a 200 kV field emission analytical electron microscope employing high-angle annular dark-field (HAADF) imaging in scanning transmission mode (STEM) mode. [4] The average diameter of metal nanoparticles (± SD), measured by EsB detector in HRFESEM (1–3% Pd-Fe_3O_4) or HAADF imaging in STEM (3% AgPd-Fe_3O_4).

For the bimetallic 3% AgPd-Fe_3O_4 sample, the Z-contrast microscopy studies were carried out differently, employing high-angle annular dark-field (HAADF) imaging in scanning transmission electron mode (STEM), using an analytical electron microscope equipped with a field emission electron gun (FEG). In a similar way to the EsB detector from the high-resolution field emission scanning electron microscopy (HRFESEM) instrument, the HAADF in STEM senses a greater signal from atoms with a higher atomic number (i.e., causing the metal atoms to shine brighter in the resulting image). Indeed, HAADF-STEM micrographs of 3% AgPd-Fe_3O_4, as the example displayed in Figure 5, show dull grey spheres within the diameter range of 11–20 nm, accordingly with Fe_3O_4 nanospheres, underneath some brighter and smaller metal particles with an average diameter of 4.3 nm (Table 2). The chemical analysis pointed to these metal particles (Figure 5) using selected area X-ray energy-dispersive spectroscopy (SA-EDS) with 0.5–2.4 nm spot size range (Table 2) and confirmed the coexistence of Ag and Pd with an averaged Ag:Pd stoichiometry of 0.2:0.8 (±0.1), consistent with the $Ag_{0.2}Pd_{0.8}$ alloy structure observed by XRD (Figure 3).

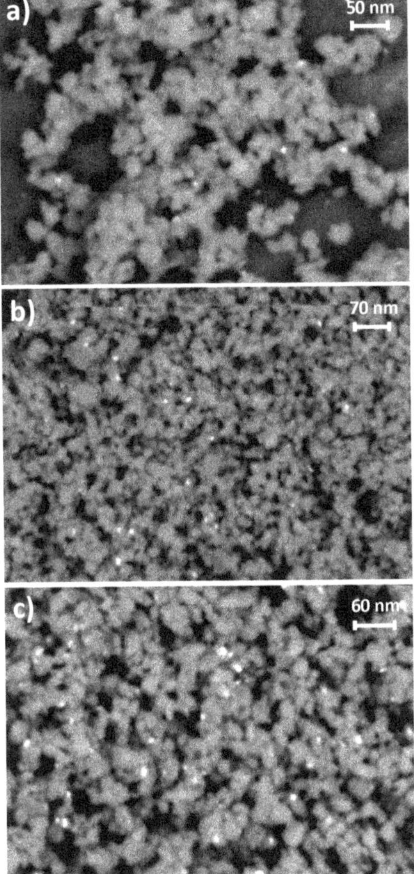

Figure 4. Z-contrast micrographs obtained by high-resolution field emission scanning electron microscopy (HRFESEM) using an energy selective backscattered (EsB) electron in-lens detector for the Fe_3O_4-based catalysts doped with 1% Pd (**a**), 2% Pd (**b**), and 3% Pd (**c**).

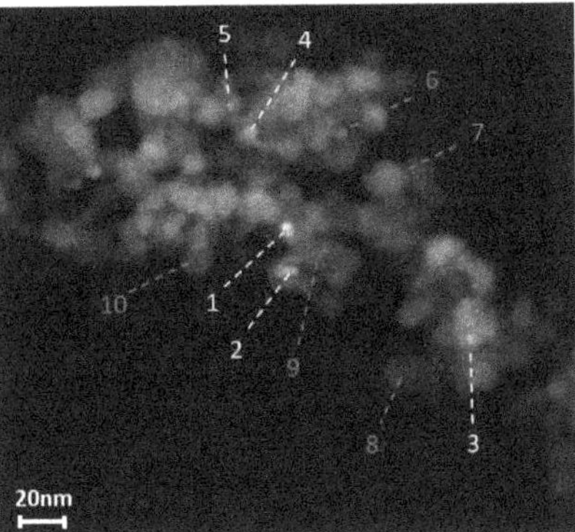

Figure 5. HAADF-STEM micrograph of 3% AgPd-Fe_3O_4 sample, with SA-EDS analysis spots marked and numbered: chemical analyses in brighter particles (1–5 spots) showed AgPd wt.% >12, with an averaged Ag:Pd atomic ratio of 0.2:0.8 (0.1 SD), while in 6–10 spots AgPd wt.% <0.9.

3.3. Catalytic Tests for Methane Partial Oxidation

The catalytic properties of the developed nanocomposite (Ag)Pd-Fe$_3$O$_4$ materials were tested in gas phase heterogeneous catalysis for the partial oxidation of methane at moderate reaction conditions (i.e., ambient pressure and temperatures up to 250 °C). Interestingly, using these mild conditions, methane gets activated although with low reaction rates. The catalytic results, summarized in Figure 6, show methane conversion at 200 °C for all the catalysts tested, leading to the formation of formaldehyde as the main reaction product (>74%), with CO$_2$ as the only secondary product detected. In general, the bimetallic 3% AgPd-Fe$_3$O$_4$ catalyst shows a catalytic activity significantly higher than the monometallic Pd-Fe$_3$O$_4$ catalysts; the latter ones presenting a maximum activity for the 2% Pd-Fe$_3$O$_4$ sample (Figure 6a). Therefore, no linear correlation is observed between the catalytic activity and the formal content of Pd which could be initially considered as the major source of methane activation at such a low temperature, according to the literature [30,31]. On the other hand, the selectivity to formaldehyde rises with the increase in the Pd load, showing a maximum above 97% for both the monometallic 3% Pd-Fe$_3$O$_4$ catalyst and the bimetallic 3% AgPd-Fe$_3$O$_4$ one at 200 °C (Figure 6b). At 250 °C the methane conversion slightly increases, while the selectivity to the partial oxidation product proportionally drops for all the monometallic Pd-Fe$_3$O$_4$ catalysts, with no absolute inverse correlation between conversion and selectivity observed. On the contrary, the high selectivity to formaldehyde hardly changes for the bimetallic 3% AgPd-Fe$_3$O$_4$ catalyst from 200 to 250 °C at which a formaldehyde productivity of 43 g$_{CH2O}$ kg$_{cat}^{-1}$ h^{-1} is reached with a turnover frequency (TOF) of 24.7 h^{-1} (reaction conditions are shown in the footnote of Figure 6).

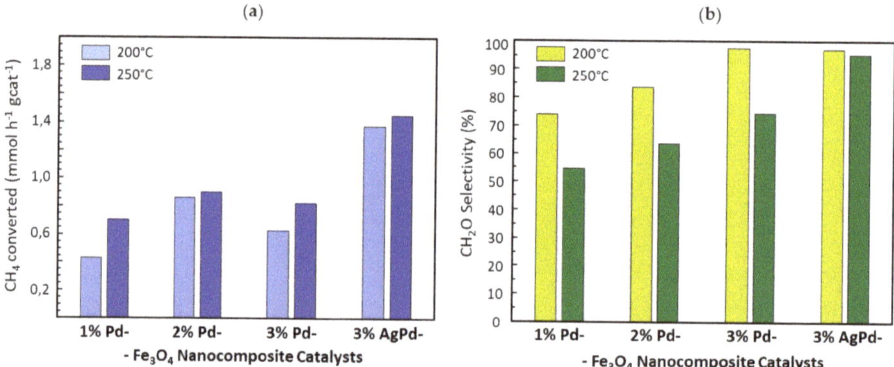

Figure 6. Catalytic activity (**a**) and selectivity to formaldehyde (**b**) for methane partial oxidation over the (Ag)Pd-Fe$_3$O$_4$ nanocomposite catalysts at 200 °C (light color bars) and 250 °C (dark color bars). Reaction conditions: 100 mg catalyst mass, CH$_4$/O$_2$/He molar ratio of 32/4.3/63.7, and contact time (W/F, at standard conditions) of 2.6 g$_{cat.}$ h mol$_{CH4}^{-1}$.

The pristine metal-free iron oxide substrate was comparatively tested for the partial methane oxidation upon identical reaction conditions used in the (Ag)Pd-Fe$_3$O$_4$ nanocomposite catalysts, detecting no methane conversion, likewise the blank results obtained no catalyst. Moreover, the catalytic results were confirmed by testing twice, under equivalent reaction conditions, all the nanocomposite catalysts displayed in Figure 6. The catalysts used in the first test were employed in a second catalytic cycle showing no apparent change in their performance (i.e., similar catalytic activity and selectivity).

4. Discussion

According to the results presented, the experimental research of this work consisted of three basic stages: (i) the development of a high surface nanostructured Fe$_3$O$_4$ substrate; (ii) its coating with noble metal nanoparticles to form mono- or bi-metallic (Ag)Pd-Fe$_3$O$_4$ nanocomposites; (iii) to assess the potential of the noble-metal/iron-oxide nanocomposites developed as catalysts for gas-phase partial oxidation of methane at ambient pressure and mild temperatures.

For the first stage, a reflux-assisted coprecipitation was the chosen method to carry out the preparation of high surface optimized magnetite. In this sense, a systematic study is presented here on the influence of both the reflux temperature (80–110 °C range) and the aqueous Fe^{3+}/Fe^{2+} molar ratio (2 or 4) in the specific surface area of the resulting magnetite precipitate. The data analysis confirmed a higher Fe_3O_4 surface area is strongly favored for Fe^{3+}/Fe^{2+} molar ratio of 4, rather than 2, in the aqueous synthesis solution (Table 2). Regardless, the magnetite surface area was also observed to increase as the reflux temperature was lowered within the 110–80 °C range, for both nominal ratios assayed (Table 2). At this point, it should be mentioned that an aqueous Fe^{3+}/Fe^{2+} molar ratio of 2, matching the Fe_3O_4 stoichiometry, has been traditionally employed in the coprecipitation-based methods reported for magnetite preparation [37,46–49]. In fact, no peer-reviewed publication could be found, with which to compare our results, that claimed the use of $Fe^{3+}/Fe^{2+} = 4$ for Fe_3O_4 synthesis. Nevertheless, the gain in the specific surface area achieved by the Fe_3O_4-based solid prepared by the higher nominal Fe^{3+}/Fe^{2+} ratio in synthesis (Table 2) is in good agreement with a size decrease reported to be observed for Fe_3O_4 particles in suspension while increasing Fe^{3+}/Fe^2 ratios [49]. However, along with this statement, in the first known work employing a coprecipitation method for magnetite preparation, no numeric values for Fe^{3+}/Fe^{2+} ratios larger than 2 are overtly mentioned.

Notwithstanding the excess of Fe^{3+} present in the synthesis media when using an Fe^{3+}/Fe^{2+} ratio of 4, compared with the Fe_3O_4 stoichiometry, the elemental and structural analyses of the higher surface area precipitates are consistent with a major presence of the mixed iron oxide bulk phase (just like the solids from stoichiometric Fe^{3+}/Fe^2+ ratio synthesis). On the one hand, an averaged Fe:O stoichiometry of 3.0:4.1 ($\pm$ 0.3), especially consistent with magnetite, was experimentally determined by EDS microanalyses on nanospheres, with 16.2 nm average diameter (Figure 2), to make up the ca. 140 $m^2 \cdot g^{-1}$ solid (Table 2). On the other hand, the 2θ peaks at 30.2, 35.7, 43.4, 53.7, 57.3, and 62.9° present in the corresponding XRD patterns (Figure 1d–f) are consistent with the (2 2 0), (3 1 1), (4 0 0), (4 2 2), (5 1 1), and (4 4 0) planes of standard magnetite structure (Joint Committee on Powder Diffraction Standards, JCPDS, file No.: 19-0629), respectively. These diffractions are also common to the γ-Fe_2O_3 structure (JCPDS file No.: 04-0755) whose assignment as major bulk phase was discarded by the absence of its additionally associated peaks from (210), (300), and (320) planes of crystalline maghemite (γ-Fe_2O_3) [39]. Nevertheless, a minor presence of maghemite cannot be completely ruled out. In fact, many studies claim that it is not straightforward to observe pure magnetite or maghemite, but an intermediate defective structure, $Fe_{3-x}O_4$, between both [50–53]. This has been especially proven for particles within the nanometric scale [52–56], for which a magnetite core with a maghemite shell structure is widely reported [48,54–56]. For the most oxidized iron oxide layer, in the cited core-shell structures, a varying thickness has been found depending on different factors such as the accurate nanoparticle size [54] and/or the reactants used and other preparation conditions [48,57].

Due to the inherent short-range periodic order that maghemite would exhibit in such a scenario, our XRD data would not be useful to directly reveal or deny its minor coexistence together with the magnetite structure. In this respect, Raman spectroscopy was complementarily employed to analyze some representative samples whose spectra are displayed in Figure 7. Thus, the Raman spectrum of the highest surface Fe_3O_4-based substrate used to incorporate the noble metals (Figure 7, spectrum a1) does not present a typical pure magnetite profile which is considered to be characterized by a single narrow and intense band reported to be centered within the 660–700 cm^{-1} range, together with a weak one between 534 and 560 cm^{-1} [33,34,58–63]. Instead, the Raman spectrum for the metal-free Fe_3O_4-based substrate is mainly characterized by the features assigned to maghemite (γ-Fe_2O_3) in the scientific literature [33,61–65]; i.e., a broad and intense asymmetric band from ca. 600 to 800 cm^{-1}, assigned to A_{1g} vibrational mode, and two more bands also asymmetric but weaker: one with two maxima reported to be centered at 345–365 cm^{-1} (E_g mode) and 380–395 cm^{-1}, and another one centered at 505–515 cm^{-1} (T_{1g} mode) [62,65,66]. In addition, two narrow peaks with relatively low intensity appear at 236 and 302 cm^{-1}, matching with the strongest vibrational bands (A_{1g} and E_g modes, respectively) related to hematite [35,62–65], along with the broadband centered at 1320 cm^{-1} [35,67,68]. In this sense, it must be said that Raman spectroscopy is extremely sensitive to the presence of hematite. Even traces of hematite impurities contained within other major iron oxides (as it is known in our case from XRD results) are able to provide narrow peaks with spurious intensity due to the large scattering power for Raman radiation inherent to hematite [60], inducing wrong conclusions about its relative content unless complementary characterization techniques are also employed.

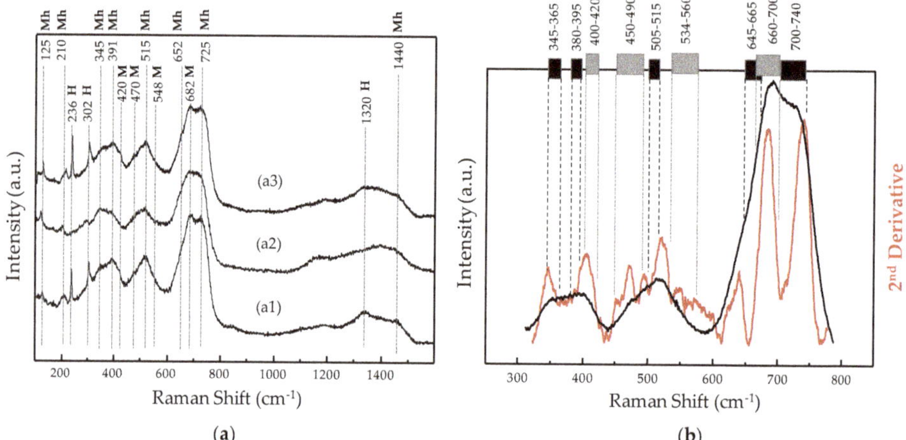

Figure 7. (**a**) Raman spectra of pristine Fe$_3$O$_4$-based substrate (a1), 3% Pd-Fe$_3$O$_4$ (a2), and 3% AgPd-Fe$_3$O$_4$ (a3) nanocomposite catalysts, collected at 0.6 mW. (**b**) High-resolution Raman spectrum and its second derivative by the Savitzky–Golay algorithm, using 81 convolution points of the Fe$_3$O$_4$-based substrate submitted to identical reduction treatment (250 °C/2 h in H$_2$) as the noble metal-containing catalysts. Acronyms from (a): *Mh* (maghemite), *H* (hematite), and *M* (magnetite) label the dotted lines according to the assignment ranges found in the literature (references at the end of the footnote). Symbols from (b): numerical ranges together with the black and grey labels define the regions where maxima have been reported in the literature for maghemite [33,61–65] and magnetite [33,34,58–63].

Regarding maghemite, its strongest band is usually described as consisting of two components whose maxima are reported within 645–665 cm^{-1} and 700–740 cm^{-1} ranges [48]. Additionally, a magnon band centered at ca. 1440 cm^{-1} and two low-intensity peaks at 125 and 210 cm^{-1} are also typical of maghemite structure [66,69]. Unfortunately, the characteristic Raman features linked to magnetite (at ca. 547 and 680 cm^{-1}) overlap with the three major bands cited for maghemite, which make difficult unambiguous statements about magnetite just based on the Raman spectrum when maghemite is present as well. A tool that might help to clarify this is the second derivative of the Raman spectrum. For this purpose, a high-resolution Raman spectrum within the region of interest was collected for the Fe$_3$O$_4$-based substrate, this time after being submitted to the same reduction treatment (250 °C/2 h in H$_2$) as the noble metal-containing samples. The spectrum obtained and its second derivative are displayed in Figure 7b. Regardless of the higher resolution, the Raman spectrum is very similar to that obtained from the pristine substrate within the same region (Figure 7a1) which leads to identical conclusions. Nevertheless, the second derivative shows a resolved accurate position for the maxima associated to the vibrational Raman modes among which it is possible to distinguish a maximum within the 534–560 cm^{-1} region (Figure 7b), where no maximum but the one from a weak vibrational mode of magnetite is reported [63,66].

Therefore, Raman results represented by the spectra in Figure 7, provide unambiguous evidence for the presence of maghemite. Unfortunately, further conclusions about either the structural configuration or the phases distribution of the given iron oxide mixture cannot be done from Raman data, due to the intimate overlapping between the main Raman band of magnetite and one of the strongest bands of maghemite (Figure 7). Furthermore, the ultrathin maghemite layer derived from a minor magnetite surface oxidation cannot be completely ruled out, although laser power was strictly kept far below the limits reported for the laser-induced magnetite oxidation [33,34]. In this respect, transformation into maghemite at temperature values as low as 109 °C has been reported for magnetite nanoparticles presenting 13–16 nm diameters (i.e., within the range measured for our Fe_3O_4 substrate) (Figure 2).

Moving forward to the discussion from the second experimental stage, the incorporation of noble metal nanoparticles to the magnetite-based substrate (the one with the highest surface area), followed by the given selective reduction treatment, leads to the corresponding mono- or bi-metallic (Ag)Pd-Fe_3O_4 nanocomposites (Table 2). Raman spectra of the representative catalysts 3% Pd-Fe_3O_4 and 3% AgPd-Fe_3O_4 are displayed as spectrum a2 and a3, respectively, in Figure 7a. These spectra show meaningless differences for the mixed iron oxide vibrational features compared with the pristine metal-free substrate, apart from the greater or lesser intensity of the narrow peaks at 236 and 302 cm^{-1} from the hematite impurity traces. Therefore, identical conclusions as those discussed above apply here.

With respect to the crystalline phases from the XRD results, new features appear for metal Pd or AgPd alloy structures, along with those remaining from the Fe_3O_4 substrate (Figure 3). On the other hand, no intensity is detected for the main diffraction of either metal Fe [70] or PdO [71–73] at 44.7 and 33.9° 2θ, respectively, which is consistent with the selective and complete noble metal reduction. The average diameter for the metal clusters was found to increase from ca. 3.3–3.4 nm for 1% Pd- and 2% Pd-Fe_3O_4 samples, to ca. 4.3–4.4 nm for the 3% Pd- and 3% AgPd-Fe_3O_4 ones. In the specific case of the bimetallic nanocomposite sample, the stoichiometry determined in the solid by structural and elemental analysis (i.e., by XRD (Figure 3) and by SA-EDS from HAADF-STEM imaging (Figure 5 and Table 2), respectively) was consistent with the nominal Ag:Pd stoichiometry used (0.2:0.8) during the impregnation treatment. Although we did not employ surfactant or any other sophisticated method to control the homogeneous size of the noble metal particles during the incorporation treatment, the standard deviation determined for the size distribution of (Ag)Pd nanoparticles on the magnetite-based substrate was acceptable (ca. 33%) in both mono- and bi-metallic nanocomposite samples (Table 2).

Finally, with respect to the stage of catalytic reactivity evaluation, promising results were obtained for the gas-phase methane partial oxidation into formaldehyde (Figure 6), for which the catalytic system (Ag)Pd-Fe_3O_4 was never reported, to the best of our knowledge. The highest formaldehyde productivity was shown by the 3% AgPd-Fe_3O_4 catalyst, yielding ca. 45 g_{CH2O} kg_{cat}^{-1} h^{-1} with a turnover frequency (TOF) of 24.7 h^{-1} (calculated for a 26% Pd dispersion) at 250 °C, far below the reaction temperatures reported so far [8,12]. In this sense, oxidation of methane has been hardly reported to occur below 600 °C [8,10,12]. Thus, the catalytic data provided by most of the studies on methane selective oxidation to formaldehyde correspond to reaction temperatures within the 600–800 °C range [8,12]. Ironically, formaldehyde decomposing into methanol and carbon monoxide starts at temperatures above 150 °C, although the kinetics for the uncatalyzed reaction is slow below 300 °C [74]. As a matter of fact, 450 °C appears as the lowest temperature reported so far, providing heterogeneous catalysis data of formaldehyde yielded by methane oxidation using molecular oxygen [75]. Nonetheless, the use of N_2O as a more reactive oxygen donor has been reported to run methane partial oxidation at 350 °C on several Fe-doped zeolites, although no selectivity to formaldehyde above 8% is claimed in that work [9]. Instead, a dispersion of partial oxidation products (CH_3OH, CH_2O, and C_2H_4) with low total selectivity was obtained, close to 40% in the best case [9]. On the contrary, selectivity to formaldehyde above 95% is achieved at 200–250 °C with the 3% AgPd-Fe_3O_4 catalyst developed in the current work (Figure 6). In addition to the major formation of formaldehyde, CO_2 was the only byproduct obtained in our catalytic tests for methane partial oxidation. No trace of methanol or CO was detected as byproducts, suggesting no formaldehyde decomposition. Moreover, reported pure radical mechanisms for aldehyde formation via methanol intermediate and/or consecutive decomposition involving CO formation, before or along with CO_2 production [8], can be discarded as plausible mechanisms being held on our (Ag)Pd-Fe_3O_4 catalysts. Instead, the so-called ion-radical mechanism, reported to involve electron transfers from the catalyst lattice oxygens, appears the most consistent with our catalytic results. Thus, this mechanism justifies the exclusive formation of formaldehyde and/or CO_2, by single-electron transfer from O^- centers and/or two-electron transfer from O_2^{2-} centers, respectively, both at the catalyst surface [8]. The ion-radical mechanism has been reported for catalysts based on metal oxides containing ambivalent metals that are able to easily change their oxidation states, such as MoO_3 or V_2O_5. As a matter of fact, these two oxides supported on SiO_2 are the catalytic systems grabbing a majority of studies concerning the oxidation of methane into formaldehyde [4–7]. Similarly, the cited ambivalence is also present in the mixed iron oxide from magnetite, the major component in our catalytic system. In fact, meaningful results in methane partial oxidation to formaldehyde have been reported for several catalysts containing iron oxide species [3,8–14]. The most encouraging finding to keep deploying resources in CH_4 conversion technology is, indeed, nature's iron-containing active center located in the methane monooxygenase enzyme able to selectively transform CH_4 into methanol, formaldehyde, and formic acid at ambient conditions [76,77]. Moreover, along with Mo and V, Fe was in the six elements list whose SiO_2-supported oxides were concluded to be the most suitable active phases for methane partial oxidation based on the screening of 43 elements [78]. However, no specific mention to the use of Fe_3O_4 in methane direct conversion reactions has been found in the literature, neither as an active phase nor support. With respect to the latter, in addition to SiO_2, other high surface substrates such as Al_2O_3, TiO_2, SnO_2, and zeolites are among the most used for supporting the active phases for partial oxidation of methane.

Noble metals such as Pd or Pt have also been briefly studied as active phases for methane activation, compared with transition metal oxides. Particularly, nanosized Pd has been reported to be highly reactive at temperatures as low as 200 °C, although toward complete methane oxidation [79]. Despite some interesting proposals for inhibiting the uncontrolled Pd activity leading to total combustion products [12,79], noble metals were practically left out in research on methane partial oxidation. Within this context, the nanocomposite material addressed in the current work, based on Fe_3O_4-supported (Ag)Pd, appears as a relatively known system, but a novel type of catalyst for the methane direct conversion into oxygenated partial oxidation products. The initial working hypothesis was to comparatively incorporate Ag with the purpose to downgrade the expected overreactive nanostructured Pd. In this regard, Ag was reported to selectively segregate at the surface of nanostructured Pd, suppressing the sites with the strongest oxidizing reactivity [30,31]. Indeed, the highest catalytic performance is achieved with the Fe_3O_4-based catalyst containing the bimetal AgPd alloy (Figure 6). Thus, the results settle that Ag improves selectivity toward the partial oxidation product at higher catalytic activity, which suggests confirmation of the working hypothesis about a softening effect on Pd reactivity. However, in general, the catalysts consisting of pure Pd supported over Fe_3O_4 have surprisingly shown high formaldehyde selectivity as well, rather than the total combustion products expected according to the published literature on Pd predisposition toward complete methane oxidation [79]. Even more, the selectivity to formaldehyde rises with increasing Pd load (from 1 to 3 wt.%). Regarding the assessment of catalytic activity from the pristine Fe_3O_4-based substrate itself, it could be dismissed by getting no conversion at all for the methane oxidation at identical reaction conditions used for the (Ag)Pd-containing catalysts. The latter statements might suggest a kind of synergy effect between Pd and the magnetite substrate. In this sense, by means of thermogravimetry analyses under reducing or oxidizing conditions, a relationship was found between the redox properties of the iron cations and the Pd content added to the mixed iron oxide. Thus, Figure 8b displays the derivative of thermogravimetry curves (DTG) collected under a H_2 stream (i.e., reducing conditions), with the positive peak maximums in the graph representing the greatest rates for each weight loss step which correspond to each dehydration accompanying a reduction. For the metal-free mixed iron oxide substrate, two separated reduction stages can be distinguished, namely, a small peak at low temperatures and a large one at a higher temperature range, below and above 360 °C, respectively. The first region between 200 and 360 °C corresponds to the reduction of minority pure Fe (III) oxides to Fe_3O_4. This reduction is clearly comprised of at least two overlapped peaks with maxima at 284 °C and 334 °C which, according to the reported literature [80], should correspond to the reduction of maghemite (γ-Fe_2O_3) to Fe_3O_4 and hematite (α-Fe_2O_3) to Fe_3O_4, respectively. This assignment is also consistent with the hematite impurity observed by XRD (Figure 1) and the homogeneous minor presence of maghemite proved to coexist with magnetite by comparing Raman spectroscopy (Figure 7) and XRD results (Figure 1). According to the following stoichiometry: $3Fe_2O_3 + H_2 \rightarrow 2Fe_3O_4 + H_2O$; a Fe_2O_3 content (γ-Fe_2O_3 + α-Fe_2O_3) of ca. 2% was calculated from the weight loss within the 200–360 °C region during thermogravimetry (TG) analysis in H_2 of the metal-free Fe_3O_4-based substrate (Figure 8a). On the other hand, the second region, from 360 °C to 550 °C, reflects major reduction consisting of a two-step Fe_3O_4 reduction sequence to Fe^0 through wüstite (FeO) intermediate formation [81].

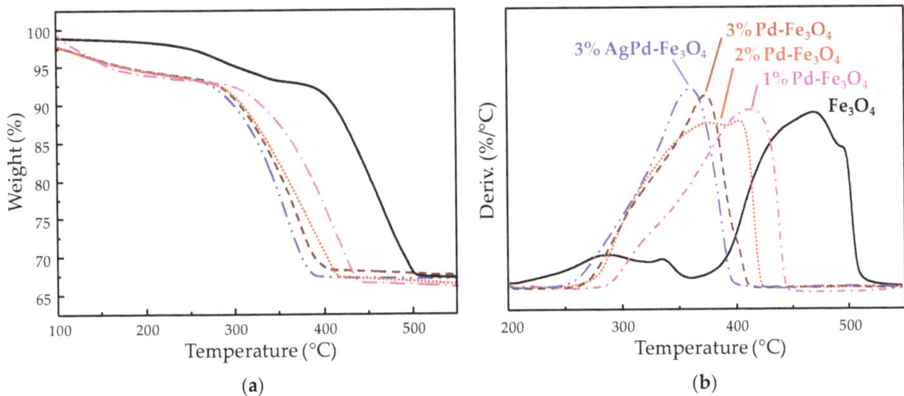

Figure 8. Thermogravimetry analysis curves collected in H_2 for the different (Ag)Pd-Fe_3O_4 catalysts prepared and for the pristine metal-free Fe_3O_4 substrate (**a**), and their first derivative (**b**).

The Pd incorporation on Fe_3O_4 leads to nanocomposite materials presenting a reduction temperature lower than the pristine magnetite-based substrate, either for the one-step reduction from minority Fe_2O_3 phases to Fe_3O_4 or the two-step reduction from Fe_3O_4 to FeO and Fe^0. Thus, the reduction temperature for the minor Fe_2O_3 phases moves bellow 200 °C, in agreement with previous publications employing Pd or other noble metals, such as Au, on Fe_2O_3 [80,82]. In Figure 8a, it can be observed that the corresponding weight loss for (Ag)Pd-Fe_3O_4 catalysts starts already at 100 °C, while the beginning of the equivalent loss for the pristine substrate delays until 200 °C.

The derivative TG curve, within the major reduction region, is displayed in Figure 8b, where the maximum of the major peaks correspond to the temperature for the maximum rate of Fe_3O_4 reduction which decreases following the next trend: Fe_3O_4 (468 °C) > 1% Pd-Fe_3O_4 (412 °C) > 2% Pd-Fe_3O_4 (376 °C) > 3% Pd-Fe_3O_4 (370 °C) > 3% AgPd-Fe_3O_4 (358 °C). According to this sequence, the Fe_3O_4 reducibility seems to be enhanced as the Pd load increases (from 1 to 3 wt.%) in the monometallic catalysts. Nevertheless, the tendency breaks for the 2.4 wt.% Pd load in the bimetallic 3% AgPd-Fe_3O_4 catalyst for which the doping with Ag forming the $Ag_{0.2}Pd_{0.8}$ alloy appears to favor the lowest temperature for Fe_3O_4 reduction at 358 °C (i.e., the highest reducibility).

A thermogravimetry (TG) analysis under oxidizing conditions (dry air stream) was also performed in three representative samples, in order to distinguish whether the effect on the redox properties of iron species, observed by the drop in their reduction temperature, goes further than the spillover effect of atomic hydrogen produced by catalyzed H_2 dissociative adsorption on Pd [83,84]. Thus, Figure 9 displays the first derivative of the TG curve collected under synthetic air flow for 3% Pd-Fe_3O_4 and 3% $AgPd$-Fe_3O_4 catalysts, and their pristine Fe_3O_4 substrate. The first negative peaks observed in Figure 9 within the 150–200 °C range correspond to the maximum rate of weight gain produced by oxygen insertion causing Fe^{2+} oxidation to Fe^{3+}, for which the corresponding temperature, appearing at 164 °C in the metal-free Fe_3O_4 increases in the noble metal-containing catalysts, especially in the bimetallic 3% $AgPd$-Fe_3O_4 with the maximum at 191 °C. Therefore, the redox properties of iron species undergo generalized changes induced by the Ag and/or Pd nanostructures supported, whereby the Fe^{3+} species become easier to reduce (i.e., present higher oxidizing power), while the Fe^{2+} ones become harder to oxidize. The strong correlation, existing between the trend in the change of the mixed iron oxide redox properties and the increased productivity to formaldehyde, points to a significant metal–support interaction that seems to be responsible for the tendency observed in the catalytic behavior of the different nanocomposite catalysts.

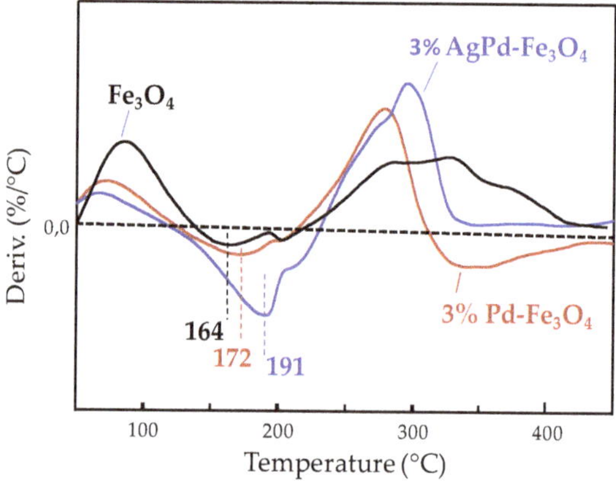

Figure 9. The first derivative of the thermogravimetry (TG) curves collected under synthetic air flow for the monometallic 3% Pd-Fe_3O_4 and the bimetallic 3% $AgPd$-Fe_3O_4 catalysts, as well as for the pristine metal-free Fe_3O_4-based substrate.

5. Conclusions

From the systematic study on the synthesis of Fe_3O_4-based solid with an improved specific surface area, an enhancement from 80 to 140 m^2 g^{-1} was achieved by doubling the nominal Fe^{3+}/Fe^{2+} ratio from 2 to 4 and decreasing the synthesis temperature from 110 to 80 °C. (Ag)Pd-Fe_3O_4 nanocomposite materials were developed through solvent-assisted (Ag)Pd impregnation on the Fe_3O_4-based substrate with optimized surface area, followed by a selective reduction treatment. The resulting nanocomposites can be defined as Pd metal or AgPd alloy nanoparticles with ca. 3.3–4.4 nm diameter supported on Fe_3O_4-based nanoparticles of 16.2 nm average diameter. The coexistence of maghemite was proven by Raman, although in a minor amount (<3 wt.%) according to XRD and TG results. Further research must be done in order to establish the accurate structural configuration linking the major Fe_3O_4 and the minor γ-Fe_2O_3 phases within the nanostructures developed.

Catalytic activity for methane conversion appears in all (Ag)Pd-Fe_3O_4 nanocomposite catalysts tested, at temperatures as low as 200 °C. The reported propensity toward methane combustion by nanostructured Pd has shown to be suppressed within the Pd-Fe_3O_4 nanocomposite systems (1–3 wt.% Pd load), which present a major selectivity to formaldehyde. The higher Pd load, within the studied range, increases the catalyst selectivity to formaldehyde which reaches above 95% for the catalysts containing above 2 wt.% Pd. On the other hand, a correlation was also observed between the increased Pd load and a gradually enhanced reducibility of the Fe^{3+} species in magnetite, which confirms a significant metal–support interaction. This reducibility in the Fe_3O_4 phase improves even further with the Ag incorporation forming the $Ag_{0.2}Pd_{0.8}$ alloy. These results suggest that the iron oxide substrate is most likely involved in the catalytic partial oxidation of methane into formaldehyde. This assumption is also supported by the CO_2 formation as the only secondary product which has been reported to occur by the ion-radical mechanism involving electron transfers from the catalyst lattice oxygens in catalysts containing oxides from ambivalent metals.

The nanocomposites have proven worthy for further research in gas-phase methane partial oxidation at relatively low reaction temperatures, compared with those reported until now using other catalysts. Forthcoming works will increase the low catalytic activity observed and shed more light on the mechanisms, active sites involved, and the role of the metal–support interaction.

Author Contributions: Conceptualization, methodology, writing—original draft preparation, project administration, F.I.-B.; validation, formal analysis, investigation, writing—review and editing, B.M.-N., E.A.-N., F.I.-B., R.S., and B.S.; resources, F.I.-B. and B.S.; data curation, B.M.-N., E.A.-N., F.I.-B., and R.S.; visualization, B.M.-N., E.A.-N., and F.I.-B.; supervision and funding acquisition, F.I.-B. and B.S. All authors have read and agreed to the published version of the manuscript.

Funding: This research and the APC were funded by *Comunidad de Madrid*, grant number 2017-T1/IND-6025 project within the program *"Atracción y Retención de Talento Investigador"* of the V PRICIT. The catalytic test experiments were funded by *Ministerio de Ciencia, Innovación y Universidades de España*, grant number MAT2017-84118-C2-1-R.

Acknowledgments: The authors gratefully acknowledge José María Gavira Vallejo (Dept. Ciencias y Técnicas Fisicoquímicas, UNED) for the Raman measurements; Inmaculada Rodríguez Ramos (Instituto de Catálisis y Petroleoquímica, ICP-CSIC), and Antonio R. Guerrero Ruiz (Dept. Química Inorgánica y Química Técnica, UNED) for their support in infrastructure and technical resources; as well as the administrative support from Mª Dolores Fernández, Cristina Ramos, and Mª Elisa Estébanez (PAS, UNED) and Rosa Mª Martín Aranda (Vicerrectora Investigación, UNED).

Conflicts of Interest: The authors declare no conflicts of interest. The funders had no role in the design of the study; in the collection, analyses, or interpretation of data; in the writing of the manuscript, or in the decision to publish the results.

References

1. Hargreaves, J.S.J.; Hutchings, G.J.; Joyner, R.W. Control of product selectivity in the partial oxidation of methane. *Nature* **1990**, *348*, 428–429. [CrossRef]
2. Zhao, G.; Drewery, M.; Mackie, J.; Oliver, T.; Kennedy, E.M.; Stockenhuber, M. The Catalyzed Conversion of Methane to Value-Added Products. *Energy Technol.* **2019**, 1900665. [CrossRef]
3. Fait, M.J.G.; Ricci, A.; Holena, M.; Rabeah, J.; Pohl, M.M.; Linke, D.; Kondratenko, E.V. Understanding trends in methane oxidation to formaldehyde: Statistical analysis of literature data and based hereon experiments. *Catal. Sci. Technol.* **2019**, *9*, 5111–5121. [CrossRef]
4. Smith, M.R.; Ozkan, U.S. The Partial Oxidation of Methane to Formaldehyde: Role of Different Crystal Planes of MoO_3. *J. Catal.* **1993**, *141*, 124–139. [CrossRef]
5. Amiridis, M.D.; Rekoske, J.E.; Dumesic, J.A.; Rudd, D.F.; Spencer, N.D.; Pereira, C.J. Simulation of methane partial oxidation over silica-supported MoO_3 and V_2O_5. *Aiche J.* **1991**, *37*, 87–97. [CrossRef]
6. Mac Giolla Coda, E.; Kennedy, M.; McMonagle, J.B.; Hodnett, B.K. Oxidation of methane to formaldehyde over supported molybdena catalysts at ambient pressure: Isolation of the selective oxidation product. *Catal. Today* **1990**, *6*, 559–566. [CrossRef]
7. Smith, M.R.; Zhang, L.; Driscoll, S.A.; Ozkan, U.S. Effect of surface species on activity and selectivity of MoO_3/SiO_2 catalysts in partial oxidation of methane to formaldehyde. *Catal. Lett.* **1993**, *19*, 1–15. [CrossRef]
8. de Vekki, A.V.; Marakaev, S.T. Catalytic partial oxidation of methane to formaldehyde. *Russ. J. Appl. Chem.* **2009**, *82*, 521–536. [CrossRef]
9. Zhao, G.; Benhelal, E.; Adesina, A.; Kennedy, E.; Stockenhuber, M. Comparison of Direct, Selective Oxidation of Methane by N_2O over Fe-ZSM-5, Fe-βeta, and Fe-FER Catalysts. *J. Phys. Chem. C* **2019**, *123*, 27436–27447. [CrossRef]
10. Fajardo, C.A.G.; Niznansky, D.; N'Guyen, Y.; Courson, C.; Roger, A.-C. Methane selective oxidation to formaldehyde with Fe-catalysts supported on silica or incorporated into the support. *Catal. Commun.* **2008**, *9*, 864–869. [CrossRef]

11. Kobayashi, T.; Guilhaume, N.; Miki, J.; Kitamura, N.; Haruta, M. Oxidation of methane to formaldehyde over FeSiO$_2$ and Sn-W mixed oxides. *Catal. Today* **1996**, *32*, 171–175. [CrossRef]

12. Brown, M.J.; Parkyns, N.D. Progress in the partial oxidation of methane to methanol and formaldehyde. *Catal. Today* **1991**, *8*, 305–335. [CrossRef]

13. Parmaliana, A.; Sokolovskii, V.; Miceli, D.; Arena, F.; Giordano, N. Silica-Supported MoO$_3$ and V$_2$O$_5$ Catalysts in Partial Oxidation of Methane to Formaldehyde. In *Catalytic Selective Oxidation*; American Chemical Society: Washington, DC, USA, 1993; Volume 523, pp. 43–57.

14. He, J.; Li, Y.; An, D.; Zhang, Q.; Wang, Y. Selective oxidation of methane to formaldehyde by oxygen over silica-supported iron catalysts. *J. Nat. Gas Chem.* **2009**, *18*, 288–294. [CrossRef]

15. Parmaliana, A.; Arena, F.; Frusteri, F.; Martínez-Arias, A.; Granados, M.; Fierro, J.L.G. Effect of Fe-addition on the catalytic activity of silicas in the partial oxidation of methane to formaldehyde. *Appl. Catal. A Gen.* **2002**, *226*, 163–174. [CrossRef]

16. Arena, F.; Gatti, G.; Martra, G.; Coluccia, S.; Stievano, L.; Spadaro, L.; Famulari, P.; Parmaliana, A. Structure and reactivity in the selective oxidation of methane to formaldehyde of low-loaded FeO$_x$/SiO$_2$ catalysts. *J. Catal.* **2005**, *231*, 365–380. [CrossRef]

17. Wang, Y.; Yang, W.; Yang, L.; Wang, X.; Zhang, Q. Iron-containing heterogeneous catalysts for partial oxidation of methane and epoxidation of propylene. *Catal. Today* **2006**, *117*, 156–162. [CrossRef]

18. Shimura, K.; Fujitani, T. Effects of promoters on the performance of a VO$_x$/SiO$_2$ catalyst for the oxidation of methane to formaldehyde. *Appl. Catal. A Gen.* **2019**, *577*, 44–51. [CrossRef]

19. Aoki, K.; Ohmae, M.; Nanba, T.; Takeishi, K.; Azuma, N.; Ueno, A.; Ohfune, H.; Hayashi, H.; Udagawa, Y. Direct conversion of methane into methanol over MoO$_3$/SiO$_2$ catalyst in an excess amount of water vapor. *Catal. Today* **1998**, *45*, 29–33. [CrossRef]

20. Koranne, M.M.; Goodwin, J.G.; Marcelin, G. Carbon pathways for the partial oxidation of methane. *J. Phys. Chem.* **1993**, *97*, 673–678. [CrossRef]

21. Yamada, Y.; Ichihashi, Y.; Ando, H.; Ueda, A.; Shioyama, H.; Kobayashi, T. Simple Preparation Method of Isolated Iron (III) Species on Silica Surface. *Chem. Lett.* **2003**, *32*, 208–209. [CrossRef]

22. Kobayashi, T. Selective oxidation of light alkanes to aldehydes over silica catalysts supporting mononuclear active sites - Acrolein formation from ethane. *Catal. Today* **2001**, *71*, 69–76. [CrossRef]

23. Kado, S.; Urasaki, K.; Sekine, Y.; Fujimoto, K. Direct conversion of methane to acetylene or syngas at room temperature using non-equilibrium pulsed discharge. *Fuel* **2003**, *82*, 1377–1385. [CrossRef]

24. Hammond, C.; Forde, M.M.; Ab Rahim, M.H.; Thetford, A.; He, Q.; Jenkins, R.L.; Dimitratos, N.; Lopez-Sanchez, J.A.; Dummer, N.F.; Murphy, D.M.; et al. Direct Catalytic Conversion of Methane to Methanol in an Aqueous Medium by using Copper-Promoted Fe-ZSM-5. *Angew. Chem. Int. Ed.* **2012**, *51*, 5129–5133. [CrossRef] [PubMed]

25. Wang, Z.-C.; Dietl, N.; Kretschmer, R.; Ma, J.-B.; Weiske, T.; Schlangen, M.; Schwarz, H. Direct Conversion of Methane into Formaldehyde Mediated by [Al$_2$O$_3$]$^{.+}$ at Room Temperature. *Angew. Chem. Int. Ed.* **2012**, *51*, 3703–3707. [CrossRef] [PubMed]

26. Grunwaldt, J.-D.; van Vegten, N.; Baiker, A. Insight into the structure of supported palladium catalysts during the total oxidation of methane. *Chem. Commun.* **2007**, 4635–4637. [CrossRef] [PubMed]

27. Lin, W.; Zhu, Y.X.; Wu, N.Z.; Xie, Y.C.; Murwani, I.; Kemnitz, E. Total oxidation of methane at low temperature over Pd/TiO$_2$/Al$_2$O$_3$: Effects of the support and residual chlorine ions. *Appl. Catal. B Environ.* **2004**, *50*, 59–66. [CrossRef]

28. Petrov, A.W.; Ferri, D.; Kröcher, O.; van Bokhoven, J.A. Design of Stable Palladium-Based Zeolite Catalysts for Complete Methane Oxidation by Postsynthesis Zeolite Modification. *Acs Catal.* **2019**, *9*, 2303–2312. [CrossRef]

29. Schwartz, W.R.; Ciuparu, D.; Pfefferle, L.D. Combustion of Methane over Palladium-Based Catalysts: Catalytic Deactivation and Role of the Support. *J. Phys. Chem. C* **2012**, *116*, 8587–8593. [CrossRef]

30. Khan, N.A.; Uhl, A.; Shaikhutdinov, S.; Freund, H.J. Alumina supported model Pd-Ag catalysts: A combined STM, XPS, TPD and IRAS study. *Surf. Sci.* **2006**, *600*, 1849–1853. [CrossRef]

31. González, S.; Neyman, K.M.; Shaikhutdinov, S.; Freund, H.-J.; Illas, F. On the Promoting Role of Ag in Selective Hydrogenation Reactions over Pd–Ag Bimetallic Catalysts: A Theoretical Study. *J. Phys. Chem. C* **2007**, *111*, 6852–6856. [CrossRef]

32. Schneider, C.A.; Rasband, W.S.; Eliceiri, K.W. NIH Image to ImageJ: 25 years of image analysis. *Nat. Methods* **2012**, *9*, 671–675. [CrossRef] [PubMed]

33. de Faria, D.L.A.; Venâncio Silva, S.; de Oliveira, M.T. Raman microspectroscopy of some iron oxides and oxyhydroxides. *J. Raman Spectrosc.* **1997**, *28*, 873–878. [CrossRef]

34. Shebanova, O.N.; Lazor, P. Raman study of magnetite (Fe_3O_4): Laser-induced thermal effects and oxidation. *J. Raman Spectrosc.* **2003**, *34*, 845–852. [CrossRef]

35. de Faria, D.L.A.; Lopes, F.N. Heated goethite and natural hematite: Can Raman spectroscopy be used to differentiate them? *Vib. Spectrosc.* **2007**, *45*, 117–121. [CrossRef]

36. Li, Y.-S.; Church, J.S.; Woodhead, A.L. Infrared and Raman spectroscopic studies on iron oxide magnetic nano-particles and their surface modifications. *J. Magn. Magn. Mater.* **2012**, *324*, 1543–1550. [CrossRef]

37. Matei, E.; Predescu, A.; Vasile, E.; Predescu, A. Properties of magnetic iron oxides used as materials for wastewater treatment. *J. Phys. Conf. Ser.* **2011**, *304*, 012022. [CrossRef]

38. van Oorschot, I.H.M.; Dekkers, M.J. Dissolution behaviour of fine-grained magnetite and maghemite in the citrate–bicarbonate–dithionite extraction method. *Earth Planet. Sci. Lett.* **1999**, *167*, 283–295. [CrossRef]

39. Yu, B.Y.; Kwak, S.-Y. Assembly of magnetite nanocrystals into spherical mesoporous aggregates with a 3-D wormhole-like pore structure. *J. Mater. Chem.* **2010**, *20*, 8320–8328. [CrossRef]

40. Pecharromán, C.; González-Carreño, T.; Iglesias, J.E. The infrared dielectric properties of maghemite, γ-Fe_2O_3, from reflectance measurement on pressed powders. *Phys. Chem. Miner.* **1995**, *22*, 21–29. [CrossRef]

41. Sun, D.; Li, P.; Yang, B.; Xu, Y.; Huang, J.; Li, Q. Monodisperse AgPd alloy nanoparticles as a highly active catalyst towards the methanolysis of ammonia borane for hydrogen generation. *Rsc Adv.* **2016**, *6*, 105940–105947. [CrossRef]

42. Wang, Q.; Chen, F.; Guo, L.; Jin, T.; Liu, H.; Wang, X.; Gong, X.; Liu, Y. Nanoalloying effects on the catalytic activity of the formate oxidation reaction over AgPd and AgCuPd aerogels. *J. Mater. Chem. A* **2019**, *7*, 16122–16135. [CrossRef]

43. Lu, F.; Sun, D.; Huang, J.; Du, M.; Yang, F.; Chen, H.; Hong, Y.; Li, Q. Plant-Mediated Synthesis of Ag–Pd Alloy Nanoparticles and Their Application as Catalyst toward Selective Hydrogenation. *ACS Sustain. Chem. Eng.* **2014**, *2*, 1212–1218. [CrossRef]

44. Lu, F.; Sun, D.; Jiang, X. Plant-mediated synthesis of AgPd/γ-Al_2O_3 catalysts for selective hydrogenation of 1,3-butadiene at low temperature. *New J. Chem.* **2019**, *43*, 13891–13898. [CrossRef]

45. Bish, D.L.; Post, J.E. Modern powder diffraction. In *Reviews in Mineralogy & Geochemistry*; America, I.S.M.S.O., Ed.; De Gruyter: Washington, DC, USA, 1989; Volume 20.

46. Veisi, H.; Najafi, S.; Hemmati, S. Pd(II)/Pd(0) anchored to magnetic nanoparticles (Fe_3O_4) modified with biguanidine-chitosan polymer as a novel nanocatalyst for Suzuki-Miyaura coupling reactions. *Int. J. Biol. Macromol.* **2018**, *113*, 186–194. [CrossRef] [PubMed]

47. Bristy, S.S.; Rahman, M.A.; Tauer, K.; Minami, H.; Ahmad, H. Preparation and characterization of magnetic γ-Al_2O_3 ceramic nanocomposite particles with variable Fe_3O_4 content and modification with epoxide functional polymer. *Ceram. Int.* **2018**, *44*, 3951–3959. [CrossRef]

48. Schwaminger, S.P.; Bauer, D.; Fraga-García, P.; Wagner, F.E.; Berensmeier, S. Oxidation of magnetite nanoparticles: Impact on surface and crystal properties. *CrystEngComm* **2017**, *19*, 246–255. [CrossRef]

49. Massart, R. Preparation of aqueous magnetic liquids in alkaline and acidic media. *IEEE Trans. Magn.* **1981**, *17*, 1247–1248. [CrossRef]

50. Boucherit, N.; Hugot-Le Goff, A.; Joiret, S. Raman studies of corrosion films grown on Fe and Fe-6Mo in pitting conditions. *Corros. Sci.* **1991**, *32*, 497–507. [CrossRef]

51. Chourpa, I.; Douziech-Eyrolles, L.; Ngaboni-Okassa, L.; Fouquenet, J.-F.; Cohen-Jonathan, S.; Soucé, M.; Marchais, H.; Dubois, P. Molecular composition of iron oxide nanoparticles, precursors for magnetic drug targeting, as characterized by confocal Raman microspectroscopy. *Analyst* **2005**, *130*, 1395–1403. [CrossRef]

52. Araujo, J.F.D.F.; Tahir; Arsalani, S.; Freire, F.L.; Mariotto, G.; Cremona, M.; Mendoza, L.A.F.; Luz-Lima, C.; Zaman, Q.; Del Rosso, T.; et al. Novel scanning magnetic microscopy method for the characterization of magnetic nanoparticles. *J. Magn. Magn. Mater.* **2020**, *499*, 166300. [CrossRef]

53. Hu, L.; Hach, D.; Chaumont, D.; Brachais, C.H.; Couvercelle, J.P. One step grafting of monomethoxy poly(ethylene glycol) during synthesis of maghemite nanoparticles in aqueous medium. *Colloids Surf. A Physicochem. Eng. Asp.* **2008**, *330*, 1–7. [CrossRef]

54. Demortière, A.; Panissod, P.; Pichon, B.P.; Pourroy, G.; Guillon, D.; Donnio, B.; Bégin-Colin, S. Size-dependent properties of magnetic iron oxide nanocrystals. *Nanoscale* **2011**, *3*, 225–232. [CrossRef] [PubMed]

55. Iyengar, S.J.; Joy, M.; Ghosh, C.K.; Dey, S.; Kotnala, R.K.; Ghosh, S. Magnetic, X-ray and Mössbauer studies on magnetite/maghemite core-shell nanostructures fabricated through an aqueous route. *RSC Adv.* **2014**, *4*, 64919–64929. [CrossRef]

56. Daou, T.J.; Grenèche, J.M.; Pourroy, G.; Buathong, S.; Derory, A.; Ulhaq-Bouillet, C.; Donnio, B.; Guillon, D.; Begin-Colin, S. Coupling Agent Effect on Magnetic Properties of Functionalized Magnetite-Based Nanoparticles. *Chem. Mater.* **2008**, *20*, 5869–5875. [CrossRef]

57. Rečnik, A.; Nyirő-Kósa, I.; Dódony, I.; Pósfai, M. Growth defects and epitaxy in Fe_3O_4 and γ-Fe_2O_3 nanocrystals. *CrystEngComm* **2013**, *15*, 7539–7547. [CrossRef]

58. Oblonsky, L.J.; Devine, T.M. A surface-enhanced Raman spectroscopic study of the passive films formed in borate buffer on iron, nickel, chromium and stainless steel. *Corros. Sci.* **1995**, *37*, 17–41. [CrossRef]

59. Kozlova, A.P.; Sugiyama, S.; Kozlov, A.I.; Asakura, K.; Iwasawa, Y. Iron-Oxide Supported Gold Catalysts Derived from Gold-Phosphine Complex $Au(PPh_3)(NO_3)$: State and Structure of the Support. *J. Catal.* **1998**, *176*, 426–438. [CrossRef]

60. Shebanova, O.N.; Lazor, P. Raman spectroscopic study of magnetite ($FeFe_2O_4$): A new assignment for the vibrational spectrum. *J. Solid State Chem.* **2003**, *174*, 424–430. [CrossRef]

61. Hanesch, M. Raman spectroscopy of iron oxides and (oxy)hydroxides at low laser power and possible applications in environmental magnetic studies. *Geophys. J. Int.* **2009**, *177*, 941–948. [CrossRef]

62. Chamritski, I.; Burns, G. Infrared- and Raman-Active Phonons of Magnetite, Maghemite, and Hematite: A Computer Simulation and Spectroscopic Study. *J. Phys. Chem. B* **2005**, *109*, 4965–4968. [CrossRef]

63. Chernyshova, I.V.; Hochella, M.F., Jr.; Madden, A.S. Size-dependent structural transformations of hematite nanoparticles. 1. Phase transition. *Phys. Chem. Chem. Phys.* **2007**, *9*, 1736–1750. [CrossRef] [PubMed]

64. Ohtsuka, T.; Kubo, K.; Sato, N. Raman Spectroscopy of Thin Corrosion Films on Iron at 100 to 150 ≡C in Air. *Corrosion* **1986**, *42*, 476–481. [CrossRef]

65. Testa-Anta, M.; Ramos-Docampo, M.A.; Comesaña-Hermo, M.; Rivas-Murias, B.; Salgueiriño, V. Raman spectroscopy to unravel the magnetic properties of iron oxide nanocrystals for bio-related applications. *Nanoscale Adv.* **2019**, *1*, 2086–2103. [CrossRef]

66. Jubb, A.M.; Allen, H.C. Vibrational Spectroscopic Characterization of Hematite, Maghemite, and Magnetite Thin Films Produced by Vapor Deposition. *Acs Appl. Mater. Interfaces* **2010**, *2*, 2804–2812. [CrossRef]

67. Bersani, D.; Lottici, P.P.; Montenero, A. Micro-Raman investigation of iron oxide films and powders produced by sol-gel syntheses. *J. Raman Spectrosc.* **1999**, *30*, 355–360. [CrossRef]

68. Marshall, C.P.; Dufresne, W.J.B.; Rufledt, C.J. Polarized Raman spectra of hematite and assignment of external modes. *J. Raman Spectrosc.* **2020**. [CrossRef]

69. El Mendili, P.Y.; Grasset, F.; Randrianantoandro, N.; Nerambourg, N.; Greneche, J.-M.; Bardeau, J.F. Improvement of Thermal Stability of Maghemite Nanoparticles Coated With Oleic Acid and Oleylamine Molecules: Investigations Under Laser Irradiation. *J. Phys. Chem. C* **2015**, *119*, 10662–10668. [CrossRef]

70. Singh, R.; Misra, V.; Singh, R. Synthesis, characterization and role of zero-valent iron nanoparticle in removal of hexavalent chromium from chromium-spiked soil. *J. Nanopart. Res.* **2011**, *13*, 4063–4073. [CrossRef]

71. Waser, J.; Levy, H.A.; Peterson, S.W. The structure of PdO. *Acta Crystallogr.* **1953**, *6*, 661–663. [CrossRef]

72. Baylet, A.; Marécot, P.; Duprez, D.; Castellazzi, P.; Groppi, G.; Forzatti, P. In situ Raman and in situ XRD analysis of PdO reduction and Pd° oxidation supported on γ-Al_2O_3 catalyst under different atmospheres. *Phys. Chem. Chem. Phys. PCCP* **2011**, *13*, 4607–4613. [CrossRef]

73. Sekiguchi, Y.; Hayashi, Y.; Takizawa, H. Synthesis of Palladium Nanoparticles and Palladium/Spherical Carbon Composite Particles in the Solid-Liquid System of Palladium Oxide-Alcohol by Microwave Irradiation. *Mater. Trans.* **2011**, *52*, 1048–1052. [CrossRef]

74. *Formaldehyde Health and Safety Guide*; IPCS International Programme on Chemical Safety Health and Safety Guide No. 57; World Health Organization: Geneva, Switzerland, 1991.

75. Yang, X.; Jung, K.D.; Cho, S.H.; Joo, O.S.; Uhm, S.J.; Han, S.H. Low-temperature oxidation of methane to form formaldehyde: Role of Fe and Mo on Fe–Mo/SiO_2 catalysts, and their synergistic effects. *Catal. Lett.* **2000**, *64*, 185–190. [CrossRef]

76. Sorokin, A.B.; Kudrik, E.V.; Alvarez, L.X.; Afanasiev, P.; Millet, J.M.M.; Bouchu, D. Oxidation of methane and ethylene in water at ambient conditions. *Catal. Today* **2010**, *157*, 149–154. [CrossRef]

77. Merkx, M.; Kopp, D.A.; Sazinsky, M.H.; Blazyk, J.L.; Müller, J.; Lippard, S.J. Dioxygen Activation and Methane Hydroxylation by Soluble Methane Monooxygenase: A Tale of Two Irons and Three Proteins. *Angew. Chem. Int. Ed.* **2001**, *40*, 2782–2807. [CrossRef]

78. Yamada, Y.; Ueda, A.; Shioyama, H.; Kobayashi, T. High throughput experiments on methane partial oxidation using molecular oxygen over silica doped with various elements. *Appl. Catal. A Gen.* **2003**, *254*, 45–58. [CrossRef]

79. Pitchai, R.; Klier, K. Partial Oxidation of Methane. *Catal. Rev.* **1986**, *28*, 13–88. [CrossRef]

80. Milone, C.; Ingoglia, R.; Schipilliti, L.; Crisafulli, C.; Neri, G.; Galvagno, S. Selective hydrogenation of α,β-unsaturated ketone to α,β-unsaturated alcohol on gold-supported iron oxide catalysts: Role of the support. *J. Catal.* **2005**, *236*, 80–90. [CrossRef]

81. Jozwiak, W.K.; Kaczmarek, E.; Maniecki, T.P.; Ignaczak, W.; Maniukiewicz, W. Reduction behavior of iron oxides in hydrogen and carbon monoxide atmospheres. *Appl. Catal. A Gen.* **2007**, *326*, 17–27. [CrossRef]

82. Wang, F.; Xu, Y.; Zhao, K.; He, D. Preparation of Palladium Supported on Ferric Oxide Nano-catalysts for Carbon Monoxide Oxidation in Low Temperature. *Nano-Micro Lett.* **2014**, *6*, 233–241. [CrossRef]

83. Mitsui, T.; Rose, M.K.; Fomin, E.; Ogletree, D.F.; Salmeron, M. Dissociative hydrogen adsorption on palladium requires aggregates of three or more vacancies. *Nature* **2003**, *422*, 705–707. [CrossRef]

84. Chen, L.; Zhou, C.-G.; Wu, J.-P.; Cheng, H.-S. Hydrogen adsorption and desorption on the Pt and Pd subnano clusters—A review. *Front. Phys. China* **2009**, *4*, 356–366. [CrossRef]

MDPI
St. Alban-Anlage 66
4052 Basel
Switzerland
Tel. +41 61 683 77 34
Fax +41 61 302 89 18
www.mdpi.com

Nanomaterials Editorial Office
E-mail: nanomaterials@mdpi.com
www.mdpi.com/journal/nanomaterials